The Three Deaths of Cerro de San Pedro

The Three Deaths of Cerro de San Pedro

Four Centuries of Extractivism
in a Small Mexican Mining Town

· ·

DAVIKEN STUDNICKI-GIZBERT

The University of North Carolina Press Chapel Hilll

This book was published with the assistance of the Wells Fargo Fund for Excellence of the University of North Carolina Press.

Set in Charis by Westchester Publishing Services
Manufactured in the United States of America

Library of Congress Cataloging-in-Publication Data
Names: Studnicki-Gizbert, Daviken, author.
Title: The three deaths of Cerro de San Pedro : four centuries of extractivism
    in a small Mexican mining town / Daviken Studnicki-Gizbert.
Description: Chapel Hill : The University of North Carolina Press, [2022] |
    Includes bibliographical references and index.
Identifiers: LCCN 2022031520 | ISBN 9781469671093 (cloth ; alk. paper) |
    ISBN 9781469671109 (pbk. ; alk. paper) | ISBN 9781469671116 (ebook)
Subjects: LCSH: Precious metal industries—Mexico—Cerro de San Pedro
    (San Luis Potosí)—History. | Gold mines and mining—Mexico—Cerro
    de San Pedro (San Luis Potosí)—History. | Silver mines and mining—
    Mexico—Cerro de San Pedro (San Luis Potosí)—History. | Mineral
    industries—Social aspects—Mexico—Cerro de San Pedro (San Luis
    Potosí)—History. | Mineral industries—Environmental aspects—Mexico—
    Cerro de San Pedro (San Luis Potosí)—History. | Cerro de San Pedro
    (San Luis Potosí, Mexico)—History.
Classification: LCC HD9535.M63 C478 2022 | DDC 332.64/42097244—dc23/
    eng/20220805
LC record available at https://lccn.loc.gov/2022031520

Cover illustration: A. Stoicheff, *Blasting at Cerro de San Pedro, February 2008.* Courtesy of A. Stoicheff.

Para todos aquellos que han luchado
Por la vida en Cerro de San Pedro
y
al Coyote que me llevó allí.

Contents

Part III
Extractivism, Again

Illustrations

Figures

Maps

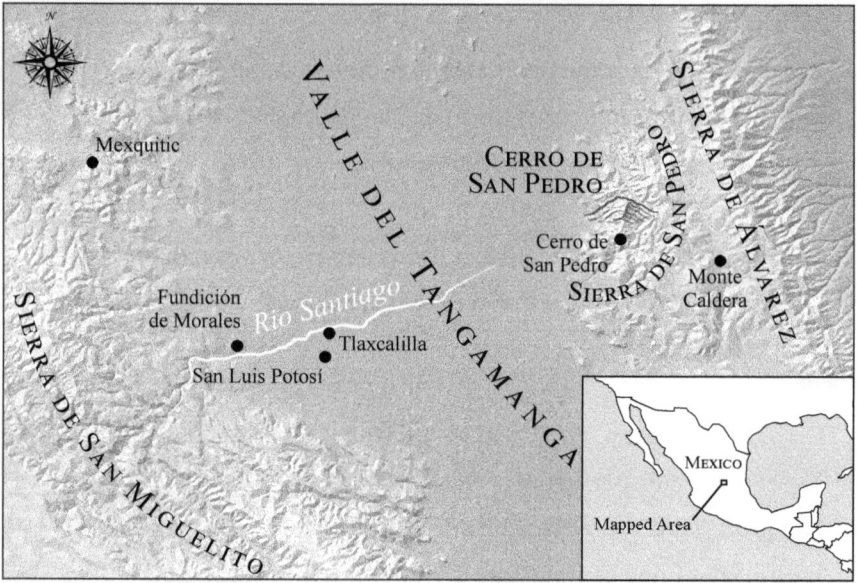

MAP 1 Cerro de San Pedro, S.L.P. and environs (© G. Wallace Cartography & GIS. Map by Geoffrey Wallace, 2021).

Preface

The following book was seeded in a conversation with my close friend and colleague Juan Carlos Ruiz Guadalajara in early December 2001. We were sitting on the steps in front of the Church of San Nicolás, enjoying the view across its plaza and the small village of Cerro de San Pedro. It was my first visit. The day had been spent wandering around, taking in the quiet scenes of life within the semi-ruins of this old Mexican mining town, absorbing its atmosphere. Evening was coming on. The winds were picking up and rustling the mesquites that seemed to be everywhere colonizing the village's abandoned places. So there we sat, looking out over the plaza to the tower and dome of the parish church of San Pedro, and behind it two Cerros, two mounts (not quite mountains but far more than hills). Looming before us was the Cerro Barreno, its shaded face murky and dun-green, its skyline a smooth bend against the blue desert sky, broken up, here and there, by the black silhouettes of some yuccas. To the immediate right of the Barreno and still catching some of that setting sun was the Cerro de San Pedro, loftier, broader. To the left, now looking down-valley, the distant lights of San Luis Potosí are coming on through the city haze.

Juan Carlos had moved to San Luis Potosí a year or so earlier, and he began to fill me in. Cerro San Pedro was the center and pivot of the local story. It had been settled in the late sixteenth century at the end of the war between the Chichimecas and Spanish. Thousands of people had poured into the area: Tarascos, Otomí, Nahuas, Kongolese, Basques, Portuguese, Castilians, Andalusians. They had all come for the mines of Cerro de San Pedro. The mining boom gave rise to the city of San Luis Potosí at the center of the valley, a city of indigenous *barrios*, a city of churches and convents, and now, over four centuries later, a city close to a million strong, sprawling across the Tangamanga valley, beautiful but also dusty, polluted, saturnine.

He also filled me in on the contemporary scene. A Canadian mining corporation had arrived to develop an open-pit gold and silver mine. It planned to raze the mountain of Cerro de San Pedro, engulf the two churches, and bury whatever remained of the village. That helped explain the graffiti spotted on the way in—"¡Fuera Minera San Xavier!"; "¡Que Viva Cerro de San Pedro!"; "Neo-colonialismo." For Juan Carlos it was a provocation and an outrage. All this history destroyed. The death of Cerro de San Pedro, the cradle of the city and of the entire region, all so that some small group of foreign investors could reap a decades' worth of dividends. And there was worse. The mine itself would consume massive amounts of cyanide and water, the former deadly, and the second, vital. In a dry city already facing failing supplies, he thought it unconscionable that a foreign corporation would put the valley's water at such risk. People had been fighting the project for years now. Up to that moment in late 2001, the company had still not been able to break ground and was running five years late. How long this stalemate would last was uncertain. The company's officers had recently met with Mexican president Zedillo in Toronto to work things out. If the Mexican government green-lighted operations, Juan Carlos and others felt that this would be the last mine. Given the scale and intensity of its devastation, the open-pit mine would finish the place off for good. Looking over to the Cerro de San Pedro, quiet and somehow watchful, the mountain seemed to pose itself as a riddle, one about how mining could give life to a place only to destroy it.

· · · · · ·

In less time than it has taken to write this book, the Cerro de San Pedro has been removed. It was shattered, dismantled, and then carted some five kilometers away to a leaching pad and refining plant that filched out the microscopic particles of gold and silver that it once contained. Half of the neighboring Cerro Barreno is also gone now. Its once clean arc of a skyline has been cut down to a ragged edge. Most of the town of Cerro de San Pedro has survived, barely. It currently perches on the edge of a steep pit hundreds of meters deep.

It took Minera San Xavier—the Mexican public limited company held by a succession of Canadian corporations—twelve years to reduce the mountain and gain its treasures. This is a remarkably short period of time given the scale of the task, but in the contemporary world of transnational mining, Minera San Xavier was well behind schedule. It had once announced the beginning of operations for 1996 and promised to be done within ten years. As it turned out, the social resistance to its operations stopped its advance for a decade. Work only began (June 2006) when it should have ended.

The struggle to defend the Cerro de San Pedro, both the mountain and the town, was one of the first of what would bloom into hundreds of social and ecological conflicts around sites of large-scale open-pit mining across Latin America. If the town escaped the destruction allotted to its namesake mountain, it is because of the fight and disputation of so many people over all those years. None of them, as far as I know, feel this as a victory. The extent of devastation to the local landscape is too great. The wake of harm churned up by this last mine is still in motion—the chronically slow contamination of the regional aquifer, the steady blow of heavy metal dust down-valley—and is set to continue for millennia. And of course, there is the loss of the mountain itself, a piece of the local geology that became a landmark, an icon, a mountain imbued with history, with stories, with generations of lives. Its destruction is a loss that escapes measure.

The Three Deaths of Cerro de San Pedro

Introduction

· ·

This book is an effort to understand the contemporary destruction of Cerro de San Pedro in light of its past. It treats two distinct but inseparable subjects. The first is the long history of gold and silver mining at Cerro de San Pedro, from the first bonanza, began in 1592 under the Spanish, to the last pass of open-pit mining organized by a transnational mining corporation. The second is the long history of what has come to be known as extractivism, an assemblage of material, political, and ideological operations that characterize capitalist forms of resource extraction. The broad history of extractivism and the local history of Cerro de San Pedro support one another. The former provides context. The latter shows how the contingencies of place modulated the logics of extractivism in the domain of precious metal mining.

The long view brings to light a pattern of recurring cycles of boom and bust. It's not what one expects from the extraction of a nonrenewable resource like a metal deposit. Boom and then bust, yes, but not a repetition of that cycle. Yet this is what the more than four-hundred-year history of Cerro de San Pedro shows. The main concerns of this book are all articulated around this basic observation. They seek to understand how serial extraction was orchestrated; with what ecological and social consequences; and with what bearing on scenarios of extractivist governance and socioecological struggle.

In brief, the book argues that the repeating cycles of extraction witnessed at Cerro de San Pedro were produced by capitalist logics operating in the field of resource extraction. Of particular interest is how capital was able to transmute busts back into boom. This happened twice in the history of Cerro de San Pedro: first in the late nineteenth century, when American capital transformed what was then a quasi-derelict hamlet of miners and peasants into Mexico's first industrial mining and smelting complex; and then again, after a fifty-year period of quietude, with the Canadian open-pit project. From a material perspective, resolving states of resource exhaustion, of turning bust into boom, was accomplished by increasing the flows of energy and increasing the mass of earth being extracted and processed.

Raising throughput compensated for the progressive decline in ore grades and other challenges associated with extraction and processing. Speeding up, scaling up, intensifying: each revival of capitalist mining at Cerro de San Pedro enabled profits to be harvested, once more, from a nonrenewable resource.

The consequences of doing so were translated directly to the ecologies and human bodies enmeshed into the work of mining. Forests, watercourses, fuel, and mineral deposits were tapped to feed the extractive metabolism. The lands, waters, and airs of the surrounding region of San Luis Potosí were forced to receive the ever-increasing mass of wastes voided by the smelters. Similar relations imposed themselves on those who worked and lived in the mining districts. They expended their vitality, health, and bodies in the labors of extraction. They absorbed its noxious releases. For the majority of the population, the cyclical efforts to revive the mines of Cerro de San Pedro only served to deepen mining's tolls on nature and society.

The social and ecological inequities created by the cycles of extraction at Cerro de San Pedro marked the political history of the district. Over four hundred years, workers and residents struggled to protect their lives and lands from the exactions forced by high-intensity mining. From the earliest days of colonial mining they appropriated the best ores of the mines as their fair compensation. They rebelled against the Bourbon monarchy and declared a republic. Their struggle for healthier working and living environments led into and past the Mexican Revolution of 1910–1917. They formed one of the first contemporary anti-extractivist movements in Latin America. Throughout, one discerns a moral ecology at work—an insistence on the equitable arrangement of social and ecological relations.

The history of mining at Cerro de San Pedro is thus a cyclical history—three booms; three deaths—playing itself out in the field of relations—material, ecological, social, political—created by capitalist forms of extraction. It is the history of a small place, but it is one linked to a history of much larger, indeed continental, scales. In Mexico and Latin America, the mining of gold and silver is both widely distributed and of long duration. Like Cerro de San Pedro, mining has developed over a series of cycles, though, unlike Cerro de San Pedro, there are districts that have passed through two, four, or even five cycles of activity and decline. Beyond these variations, it is the widely shared pattern of cycling that is of historical interest. The broader perspective shows that, since the period of the Spanish Conquest, the logics of commodification, material intensification, and exhaustion documented for Cerro de San Pedro were at play across other gold and silver mining districts. So too were the patterns of socio-ecological harm and political contestation. These are all defining elements of the cycles of extractivism that marked the tempo of the long history of precious metal mining across the continent. In this sense, the history of Cerro de San Pedro provides a closer view of a larger phenomenon.

Extractivism is the term used to define high-intensity and capital-driven forms of resource extraction. It also refers to the political dispositions that have structured regimes of resource governance in Latin America since the conquest. This, too, has a cyclical history. First established under the Catholic monarchs and the Spanish Habsburgs extractivism, as a political dispensation, has been periodically reiterated and updated under successive regimes. It made the realization of wealth from nature a pillar of the political economy, and the increase of commodity profits and rents a guiding aim of statecraft. Through regulation and administrative action, extractivist governance supported the high-intensity forms of capitalist extraction witnessed in the case of Cerro de San Pedro, and present across the mining belts of the continent. Elaborated within elite spheres of discourse and state authority, this regime weighed heavily on the conduct of gold and silver mining on the ground. In this sense, the history of Cerro de San Pedro cannot be fully understood without reference to the broader cycles of extractivist governance.

· · · · · ·

These are, in condensed form, the main concerns and arguments of this book. Woven through these are four themes that deserve further elaboration to clarify their meaning, their links with the relevant research literature, and their importance to the long-term histories of Cerro de San Pedro and extractivism. These are the cyclical dynamics of capital in the deep history of mining; the concept of social metabolism as a heuristic that links the work of extraction and refining to the social and ecological consequences it produced; the concept of a moral ecology as a framework to understand popular struggles for environmental justice in the extractive zones; and the history of extractivism as a foundational and persistent regime of resource governance.

Cycles of Extraction

The aim of the longue-durée approach is to discern the patterns of change that run past the conventional divisions of history. By following the entire course of the more than four-hundred-year history of mining at Cerro de San Pedro, what appears most prominently into view is the pattern of cyclical repetition. This was how locals saw their own history. In 2006 Marcos Rangel, the last in a long lineage of local miners, believed that the contemporary open-pit project orchestrated by the Canadian company was the return of colonial scenarios. The project, he felt, "is a pillage of our homeland. Why? Because all these riches that are being taken out, they're leaving the country wholesale. And well, this goes back a long-ways. It roots back to the Virreinato, don't you see? And now, now it resurfaces again."[1] One of his neighbors, Aristeo Gutierrez, was more sympathetic to the new project, but he too placed it within a history of repeating stages. "Mining has come here in stages," he explained, "First the Spanish, then the Mexicans, then the Americans, then the Mexican company Peñoles, and now the Canadians."[2]

The periodic return of mining structured local memories of place. But they ran against the more truncated boom-to-bust narrative that commonly frames the history of mining. The classic account is given in Homer Aschmann's essay, "The Natural History of the Mine." He describes the standard life history of a mine as it moved across a set of stages: (a) prospection and discovery, (b) rapid development, (c) stable operations, (d) decline and abandonment.[3] Generations of historians, working in multiple languages, have produced a remarkably well-furnished literature on mining in Latin America. They have documented its material organization, the particular knowledges and techniques that it entailed, its place in local, regional, and global economies, the internal relations of work, class, and gender, and, not least, the political struggles that have defined the histories of different mining districts.[4] Few studies, however, cut across the standard period divisions of professional historiography.[5] One important exception is the doctoral research undertaken by Letizia Silva Ontiveros that considers the landscape impacts of contemporary megamining in light of the colonial experience.[6] In general, however, chronological segmentation has kept most historical treatments of mining within the bounds of a single cycle as described by Aschmann. We are missing a view that carries us past the arc of boom and bust.

What the long history of mining at Cerro de San Pedro shows, and what the likes of Marcos Rangel and Artisteo Gutierrez knew as a matter of local knowledge, was that a mining bust was not a definitive sentence. The place had known abandonment, true, but also revival. Cerro de San Pedro passed through three cycles of mining bonanzas and *borrascas* (busts). The first bonanza was ushered in as part of the Spanish occupation of the region in March 1592. It lasted until the 1650s, briefly becoming the Habsburg monarchy's most important gold mine before lapsing into a long-lasting period of quasi-abandonment. Gold and silver mining at Cerro de San Pedro would boom again under an American-owned industrial operation in the 1890s, only to collapse, again, in the late 1940s. The open-pit project developed by the Canadians was thus the third bonanza, but also, now that work has again ceased, the third death of Cerro de San Pedro.

This pattern of serial exhaustion and renewal is not unique to the mines of Cerro de San Pedro. In 2011, Mexico's geological survey listed 718 active projects. All the gold and silver projects on this list were tapping deposits that had been mined before. They were, in the parlance of the industry, brown-field mines (in contrast to first-strike "green-field" mining). Tracking the same phenomenon in Canada's north, historical geographers Arn Keeling and John Sandlos call them zombie mines.[7] Given the centuries-deep, even millennial, history of gold and silver mining in many regions of Latin America, the revival of spent deposits appears as a continental pattern.

The recurring cycles of boom and bust at Cerro de San Pedro are the historical signature of capitalist forms of resource extraction. The specific case here is a gold and silver mine in Mexico, but it features aspects that can be found in other mines and in other areas of resource extraction. Capitalism in mining brings characteristic elements into the extraction and processing of ore-bound metals. The first is the commodification of the metal itself which, under other economies, does not carry monetary value.[8] Under capitalism it does. Each twenty-seven grams of minted Mexican silver was worth a Spanish peso; each 3.33 grams of gold, an escudo.[9] The Spanish *peso de a ocho reales*, the famous piece of eight, was a currency recognized across the world from the sixteen-century on. The precocious globalization of the bullion trade has been well-charted out by the likes of Dennis Flynn and Arturo Giráldez, and it has long been a central element in histories of capitalism from Marx to André Gunder Frank, Immanuel Wallerstein, John Tutino, and Jason W. Moore.[10]

At sites of extraction like Cerro de San Pedro, the global commodification of gold and silver supported a market for local mine and smelter-operators, and underwrote the investment of capital into the work of extraction and refining. These investments, tendered as credit, paid for the infrastructure, equipment, labor, and consumables (energy, food, reagents) that enabled the expansion of mining output. Mining capital concentrated in a few hands. Spanish-era bonanza at Cerro de San Pedro was dominated by a small handful of individuals, less than 1 percent of the over two hundred operators at work on the mountain. The distinction between elite mining capitalists and the rest is measurable in the vastly greater amounts of the gold and silver they rendered; of the ores they processed; of the numbers of workers they managed; of the amount of fuel and goods they consumed; and of the amount of capital they drew on to operate at such scales. The late nineteenth-century return to bonanza at Cerro de San Pedro was similarly a product of capital. The industrial scales of its operation, concentrated in the hands of a single company, were only made possible by the massive flow of investments from financiers from the US East Coast. The contemporary open-pit project, operating at mountain-removing scales, was funded from the Toronto Stock exchange, today a central hub for global investments in mining. In each boom, the same basic rationales applied. Since profits accrued with each unit of metals sold, the more mined, the better. To mine more meant increasing extraction and processing rates. To increase output, operations had to be scaled up. To scale up required capital.

The catch was that the same logics that produced the booms also produced the busts. The common view of a mining bust is that it is caused by depletion of the metals in a given deposit. This is a sensible idea since metal stocks are, effectively, nonrenewable, and thus limited. The history Cerro de San Pedro shows, however, that earlier abandonments were not solely, or even primarily, caused by physical depletion *in strictu sensu*. There remained a great deal of silver and gold to be mined after the Spanish cycle fell into decline, and again after the departure of the US company ASARCO in 1948. And it seems that even now, after the mass removal of the better part of the mountain, there remains more to be had.

The matter of exhaustion is thus more complex than measuring the quantities of metal in the ground. It is true that as each cycle mined the deposit, the average ratio of metal to ores declined. During the Spanish bonanza miners worked ores with an estimated average grade measured in percentages. Today open-pit mines process material whose gold content is measured in parts per million, or less. The challenges of mining ever more deeply into the earth also grew. Likewise, the work of refining ores whose chemistry became increasingly intractable. However, what drove the busts was the response of capitalist miners to the rising costs of meeting these challenges. As per-unit extraction and processing costs rose, the margins of profit narrowed. Eventually there was no reason to continue investing. Labor forces dropped. Activity slowed. Mines were closed. Smelters ceased operating.

Since large-scale capitalist operators accounted for the greatest part of the output from a given deposit, their retreat produced a crash in production figures.

Resolving the bust, and thereby bringing the mines of Cerro de San Pedro back into activity, was also the work of capital. It was possible to make up for dropping ore-grades and maintain profits by increasing the rate at which ores are processed. In the Spanish period, on average, Mexican mine operators had to extract 144 kilograms of ore for every kilogram of silver recovered. During the industrial cycle ten times that amount had to be processed to render the same kilogram of silver. A contemporary open-pit mine needs to churn through close to six thousand kilograms of ore per kilogram of silver produced. To produce a kilogram of gold—the principal target metal for two-thirds of these open-pit projects—requires the excavation and processing of five hundred thousand kilograms of ore. This is the historical confirmation of Lasky's Law whereby the processing rate of ores "increases geometrically as grade decreases arithmetically."[11] By jacking up processing rates, each cycle of capitalist revival was able to be more productive than the last. Mexico's maximum yearly silver output went from 4,880 tons under the Spanish, to 19,880 tons under the industrial system, to over 53,640 tons under the contemporary system of mass open-pit mining. For gold, the series has run as follows: 24 tons (colonial period), 134 tons (industrial period), and an astounding 1,102 tons under the current dispensation.[12] The trade-off has been duration, with each cycle lasting progressively less time than its predecessors (see figure 1).

When revival came to Cerro de San Pedro in the 1890s, it was thanks to the combination of the new energetics enabled by fossil fuels and the new techniques of industrial hard rock mining. These greatly increased extraction and processing rates that, in turn, enabled profits to be reaped from ores located at unprecedented depths, of low metal-to-gangue ratios, and of complex chemical composition. After some fifty years, however, the industrial mining regime met the limits of profitability and ASARCO abandoned Cerro de San Pedro once again. The Canadian revival of the last decades simply updated the basic formula. The exhaustions of the past could be transcended, and profitability thus recovered, by further accelerating the rate of ore extraction and processing. This was accomplished through the wholesale excavation of the remaining deposit through mass explosives, bulk haulage, and open-air cyanide ore leaching.

Thus the repeating booms and busts of large-scale mining at Cerro de San Pedro were not exclusively set by geology and nature. Exhaustion and

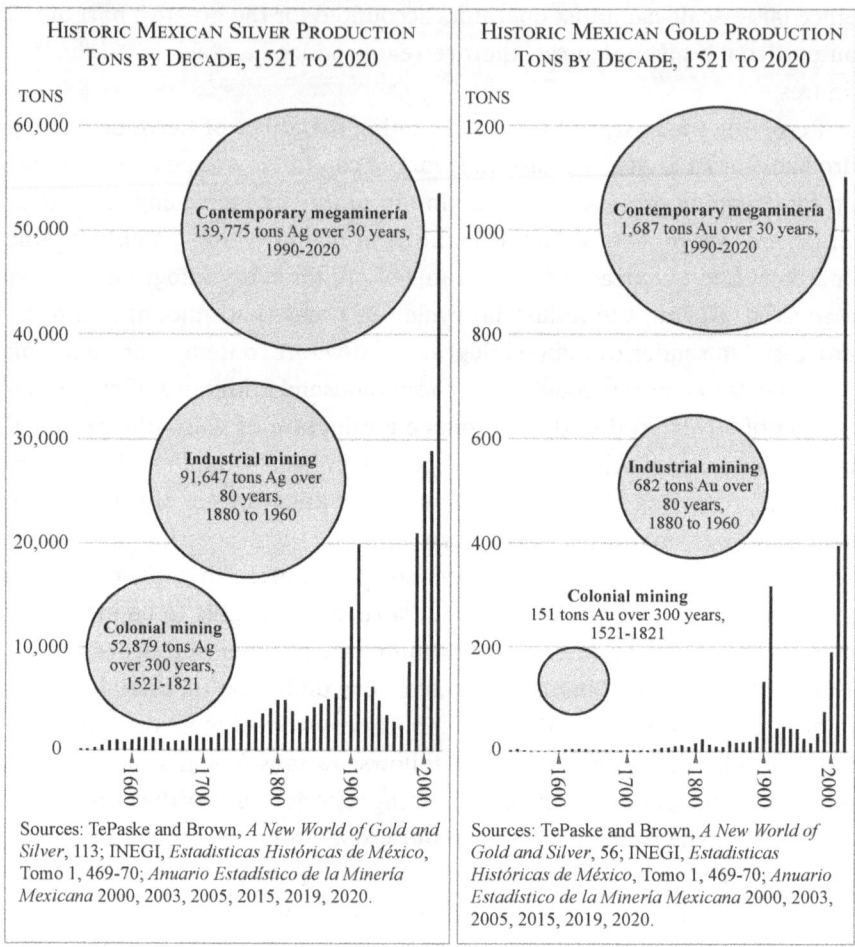

HISTORIC MEXICAN SILVER PRODUCTION
TONS BY DECADE, 1521 TO 2020

TONS

60,000

50,000

Contemporary megaminería
139,775 tons Ag over 30 years,
1990-2020

40,000

30,000

Industrial mining
91,647 tons Ag over
80 years,
1880 to 1960

20,000

Colonial mining
52,879 tons Ag
over 300 years,
1521-1821

10,000

0

1600   1700   1800   1900   2000

Sources: TePaske and Brown, *A New World of Gold and Silver*, 113; INEGI, *Estadísticas Históricas de México*, Tomo 1, 469-70; *Anuario Estadístico de la Minería Mexicana* 2000, 2003, 2005, 2015, 2019, 2020.

HISTORIC MEXICAN GOLD PRODUCTION
TONS BY DECADE, 1521 TO 2020

TONS

1200

1000

Contemporary megaminería
1,687 tons Au over 30 years,
1990-2020

800

600

Industrial mining
682 tons Au over
80 years,
1880 to 1960

Colonial mining
151 tons Au over 300 years,
1521-1821

200

0

1600   1700   1800   1900   2000

Sources: TePaske and Brown, *A New World of Gold and Silver*, 56; INEGI, *Estadísticas Históricas de México*, Tomo 1, 469-70; *Anuario Estadístico de la Minería Mexicana* 2000, 2003, 2005, 2015, 2019, 2020.

FIGURE 1 Historic bullion production in Mexico, 1521–2020.

abundance, as critical geographers David Harvey and Jason W. Moore observe, are contingent states, determined by what is possible within a given system of extraction. The historical contingency of exhaustion has also been treated by historians such as Fredrik Jonsson, Chandra Mujerji, Paul Warde, and John Wing.[13] Their research shows how the abundance, or scarcity, of timber, coal, and soil fertility was discursively construed, and how such constructions reflected the shifting play of political and economic relations over time. The long history of Cerro de San Pedro shows the play of capitalist logics in shaping the cycles of mining. It shows the tempo of extractivism.

Extractivist mining at Cerro de San Pedro appears as a series of pulses, driven by the rationales of profitability and material increase, each more productive than the last, each shorter than the last, all set against the progressive emptying of the deposit. That this was a characteristically capitalist pattern is clarified by the contrasting experience of small-scale miners. These miners have always been present in the history of Cerro de San Pedro. They arrived by the hundreds during the Spanish boom. The majority ran family-scaled operations, smelting small amounts of carefully selected ores in backyard furnaces. Unlike the larger capitalist operations, they were embedded in local, rather than global, circuits of exchange. In the wake of capitalist abandonment, they continued, well-stitched into the local subsistence economy. They were known as gambusinos and buscadores. They operated continuously for four hundred years until the beginning of the open-pit project finally evacuated them from the deposit. Against the sharp and intermittent cycles of extractivist mining, theirs was the long wave.

Social Metabolisms

As the rate and scale of precious metal mining increased with each cycle of extractivist mining, its demands on local environments and people rose proportionally. The social metabolism concept helps clarify these links. The concept entered into social theory in the nineteenth century with Justus Von Liebig's and Karl Marx's critiques of how trade and capital drove the exhaustion of land and fertility.[14] It has, since the 1990s, returned as a key heuristic in the work of Juan Martínez Alier and the Barcelona school of ecological economics and is now adopted in the fields of political ecology, and environmental history.[15] Analyzing an activity like mining as a metabolism widens the view from a narrowly economistic consideration of investments, costs, and profits. It takes in the flows of energy and matter in to, and out from, the work of extraction. It follows these flows as they drew from, and voided into, adjoining ecologies and bodies. It places importance on the intensity at which a given metabolism operates, that is, the rate and volume with which energy and matter flow through a given productive or extractive operation. All these aspects are important for historical understanding. They allow us to see the cyclical intensification of mining more clearly—as measured by accelerating rates of consumption and release— and from there assess its environmental and social repercussions.

The defining ecological pattern of the extractivist metabolism is that it simultaneously exhausts and wastes the natural and human bodies with which it interfaces.[16] The historical evidence shows that the intensity at which it operated exceeded the rates of replenishment and safe absorption for existing forms of life. In the preindustrial colonial period (1590s–1880s), the mines and smelters at Cerro de San Pedro relied on human and animal labor for mechanical power. In turn, these "human engines" (as per Fernand Braudel) depended on the cultivation of plants and the rearing of animals to keep them fed, and, in the case of the people, clothed and otherwise provisioned.[17] The metabolic needs of labor explain how the opening of mines at Cerro de San Pedro in the sixteenth century created a new agrarian landscape around it. At more individual, embodied scales, the exertions demanded of mine and smelter labor were such that workers' bodies were not regularly or fully replenished. The sources describe how mine and smelter workers were consumed by their labors, the emaciation of their bodies, their physical deterioration. The extractivist metabolism also had a voracious appetite for heat energy. Critical to achieving the high-temperature chemistry of refining, heat was produced through the combustion of charcoal derived, in turn, from existing forest stands. The take-off in smelting drove deforestation, mass soil erosion, and landscape change across the region. As for the outflows produced by the mines and smelters, these consisted of the release of heavy metals—primarily lead and mercury—through emissions, effluvia, and dust transport. They produced their own kind of wasting, undermining the healthy physiological and ecological function of local peoples, waters, and lands. The different elements that composed the new mining ecology at Cerro de San Pedro, with its flows, exhaustions, and wastings, are all amply documented across the mining belts of Spanish and Portuguese America. That is to say, the mines at Cerro de San Pedro presented the local variant of a general pattern.[18]

When the American companies arrived to revive the abandoned mines of Cerro de San Pedro in the nineteenth century, they created an industrial mining and smelting complex that greatly sped up the extractive metabolism. The stepwise increase in energy and material flows shifted the spatial configuration, scale, and form of social and ecological exhaustion. Energy consumption, for instance, exceeded what local bodies—human, animal, trees, or waters—could produce and so provisioning depended on interregional flows of train-borne fossil fuels. Whereas under the Spanish bonanza crews of miners drew out subterranean material by the hundredweight, now, with fuel-driven machinery, the rate of extraction from Cerro de San Pedro rose to the hundreds of tons. Moreover, the consolidated smelterworks established by the US company, received and processed ores from a national network of mines. Every day the smelter's banks of furnaces processed over a thousand tons of ores. And since only a tiny fraction of this material was recovered as metal, it resulted in a daily voiding of around one thousand tons of mining waste. Drawn from the underground, highly charged with toxic heavy metals such as lead and arsenic, this material was dispersed across the surface environment. It was dumped in massive piles of tailings, volatilized as smokestack emissions, and mixed with local waters. New atmospheres were created inside the working environments of the industrial mines and smelters. They were exceedingly hot, suffused with gases, and densely fogged by the dust kicked up by mills and the new pneumatic drills. New exactions on workers' bodies were added to the old. Atop the persistence of over-exertion, heat prostration, exposure to mine gases, workers now had to contend with chronic silicosis, diesel and furnace emissions, and the injurious touch of heavy machinery.[19] Above ground, families composed with water supplies that were either scarce or contaminated. A constant, barely perceptible mist of heavy metal dust fell on their barrios and coated the surface of life. Their bodies were slowly laced with heavy metals. Rivers and streams were sterilized. Mining towns, rapidly assembled to house the influx of workers and poorly provisioned with water or waste disposal, provided the optimum ecological conditions for the mass propagation of parasites, bacteria, and viruses.

The contemporary revival of mining at Cerro de San Pedro in the twenty-first century sped up the extractive metabolism yet again, by an order of magnitude, yet again. It represented the last effort to wring out the residual stocks of commodity value from the earth. The gargantuan dimensions of the wasting and exhaustion it has produced are still being absorbed. To profitably mine low-grade ores, the open-pit project operated by the Canadian corporation had to process tens of thousands of tons of material per day. This was simply not possible within an underground operation, even a highly industrialized one. Thus, the hyperkinetic metabolism of the open-pit mine operated at landscape scales. The better part of two mountains has been removed wholesale, as well as a further mountain's-worth of material from a pit that now measures hundreds of meters in depth and width. A large portion of this material, the so-called overburden, was dumped immediately beside the pit-works. It has come to fill entire drainages with a raw and highly acidic jumble of boulders. The ore-bearing material, was hauled a few kilometers away where it was doused with a solution of water and cyanide to leach out the gold and silver. The immediate ecological impacts of the open-pit operation have been to make entire swaths of the local landscape completely uninhabitable by any lifeform other than bacteria. Its daily consumption of water from the region's diminishing aquifer was comparable to that of a large neighborhood. The levels of energy consumption and greenhouse gas emissions were commensurate to the scale of operations. The daily blasting filled the valley of Cerro de San Pedro with a thick yellow smoke of nitrogen compounds and dust. Today these operations have ceased, and the extractive metabolism has quieted once more. Now, and for the foreseeable future, locals will have to compose with the long-term effects of all the toxic material the company has left behind. It is chemically active and acidifying. With every passing year, decade, century, heavy metal compounds will slowly, imperceptibly, percolate down into the aquifer beneath the city of San Luis Potosí, and blow across the surrounding area in fines and dusts. The pace and duration of these effects invert those of the extractive metabolism. These are chronic processes that will unfold according to the quasi-geological rhythms. A bit more than ten years of operation have reaped thousands of years of consequences.

A Moral Ecology

The accelerated metabolism of extractivist mining led directly to over-drafts and over-loads. These tolls marked the lands and waters of Cerro de San Pedro and the adjoining valley of San Luis Potosí. And they marked the bodies of those who worked in and lived around the extractive complex. These harms expressed the "slow violence" identified by Rob Nixon. Chronic, invisible, embodied, they were present nonetheless. These were the social and ecological costs of restoring profitability and capital accumulation on a recurring basis. The Mexican political ecologist Claudio Garibay has characterized this state of affairs as the political ecology of negative reciprocity.[20] The benefits were drawn out. The harms were laced in.

Over the centuries, the people and the lands of Cerro de San Pedro and San Luis Potosí have been made to bear the costs of a high-intensity mode of extraction. These were the consequences of the unequal relations—economic, material, social, and ecological—that operated at its core. But endurance was only half the story. There was also a response: a centuries-long struggle to right these inequities. Workers and residents of the mines and smelters, and, in time, people from across San Luis Potosí and beyond, positioned themselves against the negative reciprocities of extractivism. Their struggle matches those across the mining districts of Mexico and Latin America.

In act and expression, such struggles enacted a frame of justice in the realm of socio-ecological relations that historian Thomas Klubock has described as a moral ecology. In Klubock's study the concept describes how the Mapuche of southern Chile articulated claims to social justice, decolonization, and ecological vitality in the face of the destruction of their forest commons.[21] The concept is a reworking of E. P. Thompson's classic formulation of a moral economy, that customary frame of justice that animated claims to subsistence, most notably to bread and grains, made by English commoners. Thompson argued that it was not privation, starvation, or dearth that triggered riot and revolt, but rather the outrage at the elite abandonment of their responsibilities toward popular subsistence.[22] The moral ecology concept, which expands the terms set by Thompson to include the web of relations that link people to their environment, is likewise useful in understanding popular resistance not as a spasmodic reaction to harm, but as the positive defense of human-nature relations that are more amenable to the reproduction of health and life and more equitable in the distribution of benefits and sacrifices. It is the ideological frame that can be discerned in the early period of Spanish mining and that continues to animate what are currently known as socio-ecological or environmental justice movements.[23]

Given the colonial context of frontier war and territorial appropriation that surrounded the beginnings of mining at Cerro de San Pedro, the first line of resistance to early extractivism was the Guachichil defence of their autonomy and sovereignty. They, along with other nations of the wider Chichimecan confederacy, stalled the advance of the Spanish mining frontier for decades. It took the internal break-up of Guachichil resistance, through tactics of gifting and alliance-making, to allow colonial mining to begin in 1592. When it did, the field of contestation moved deep inside the extractive complex. Taking advantage of their effective control over the subterranean spaces of the mines, workers laid claim to the richest and most easily refined ores they found. These were reserved as the *partido* or the miners' share. The remainder of the ores were delivered to the mine owners. This appropriation was strictly speaking illegal but it was ubiquitous, not only at Cerro de San Pedro but throughout the mining districts of Spanish America. The practice was also enduring, lasting well into the twentieth century. Across these different scenarios this customary claim has been justified as the fair compensation for the consumption and harms endured from mine work. That is, it declared in practice a relation of positive reciprocity between miners and mine, each one giving to the other. This is the frame famously expressed by twentieth-century miners of Bolivia: we eat the mines and the mines eat us.[24] At Cerro de San Pedro, the popular appropriation of the mountain's mineral wealth eventually led to a regional insurrection against the Bourbon monarchy when, in 1767, miners seized the mines and helped constitute an autonomous republic in the city of San Luis Potosí. This was but one of dozens of miners' insurrections in defense of the partido that erupted across the mining districts of Brazil and Spanish America in the eighteenth century.

Industrialization redefined the moral ecologies of the districts. The mobilization of workers and their families were directed at the living and working environments of the mines, smelters, and barrios. They contested the burns they endured while washing in the local arroyo. They wanted floors for their houses, and they wanted real latrines. They created the first unions: to push for their share of the benefits generated by the mines, but also, crucially, to lower the bodily tolls they exacted: the silicosis, the lack of fresh air, the heat, the injuries from rock and machine, the repeated exhaustion of shift work. This was not a revolutionary politics. It was a moral ecology that sought to achieve a greater measure of social and ecological justice within the extractive complex.[25] And it had political potency, fuelling the famous militancy of miners' unions across Mexico and elsewhere in Latin America.

The last reconfiguration of mining at Cerro de San Pedro has once more forced a shift in moral ecology. For all its immensities, open-pit mining has only need of small crew of workers to run what are highly mechanized operations. They live in trailers and camp-dorms, separated from their surroundings by chain link fences. The contemporary resistance to extractivism comes from the other side of those fences. In Cerro de San Pedro, it was composed by the children and grandchildren of the miners who had struggled against ASARCO. In the mid-1990s, when the project was first proposed and opposed, they were *ejidatarios*, the collective owners of a regional commons secured in the wake of the Mexican Revolution. Prior to the arrival of the open-pit mine, they had built a quiet life of herding, honey production, and small-scale mining. For the mine to move forward, all of this had to be destroyed and this is why they did what no one had ever done in the history of mining in Mexico: they tried to stop a mine. At first this was a simple matter of territorial self-preservation. In time, as the deeper ecological consequences entailed by such a mine became apparent, the fight to defend Cerro de San Pedro broadened into a political movement. It became a most minimally coordinated coalition of ejidatarios, former miners, urban environmentalists, church groups, intellectuals, students, and anarchists. As it grew, this movement developed the idiom of environmental justice. They denounced the mine as a form of ecocide. In terms of their moral ecology, the mine was unacceptable because it would permanently leach toxic waste into the waters needed by over a million people so that a transnational company could take the last few billions that remained in the Cerro de San Pedro.

When it began in the mid-1990s, the struggle against the renewal of extractivist mining at Cerro de San Pedro, was one of a handful of emergent environmental justice movements in Latin America. A generation later, these are well-installed and widespread. Regional and sector-specific observatories, as well as collective cartography projects such as the EJOLT Atlas, provide an overview of the extent and scale of the phenomenon.[26] Environmental justice conflicts, also known as socio-ecological conflicts, are now counted in the hundreds. They are documented across all the different kinds of extractivist projects—dams, agro-industry, fisheries, forestry, nonrenewables—though movements against large-scale mining are, by number and geographic extent, the most prevalent. The majority are rooted in rural or peri-urban settings where they build on long-standing struggles for collective title in land and forests. They originate in the local response of small-scale *campesino* farmers, Afro-descended communities, and indigenous peoples. Many have developed broader coalitions with urban organizations, unions, environmentalists, and have scaled up into regional, national, and transnational networks.[27] The object and idiom of the struggle over the land have deepened.[28] It is now rooted in the ongoingness of life. Materially, this means assuring continued access to land and water and resisting their degradation, contamination, or wasting.[29] More politically, claims are made to sovereignty over local territories and natural resources, to collective property in "natural commons," and to justice in the realm of socioecological relations.[30] They express the political aspirations of what Ramachandra Guha and Joan Martínez Alier have characterized as the environmentalism of the poor.[31]

Running through these movements is the discursive frame of life and death. The figure of water, as a life-giving substance, is constantly mobilized.[32] So too is the figure of the *proyecto de muerte* ("the death project") when referring to a dam, a mine, or a mono-crop plantation. After fifteen years of blockades and other acts of civil resistance against the La Alumbrera open-pit mine in Argentina, the local opposition declared its intent: "we stand vigil day and night for our ultimate aim, the flows of rivers, fertility, for the land, for life itself."[33] Life versus death. Water versus gold. These starkly drawn oppositions can be quickly dismissed as hyperbole, but a closer view of the wasting effects of the extractivist metabolism suggests that they express a deeper truth.

Extractivism, Historicized

The struggle for healthy bodies and territories, with its collective actions and moral ecologies, has always faced a political regime dedicated to maximizing the commodification of nature. First established in the first years of the Española gold rush (1494–1530s), this mode of resource governance is now known as extractivism. The term is a neologism, first articulated in debates over mining in Ecuador's 2008 constituent assembly, and then described in Eduardo Gudynas's 2010 essay "Ten urgent theses about extractivism."[34] It has since been taken up, elaborated, and refined by a growing coterie of writers from Latin America (the Argentine sociologist Maristella Svampa, the Ecuadorian economist Alberto Acosta, the Chilean historian Mauricio Folchi, and the Mexican political ecologist Gian Carlo Delgado R.) and beyond (Anthony Bebbington, Gavin Bridge, Alain Deneault, Donald Kingsbury, Naomi Klein, Liisa North, Thea Riofrancos, Henry Veltmeyer, Jingzhong Ye, Anna Zalik). Taken together, their work shows the multiple and mutually supporting elements that compose extractivism.

Gudynas's essay was a critique of the resource regimes put in place by leftist governments in Argentina, Ecuador, Bolivia, Venezuela, and Chile, among others. It showed that the leftist governments of the early 2000s were no less committed to extraction than their neoliberal predecessors. The forms and ways varied but the key elements remained: state property in nature, the fiscal importance of realizing rents from nature, the political motivation to maximize extraction, and the functional position of Latin America as a resource continent in the global capitalist economy.[35] Acosta, Gudynas, Svampa, Bebbington and Bury all signaled the historical depth of this regime, rooted it in the Iberian conquest and persisting, across different iterations, to the present.[36] Fernando Coronil, writing on the contemporary boom in resources (reprimarization), saw it as a "regression to older forms of colonial control" even as new assemblages of technology and capital were bearing down on Latin American Nature.[37]

The political regime of extraction was assembled far from places like Cerro de San Pedro but it had a heavy influence on how extractivism worked itself out on the ground. From the Spanish appropriation of Cerro de San Pedro to the present, the state has organized the property relations between mine operators; it has liberally granted them concessions to timber, land, and waters; it has granted fiscal exemptions and direct subsidies to operations. When the opposition to extractivist forms of mining grew dangerous, it provided coercive force to subdue protest. All of this in order to increase the outflow of gold and silver and the inflow of revenues.

Like the local history of Cerro de San Pedro, the broader history of extractivism is one that developed in cycles. The central legal and institutional opinions of extractivism were innovated in the exceptional circumstances produced by the Spanish conquest. It began with the wholesale appropriation of Latin American nature into the royal estate. This provided the base for an administrative regime that ordered private use rights to extract this wealth and received the royalties produced from its extraction. Rents from colonial nature, most especially from silver and gold, helped make the Spanish monarchy one of the wealthiest states of the early modern Atlantic world.[38] When this flow of mineral wealth was slowed or put at risk, the Crown intervened. It reorganized subject populations to assure much-needed labor supply. It legislated on the conservation of fuel-wood. It subsidized the construction of needed infrastructure and the supply of reagents, most especially mercury. Exemptions for tribute and taxes and other protections were all extended to support mine owners. Colonial extractivism also mobilized an intellectual defense of the regime. In the face of critique by the likes of Bartolomé de Las Casas and other Dominican moralists, there emerged a doctrine of providential mission in which the exceptional natural endowment of the Americas was described as God's reward for Spanish service to the faith.[39]

The political regime of extractivism would evolve over time. It would be periodically updated: under the Bourbons in the eighteenth century, the Porfiato of the nineteenth, the postrevolutionary state of the twentieth, and lastly, under neoliberal reform in the 1990s. The precise shape and form of these different regimes of resource governance were adapted to their circumstances, but throughout, as per Fernando Coronil and others, the basic grammar has endured: state property in resources, a concessionary system of use-rights, natural rents through royalties, and state support of extraction through the provisioning of inputs and congenial fiscal policies.

Gudynas's article asked the question of how the progressivist governments of Latin America's new left could have maintained the basic features of the resource regime run by their neoliberal predecessors. The historical answer is that this was simply the latest reiteration of a centuries-old pattern.

Indeed, even today, under the populist governments of the left (Obrador in Mexico) and right (Bolsorano in Brazil), the state's commitment to extractivism remains entire.

A Note on Form and Approach

The histories of Cerro de San Pedro and extractivism in Mexico have been entwined for centuries. To encompass them, the following book is organized around the three major cycles of mining that have passed through the place: the colonial period under Habsburg and Bourbon rule, the industrial period of the Porfiriato and the early postrevolutionary state, and the last period opened with the neoliberal reforms of the 1990s. Each cycle forms a part. Each part leads with a synthetic overview of extractivism in Mexico and in Latin America more broadly to provide the overall context. Then follow shorter, more thematically organized, chapters that document the material, ecological, social, and political histories of mining at Cerro de San Pedro during a given period. The structure and sequence of these parts, like the cycles they describe, is reiterated. The hope is that this will help clarify the serial dynamics of mining and allow an evaluation of what changed and what was repeated across the different cycles of extractivist mining.

Researching and writing this history did not follow the normal path of a typical scholarly project. When I first arrived in San Luis Potosí in 2001, I had no intention of working on mining as such. The initial project was to have been a social history, researched in collaboration with my friend and colleague Juan Carlos Ruiz Guadalajara, of the early (mainly indigenous) settlement of the area during the first boom under the Spanish. It was only because of the advance of the Canadian company, and the serious consequences that would result from its open-pit mine, that we turned our attention to the contemporary situation and to the question of the harms—ecological, social, patrimonial—entailed by large-scale mining.

Having shelved the original project, my research on the Cerro de San Pedro followed two distinct lines. The first addressed the contemporary dimensions of this transnational mining operation. With a student-based research collective at McGill (MICLA), we researched the company behind the mine, the economic and political resources it drew on, and the broader context that surrounded it. We quickly came to realize that what was taking place in San Luis Potosí was but one of well over a thousand Canadian mining projects deployed across Latin America since the 1990s. Inventorying these, we documented over a hundred social and ecological conflicts involving a Canadian mining corporation, and came to work with the organizations and communities pushing back against this twenty-first century extractivist frontier. The research material generated during this period was primarily fed into more public venues of meetings, documentary films, Parliamentary testimony, and a (now-defunct) website. In this book, it comes to support what is basically a participant's account of the years between 2001 and 2018, a period when I traveled regularly in Mexico and worked with various organizations, Canadian and Mexican, opposed to the project. This forms the last part of the book. It is, inevitably, a partisan account. But it is a well-documented one.

The second line of research delved into the past. Having already begun to work on the colonial origins of the mines of Cerro de San Pedro, it was impossible to ignore the historical depth of the present-day story. And so, I began to reconstruct that past, through the more conventional forms of archival and documentary research. I started in the state archives of San Luis Potosí, which contain a remarkably large and well-organized repository for the Spanish period: hundreds of criminal and civil suits, official correspondence, and notarial records, all thoroughly cataloged. The holdings for the late-nineteenth and twentieth centuries are smaller and mainly composed of concession maps and titles. Other important primary sources for the project included the reports of officials and mining engineers kept at the Palacio de Minería in Mexico City and in various US collections. I was also able to draw on local newspapers and, for the period beginning in the 1940s, on a series of oral history interviews with residents of Cerro de San Pedro. To this primary source material was added the very rich historiography presented in theses produced at the Colegio de San Luis Potosí and other Mexican universities. This work was invaluable for fleshing out the full history of mining at Cerro de San Pedro. The broader national and continental contexts were built from the exceptionally deep historiography of mining in Mexico and Latin America, a line of publication that begins with Las Casas, Gonzales de Oviedo, and de Acosta in the sixteenth century and moves forward through Barba, Gamboa, Humboldt, the reports of English, French and US mining engineers to reach the modern research of Mexican, Latin American, and Anglo-American scholars.

When it first began to take shape, the book was originally conceived as a collection of chronicles, a narrative account that linked together important episodes in the history of Cerro de San Pedro—such as Caldera and Guichichils, the 1767 republic, the 1948 strike, and abandonment of

ASARCO—and then tied these to the contemporary scenario. Insofar as research was concerned, my initial aim was to gather the material necessary to provide a coherent account of each episode. It was only later in the process that the book was recast in more analytical terms and addressed itself to the central questions raised here: the cyclical reanimation of a nonrenewable resource, the role that state and capital played in this process, its social and ecological consequences. To answer these I drew on concepts developed in different fields: Lasky's Law, social metabolism, the body as a key nexus between the social and the ecological, moral ecology, negative reciprocity, and of course, extractivism (this last a neologism that had not yet been voiced when this project began). Each of these helped make analytical sense of these questions, often forcing a reworking of the existing material, or further research to fill in new gaps.

The following pages thus relate a history of Latin American extractivism as it wove itself into, and out from, the history of a small Mexican mining town. At its heart is Cerro de San Pedro, which is, simultaneously, a mountain, a mine, and a place. As a mountain, the Cerro de San Pedro is not especially big or small, nor striking in any spectacular way. So too with the mines: they have always been middling in size, conforming to the general pattern. But this may very well be why—beyond my own personal commitments—the story of this place is worth telling. It provides the story of an everymine, one that can resonate with others. I am not arguing that all the mines of Mexico and Latin America were identical nor that they all marched in lockstep through the same chapters of history. But they did share a great deal. Sufficiently so to allow the story of Cerro de San Pedro to shed light on wider patterns.

Writing of the irresistible atmosphere of Mexico's northern mining towns, the Mexican essayist Carlos Montemayor admitted that "the mining world absorbed me."[40] Likewise. It is a strangely compelling world: lithic, dark, situated at the precise antipodes of life on the surface. Working and living in and around that environment brought people to the extremes of what a body can bear. Their lives were remarkable in so many ways, for what they endured, for what they created, for what they struggled for. Through them, I hope to have relayed something of the historical ambiances of Cerro de San Pedro: mountain, mine, and place.

Part I   **Extractivism Assembled**
....

1   Extractivism Assembled (1494–1650s)

In the last months of 1550, Don Antonio de Mendoza, first viceroy of New Spain, interviewed with his replacement, Don Luis de Velasco, about the affairs of the land. Over the fifteen years of his term, Mendoza had brought Hernán Cortes back under royal authority, successfully prosecuted a frontier war against the northern peoples of the altiplano, and given legislative order to Mexico's first boom in gold and silver mining. In 1550, however, the challenges for royal administration remained. The kingdom was still reeling from the *cocoliztli* epidemic that had swept through central Mexico and emptied mining districts in the northwest. Mexico City wanted for drinking water, fodder, and firewood. The new silver district at Zacatecas was in bonanza but remained dependent on a slender and exposed provisioning road that crossed the unconquered territories of the Chichimecan confederacy. Mendoza himself was seriously ill, still battling a recurring fever contracted during the Mixtón war.[1] Taking stock of the situation, he was determined to impress upon Velasco a point of particular importance: above all else, tend to the mines. "What presently gives being to the land," Mendoza wrote, "and sustains it, are the mines . . . For if the mines fail, then all the other estates of the land will suffer a great decline . . . and His Majesty would lose almost all his rents, because the being of the land is in the mines."[2]

   "The being of the land is in the mines" was a statement of remarkable philosophical depth. It was metaphysical, indeed ontological. It suggests two possible readings. The first is a claim about the nature of nature in Mexico. Reading the land as nature, Mendoza's phrase posited that the essence of New Spain, its inner being, was embodied in its subterranean wealth. The second is to read land in the sense of country or society. In this rendering the mines have animating agency. Mines created and maintained the kind of territory that a Viceroy might imagine—one expected to contribute to the material fortunes of the Habsburg monarchy. Note the instrumentalist frame: the mines must be cared for because they gave life to the colonial economy and provided rents for the Royal fisc. The two readings are not incompatible. They show how a particular ontology was braided into the political economy of empire to declare the primacy of extraction.

Similar lines of thought about the nature and place of mining in the Spanish Habsburg monarchy can be pulled from other elite texts of the period. Take, for instance, the conclusions of the Junta Magna of 1568. This was the joint meeting of the Council of the Indies and the Council of Castile convened to prepare the designated viceroys of Peru, Francisco de Toledo, and New Spain, Martín Enríquez, for their forthcoming commissions. The written instructions that emerged from this meeting led with the mines: "1. Upon [the mines] depend our *quintos* and duties and the wealth and substance of those provinces [of the Indies]." The councils then elaborated as to the policy that this would require: "Those who own the mines and make them work must be assisted insofar as possible, and all the means and things necessary to mining made available so that it grows and increases as much as possible." "Do whatever you can, and plan, and provide, so that the said labors of the mines and smelters grow and continuously augment." This included: assuring an ample and constant supply of charcoal; the furnishing of mercury, expertise, tools, materials, provisions to the mines; and, of central importance, the "conservation" of indigenous labor and its application to mine and smelter work.[3]

In these passages, and others of the sort, was expressed an incipient form of extractivism, one articulated with the colonial mining of precious metals.[4] This chapter more fully documents the process of its emergence. It makes three basic arguments. The first is that extractivism was an innovation in political economy enabled by the exceptional circumstances of the Iberian conquest and colonization of the Americas. Gold and silver mining, as such, were not new, either in Iberia or in Europe. Nor was the realization of rents from the extraction or harvesting of natural wealth. What was novel was how the commodification of nature came to occupy a cornerstone position in the political economy of a modern state. This was a notable departure from medieval precedent, and it set the Spanish monarchy apart from its early modern European rivals whose land rents were rooted in cultivation, rather than extraction. The political, legal, and institutional developments that took place during the first century or so of Iberian colonial rule set important foundations for modern regimes of resource governance that are dominant in Latin America, and indeed the world, today.

The second argument is a more heuristic one. Extractivism is best understood as an assemblage. It was not a single thing but rather a grouping of elements in different fields of social life that mutually reinforced each other, together creating a regime that aimed at the limitless commodification of nature. The central elements that composed the extractivist assemblage were ideological and discursive, such as the cornucopian and providentialist readings of New World nature. They also involved new juridical frameworks and administrative practices regulating the extraction and fiscalization of precious metals. They entailed the organization of new material and labor relations to assure a constant increase in the rate of extraction.

The third argument concerns the important role that crisis played, as both cause and context, in the assembly of extractivism. These crises were of multiple kinds. They included the internal disorder and lack of revenues that characterized the first years of Spanish rule in Española; the material depletion of treatable ores; the ecological exhaustion of wood fuel; the social collapse of indigenous populations; the fiscal emergencies of state bankruptcy; and the political crises that flared around the legitimacy of Spanish dominion in the Americas. These crises motivated the royal decrees and administrative responses that built out an extractivist regime. Noteworthy are the material dimensions of these interventions. The royal administration of mining aimed at increasing revenues, but since these were levied pro-rata on the gold and silver excavated from the Americas, there was an important incentive to accelerate the rate of extraction. Scarcity crises were met by multiplying the number of mines through prospection and frontier wars, or by speeding up the extractive metabolism by increasing the inflows of labor, energy, and materials, notably mercury, needed to run the mines and smelters. Assuring a steady or growing supply of these inputs was, for the same reasons, a keystone of early Spanish extractivist governance. These were the root causes for laws and reforms relating to the *conservación* (conservation) of the subject populations or woodlands.

To take in the emergence of these different elements, and to see how they came to link with one another, the following pages revisit a number of the key episodes of the history of the period: the early colonization of Española, the expansion of the Spanish frontier, the rocky development of Habsburg imperial pretensions and fiscality, the mass compulsion of indigenous labor, and the extension of the global bullion trade.

· · · · · ·

When Columbus left the port of Palos, Andalusia, in March 1492 he sought a westward passage to the Indies. As he went, he oriented his ships along the lines of latitude where one would most expect to find gold and the other precious things of nature's creation. As Wey Gómez's careful and thorough reading of the Columbine corpus shows, Columbus's four navigations consistently tracked to the south and towards the equator. It was, Columbus wrote, "under that parallel of the world that more gold and things of value are found," explaining that "gold is generated . . . wherever the sun is strong."[5] From his early career, he knew just how rich the gold fields of sub-Saharan West Africa were. His reading and correspondence indicated that a similar bounty existed in India, along the same tropical band. The Catalán cosmographer, lapidarian, and partisan of Columbus Juan Ferrer de Blanes wrote that it was "at the turn of the equator [that] precious stones, gold, spices and medicinal plants are abundant . . . from what I have heard most often from many Indians, Arabs and Ethiopians is that most of the good merchandise comes from a very hot climate."[6]

In 1492, Columbus followed the sun and it brought him to the islands of the Caribbean. Since Humboldt, students of Columbus's writings have noted the emphasis he placed on the natural wealth of the archipelago.[7] It was the expression of the broader theory that related terrestrial fecundity to the agency of the sun. Coming upon the large island that he claimed and named as La Española, he described a mountainous land "covered with trees of a thousand kinds, of such great height that they seemed to reach the skies."[8] The coastal waters abounded in fish. There were thousands of kinds of birds, palms, and fruits. The variety overwhelmed Columbus. He regretted his inability to identify them all but surmised that therein were "many herbs and many trees that will be worth a great amount in Spain."[9] But there was no time to investigate. During his first parley with the islanders, October 13, 1492, Columbus's eye had caught sight of their golden adornments. They confirmed that he had indeed arrived into a zone of tropical bounty. "There may be many things of which I know nothing because I am reluctant to linger here, being anxious to explore numerous islands with a view to finding gold."[10]

The signs of gold abounded during Columbus first reconnaissance. He heard, or thought he heard, of islanders overladen with golden bracelets; of people harvesting gold at night by the flicker of torchlight; of golden kings and even the possible whereabouts of King Salomon's mines at Ophir.[11] Nor did he hesitate to seize or barter for whatever gold actually came into view. All told, it amounted to some twenty-thousand escudos, a bit more than sixty-four kilograms of gold.[12] Locating the source of the gold would fall to the captain of the Pinta, Martín Alonzo Pinzón who, after some time trading for gold at the harbor of Luperón, Española, traveled inland to the river placers of the Cibao.[13] Heading home, Columbus wrote: "This land is to be desired, and discovered, and never to be left."[14]

Columbus returned to the Caribbean a year later, now at the head of an expedition of fifteen hundred men, equipped for war, planting, and settlement. Within weeks of its arrival in Española, news came of "extensive mines of gold and other metals" in the island's mountainous interior. In January 1494 two parties under the command of Ojeda and Gorbalán were dispatched to investigate more closely.[15] Over fifty rivers and creeks were prospected. Each bore gold such that "they had only to seek throughout that province and they would find as much as they wished."[16] Columbus's report back to the Catholic Monarchs captures the grandeur of what he had reaped for them. "Surely," he wrote, "their Highnesses the King and Queen may henceforth regard themselves as the most prosperous and wealthy Sovereigns in the world; never yet, since the creation, has such a thing been seen or read of; for on the return of the ships from their next voyage, they will be able to carry back such a quantity of gold as will fill with amazement all who hear of it."[17]

Columbus's accounts mark the beginning of an Occidental tradition of representing the Americas as a territory of natural bounty. The tropes of exceptional fecundity, variety, and abundance can be found and elaborated in the work of the chroniclers and natural historians of the Iberian conquest and colonization of the continent—Peter Mayrtyr d'Anghiera, Oviedo, Velasco, Cardenas, Acosta. This discursive mode would be reproduced north of the tropical band, as French, English, and Dutch writers proved up the bounty of the septentrional zones of the continent with their seemingly limitless stocks of timber, fish, fur-bearing animals, and fertile soils.

These accounts signified the Americas as a reservoir of wealth. Their territories were defined by the superabundance of substances (minerals, salts, and oils), objects (pearls and gems), and lifeforms (trees, animals, and fish) whose extraction, circulation, and use contributed to human welfare. It exemplified what geographer Gavin Bridge terms a resource imaginary, a form of geographical representation in which "space is reinscribed in the image of the commodity."[18] This view of the world entwined itself with an instrumentalist argument of a religious and political order. The remarkable bounty of the Americas was read as a sign of Providence's designs for the Catholic monarchy. This was how Columbus framed things. In his more eschatological writings, he believed that the treasures of the Indies would help Ferdinand of Aragon reconquer Jerusalem and set in motion the calendar of the Second Coming.[19] He was not alone in reading the natural wealth of the Americas as the material base for the spiritual renewal of

European Christendom.[20] Linking the inexhaustible riches of the land to the higher purpose of the polity would prove to be an enduring frame.

· · · · · ·

On the strength of Ojeda's and Gorbalán's reports, Columbus led a large expedition from the north coast of Española up into the highland reaches of the Yaque River.[21] Relations between the expeditionary force and the islanders broke down as the assaults and sexual predations of the former forced the riposte of the latter. This provoked Columbus's invasion of the indigenous Kingdom of Macorix, the highland realm that concentrated an important part of the island's gold fields. Military occupation and the work of extraction advanced together. Indigenous warriors who were taken in combat were held as slaves and set to work in the river placers. To assure the control of the gold fields, and the secure movement of supplies, labor, and ores, Columbus had built a series of forts and roads. Around the gold placers, makeshift camps of huts, corrals, and gardens sprang up. Some had a thatched-roof church. All were characterized by an overwhelmingly male population, the surfeit of day-to-day violence (of Spaniards upon islanders, of Spaniard upon Spaniard), the regular visitations of hunger, and fatally, disease.[22] But it was the factional disputes within the conquering group that incited the Catholic Monarchs to act. They sent Francisco de Bobadilla, a court veteran, to resolve the disputes between Columbus's party and its opponents, which he did, but at the price of lifting the obligation to pay the 50 percent royalty on mined gold.

Since this was not the resolution that they hoped for, in 1501 the monarchs sent Nicolas de Ovando to bring royal order to the gold districts of Española. The instructions he received and the policies he put in place over the next years set the cornerstones of a new regime of mineral extraction. Leading these was the state appropriation of natural resources such as minerals, trees, salt, fish. The Catholic monarchs had already made moves in this direction. In their 1497 instructions to Columbus regarding the distribution of lands, the monarchs had reserved all those lands that contained gold, metals, and brazil wood trees "or any [other] thing that belongs to us."[23] Only those lands that were dedicated to planting or livestock could be granted to settlers. The instructions thus established a very early distinction between property in nature and property in land. Now under Ovando, Ferdinand and Isabel pushed their claims to encompass all the natural treasures of the Americas. At first they claimed half of all gold in the islands and mainland. Less than a year later they declared that all mines were their property. This was then projected to any as yet undiscovered mine, and then extended to all kinds of metals (silver, tin, copper, iron) and other forms of natural bounty such as brazil wood stands, pearl fisheries, and salt pans.[24]

Ferdinand and Isabela's sweeping appropriation of the natural wealth of the Americas represented an important break from the metropolitan scenario. On the Iberian Peninsula, the rights to what might be termed the goods of nature composed a densely layered mosaic of claims. These varied according to the natural good in question. Different rules pertained to woods, to waters, to fisheries, to pearls, and to minerals. These were further divided by morphology (for instance, commoners held rights to harvest "lops and tops," while landlords owned and disposed of the tree itself), or geography (gold deposits generally belonged to the landowner but those found on riverbanks, islands, or beaches were free to the first taker).[25] They were maintained by custom, enacted in ritual, and defended through social action. When not considered "those things that belong to all creatures" (Siete Partidas), the goods of nature were claimed by villages, municipalities, corporate bodies such as universities, religious and military orders, and aristocratic families. There was very little here for the monarchs as such, beyond what their familial estates might possess as property. In one rather exceptional circumstance did medieval Spanish law grant the monarch a radical right to natural treasure: when it was "found by enchantment, in that case it should belong to the King."[26]

With the novel circumstances afforded by the conquest of Española, a new class of property was created and just as soon appropriated. All manner of natural goods, anywhere within the imperial domain, were now integrated into the *Patrimonio Real* or *Realengo*.[27] It marked an important juncture in the development of the modern doctrine of state ownership of natural resources. State property in nature formed the base for the regime of colonial resource-rents. Under this system, private parties could not own gold fields, groves of brazil wood, or pearl fisheries. They obtained use rights. In the case of precious metal mining, these were conditional on the observation of royal regulations and on the payment of royalties, the famous *quinto* (20 percent) charge that mine owners paid for the privilege of extracting gold and silver.[28]

To assure the proper receipt of these royalties, Ovando was tasked with organizing a better control of the extraction process and of the gold that it produced. He called for a new regimentation of labor. Teams of indigenous workers were to be divided into *cuadrillas* (squads) of ten apiece. Each cuadrilla would be supervised by an official who watched to see that all the gold ore recovered from the rivers was dispatched to a royal smelter. There were only two of these, one in Concepción de la Vega, the other at Buenaventura. By law all ores had to be refined and stamped at these places. It was at that point that the monarchs received their royalties.[29]

The rents drawn from the first years of this colonial mining regime were considerable. Columbus's first two voyages brought back twenty and thirty thousand gold escudos, respectively. Materially speaking this amounted to about 140 kilograms. As indigenous miners were pushed into mine work, production exploded. By 1510 the conquered gold fields of Española were sending back enough bullion to triple all of Christian Europe's imports.[30] Over the three decades following Ovando's arrival, Española, and to a smaller extent Puerto Rico and Cuba, produced twenty-seven tons of the metal.[31] Twenty percent of this was destined to the royal exchequer as a new and much welcomed revenue stream.

The last piece of the new regime of gold mining established by Ovando concerned the administration of indigenous labor. Compared to the riverine harvesting that generally characterized indigenous gold mining, the colonial system of extraction entailed a much more concentrated and regimented effort.[32] The Spanish chronicler Gonzalo Fernández de Oviedo, described the operations. Individual Spanish mine owners secured, through obligation, dozens of laborers. They were formed into cuadrillas dedicated to the mass excavation and processing of alluvium in deep trenches. A front line of diggers advanced to break up the dense conglomerate of sand, mud, and cobble with hoes and picks. Then another crew shoveled the ores into woven baskets and hauled them to the nearest water flow where they were dumped in heaps. There the washers used water and craft to separate the gold from the dross. According to Oviedo, such washers were mainly women known as *lavadoras*, "because the work of washing requires more science . . . They handled its flow with certain craft and fluidity . . . swirling it in their bateas until it stole away the earth and left behind the heavier gold."[33]

Bartolomé de Las Casas, who was also present and chronicled these early years directly, described how colonial gold mining not only scaled up and rationalized the work of gold extraction, but how it also violently increased the intensity of labor. "There was a burning rush," he wrote, "to take out the gold from the mines and do all the other things that this required . . . They were put to hard and sharp labour, moved from one extreme to another, not bit by bit, but in a sudden acceleration. No surprise that they could not last very long." People were beaten severely to keep them moving. Las Casas recorded that "If they did not sweat, the [Spanish] gave them the stick, yelling, "You're not sweating, dogs? You're not sweating?"[34]

In Española indigenous workers conscripted into the mines were required to serve a five-month term known as *la demora*.[35] The remainder of the year was to be dedicated to farm-work to assure the food supply for the Spanish settlers, the workforce at the mines, and their families. The arrival of the Eurasian diseases destroyed that plan. The gold fields concentrated indigenous conscripts and Iberian overseers in camp-like conditions marked by overwork and inadequate food. They became the perfect foyers for contagion and death. Anywhere between a quarter to a third of the local conscripts did not survive to see the end of their six-month term. Concerned with maintaining production, Spanish officials responded by increasing the length of the corvée to eight months. This only made things worse, but the Spanish kept on, lengthening the demora to a full year. Native peoples continued to die. Scrambling to capture whatever was left of the fast-diminishing labor reserve, Governor Ovando transformed the demora into perpetual servitude.[36]

Eventually, the catastrophic effects of this regimen could no longer be ignored. The passing of the Laws of Burgos (1512), which is often presented as a reformist code compelled by the activism of Christian moralists, should be read in this light. Fully cognizant of the importance of indigenous labor for mining, these laws express a very early form of bio-political concern, a way of knowing and acting upon the indigenous population and its health in function of the labor service it could render.[37] The laws required that detailed records be kept of how many individuals died and how many were born, and that these should be updated every ten days. Whenever a miner brought his gold to the royal foundry to be smelted, tithed, and marked, the Spanish *visitador* would gather his records "so that he might have an entire understanding of the growth or diminution of said *Indios*."[38] Miners were responsible for providing adequate food to those in their service: bread, vegetables, and a pound of meat or fish per day. At the end of their term in the mines, workers were freed of service for forty days so that they might return home to rest and recover. As for women, the regulations sought to assure their reproductive labor. Pregnant women were exempted from mine labor and were ordered to remain home so that they might provide for their families and raise their children.[39]

As historian Demetrio Ramos Pérez noted, the Laws of Burgos represent the first body of state law governing mining in the Americas.[40] It is telling that it should devote such attention to conserving subject populations and thereby assure the labor supply needed to maintain mining production. By 1512, it was already too late. The consequences of the gold rush at Española proved fatal for local societies. Community after community had been wiped out, decimated by disease, unable to cultivate their fields, and harried by the brutality of Spanish arms. The population of the island collapsed from an estimated one million to thirty thousand survivors.[41] Witnessing the disaster firsthand, the Dominican Bartolomé de Las Casas wrote that the Spanish hunger for gold was the cause of the "killing and destroying of an infinite number of souls."[42] The gold fields of Española, however, had to be kept alive. New expeditions were thus dispatched to neighboring islands and to the mainland to tap new reserves of workers. In each place—Jamaica, Puerto Rico, the Lucayas—the terrible cycle churned its way towards the extinction of local peoples.[43] Coming into the 1530s the Caribbean mining cycle could no longer be sustained. Production dropped and, by 1550, the remittances of Caribbean gold to Spain ceased entirely.

· · · · · ·

The collapse of gold mining at Española proved to have little impact on the overall volume of precious metals flowing to Spain from its American possessions. By the 1530s and 1540s new gold and (increasingly) silver mines began to be exploited on the mainland. They were more numerous, were distributed across a much larger swath of territory, and were often larger than the mines of the island. The extension of the colonial mining frontier to the continent began even as the gold cycle of the Caribbean was in full swing. Spanish *entradas* (expeditions of conquest) fanned out from Española. The spoils of war seized in Panamá and the Gulf of Urrabá (1501–2), Tenochtitlán (1519), Esmeraldas (1526), Cajamarca (1532), and Sinú (1533) all signaled just how exceptionally, and ubiquitously, well-stocked the Americas were in precious metals.

Following the conquest of new territories and the division of war prizes, certain groups of Spaniards turned to the excavation of the graves and shrines where ancestors and other beings were gifted with gold and silver. The practice came to be known as *huaqueria*, huacas being the common name for Andean shrines and sites of veneration. It was conducted systematically from Panama down to Peru, meriting a distinct fiscal category in the account books of the Royal Treasury.[44] The effort required raised tomb-raiding to an almost industrial scale. In the 1530s, Spaniards marched hundreds of African and indigenous slaves up the Sinú river valley of coastal

Colombia to undertake the regimented pillage of a funerary complex that extended over tens of square kilometers. This indigenous necropolis was composed of many dozens of *mogotes*, or burial mounds, shaped into circles, hemispheres or large pyramids.[45] At the Huaca del Sol, a massive abode brick temple situated on the north Peruvian coast, Spanish *huaqueros* had the local river diverted and channeled into hydraulic jets that washed out the gold objects it contained.[46] A similar attempt was made at the Akapana pyramid at Tiwanaku, the great pre-Inca city immediately south of Lake Titicaca.[47] In sixteenth-century Quito, the ledgers of the Royal Treasury reveal that plunder, rather than extraction, provided the greatest share of the colony's early gold production.[48]

Pillage and huaqueria formed, by definition, a colonial *raubwirtshaft*.[49] It was an economy that plundered the reserves built by generations of indigenous miners and metal workers. The pillage economy could not be sustained, not at the scales witnessed during the first decades of mainland conquests. The long-term development of the extractive economy entailed a move towards the actual work of prospection and extraction. This happened quickly. By 1592, when mining began at Cerro de San Pedro, there were close to two hundred and forty colonial mines in operation across Spanish America. They composed a continental zone of extraction running belt-like down the cordillera from the northernmost mines of Santa Barbara, Todos Santos, and San Juan (Chihuahua) to the southernmost gold placers of the Valdivia (Chile), a close to ten-thousand-kilometer journey.[50]

The rapid extension of the early Spanish mining frontier mapped itself directly onto the preexisting geography of indigenous mining. At the time of the Spanish conquest, gold was mined throughout Mesoamerica, across the Chibcha territories of Panama, Colombia, and Ecuador, and down the Andes. Indigenous silver mining was less widespread. The best-known mines were worked in Andes: Carabaya, Chuquiabo, Porco, and general region around Potosí.[51] The discovery of the last, the greatest silver mine in the history of the world, was long attributed to Juan de Villaroel in 1545. It is now clear from the archaeology that it was well-settled and long-mined for centuries.[52] The silver mines of Mexico, however, were not mined at the time of Cortes's expedition and were, instead, opened by colonial prospectors or indigenous individuals who served them. Almost immediately after the fall of Tenochtitlán, Hernán Cortes dispatched "men trained in the arts of the metals" to see what they could find. They followed the local leads that brought them to the gold placers of Tehuantepec and also, significantly, to Taxco, the site of Mexico's first colonial silver mine. Over the next decades subsequent parties opened mines across the Nahua region of central Mexico (Sultepec, Zumpango, Tlupujalma, Pachuca) and then further north (Zacatecas, Guanajuato, Charcas, Santa Barbara, and, eventually, Cerro de San Pedro).[53]

The mainland advance of the Spanish mining frontier thus consisted of either seizing control of indigenous mining districts where feasible and then speeding up the rate of extraction, or it entailed the prospection and development of new mineral deposits. Both scenarios required the conquest of local peoples and this often provoked a stiff response. Mainland resistance began against Columbus, not during his first voyage but during his last, the fourth voyage of 1501–2 on the Atlantic coast of Panama. He was once again searching and raiding for gold, now in the Buglé and Ngäbé territories of Veragua. Here, however, a coalition of communities and kings destroyed the Spanish fortifications and sent survivors fleeing for the safety of their ships. The Buglé and Ngäbé would subsequently repel more than a dozen other expeditions over the next decades. The Shuar of the Amazonian piedmont similarly over-ran the Spanish at Macas and Logroño to set the scene, possibly apocryphal, wherein they poured molten gold down the throat of the governor "to see if for once he had enough gold."[54] The Mapuche surge in southern and central Chile resulted in the Crown's loss of many of the richest gold mines in the Americas, "even if" noted a Spanish observer, "they don't esteem gold any more than lead."[55] In Mexico it was the Caxcán communities that pushed back against the Spanish advance. Situated in the transition zone between the agrarian heartland of central Mexico and the semi-arid ranges and mountains of the north, they forcibly dislodged the first Spanish settlements in the area. This triggered the Mixtón war (1540–41), a bitterly fought two-year conflict that was only resolved with Viceroy Mendoza's deployment of tens of thousands of Nahua and Tlaxcalan combatants and heavy Spanish cavalry.[56]

On balance, however, the Spanish mining frontier advanced. By 1550 the Spanish had opened enough gold and silver mines to increase bullion output five-fold from the peak of the Española gold rush.[57] Royalties and other taxes on precious metals accounted for the great majority of the revenue provided by the Indies. And although land rents continued to provide most of the Habsburg monarchy's income, receipts from American treasure occupied a growing share. Under Charles I they reached close to a million ducados per year or 11 percent of the Crown's total revenues. They continued to increase under Philip II, averaging a full fifth of revenues over the course of his reign. However, as various economic historians have observed, for the Habsburgs the importance of American bullion should not only be gauged by the quantities involved. The fact that gold and silver were readily convertible into capital, that they presented a kind of "natural money," meant that they provided an important lever for the raising of financial capital. Bullion served as the surety for growing volume of loans to the Crown that not only far exceeded the value of the metal itself, but that could be easily deployed across the exchanges of Europe. It was a critical pinion of Charles's imperial pretensions and Phillip II's military campaigns against the Protestants and the Ottomans.[58]

By mid-century the extraction of gold and silver proved a central element in the political economy of the Habsburg empire. In the language of the time, they vitalized the body politic. The Laws of the Indies declared that they were the "nerve and spirit that gives vigor and being to the Royal estate."[59] But there were problems too. In the Americas, Spanish officials noted that the material and social foundations of the colonial mining regime were not fully assured. In the 1540s the Viceroy Mendoza had noted the exhaustion of woodlands in the areas around the mines of Taxco, stripped bare to provide charcoal for the smelters.[60] A similar portrait was obtained at Potosí in the Andes, where indigenous and Spanish foundry men were stripping the local landscape of its *quishar* trees, and resinous *llareta* plants.[61] Of even greater concern was, as in Española, the provisioning of labor. Just as they had in the Caribbean, repeating waves of epidemics, combined with colonial war and violence, wrecked the peoples of Central Mexico and the Andes. This social crisis, in turn, undermined the colonial capacity to conscript labor for the mines. This was felt in the mining districts of Mexico such as Compostela in Nueva Galicia, which in 1548 was completely emptied by cocoliztli.[62] Finally, the first signs of depletion began to be observed, most notably at the mines of Potosí. The heavily weathered silver ores that were excavated by trenching operations on the flanks of the Cerro Rico, and then smelted in the indigenous wind-powered guayra furnaces began to run out.[63] As they did, Andean miners left the district. Silver production dropped from sixty-four to twenty-one thousand kilograms per year over a five-year span in the 1560s.[64] In the metropole, Phillip II's increasing need for loans outpaced the flows of bullion required to service them, leading to the first in a series of state bankruptcies declared between 1557 and 1577.[65]

It was to address these different crises, and thereby assure vital inflow of rents, that the Habsburg monarchy updated and consolidated the extractive regime first limned half a century earlier in Española. In 1559 the first dedicated set of mining laws in the Americas was legislated, the *Ordenanzas sobre la minería*. The Ordenanzas began by reaffirming the Crown's radical title in the mineral wealth of the Americas: "We incorporate into Ourselves and Our royal crown and estate all the mines of gold and silver and mercury in our Realms, wherever they might be found . . . whether in Royal properties [*Realengo*] or public, or of councils or commons, or in inherited properties or lands of individuals . . . perpetually and without hindrance or condition."[66]

The guiding aim of the Ordenanzas was the overall increase in mineral extraction. A miner who ceased to work a deposit forfeited his title and the deposit reverted back to the Crown's possession until a more energetic successor could be found.[67] The discovery and claiming of new mines was opened to all subjects of the Crown. The distinctions of caste and *calidad*, so determinant in this colonial society, were here dispensed with: "the mines of gold, silver and other metals are common to all."[68] In the subsequent Royal Pragmatic of 1563, the mines were also opened to foreigners.[69] The indigenous subjects of the Crown were encouraged to exploit mineral deposits by exempting those who mined from labor corvées, such as the repartimiento or mita, and tribute. The privilege would be inherited "in perpetuity" by their descendants.[70] Mine owners received other incentives. In case of bankruptcy, their tools, machinery, buildings, livestock, and slaves could not be seized "so that the work of the mines might never cease or be impeded."[71]

As in Española, the administration of indigenous labor, so necessary for the work of extraction and smelting, became a central object of Spanish mining governance.[72] In the early 1570s, the Viceroy of Peru, Francisco de Toledo, wrote that the conservation of local populations ("*la conservación de los indios naturales de la tierra*") was imperative for the proper administration of territory ("*la disposición de la tierra*").[73] This was the motive for Toledo's organization of the Andean mita in the 1570s, the mass labor draft that forced the displacement of an estimated million and a half people across the Andes.[74] They were moved into the newly planned towns, the reducciones, where they could be more effectively Christianized and their service more closely administered.[75]

Directives were passed that aimed at the preservation of the health and vitality of working bodies. Food and nourishment were to be provided to drafted workers during the term of their service in the mines. They were not to serve in regions too far removed from their native climes for the risks that this might pose to their health. The work that they were asked to accomplish ought to be moderate. Mita laborers should not be asked to work in areas of the mines that might be dangerous to their health, or in the refining of mercury. They were to be excused from drainage works, since constant labor in water was also held to be perilous to their health.[76] These regulations, as so many historians have shown, were almost entirely observed in the breach.[77] What they demonstrate, however, are the links Habsburg administrators made between the conservation of indigenous labor, the production of the mines, and thereby, the financial welfare of the empire.

Royal officials also acted to ensure the plentiful in-flow of energy and materials consumed by extraction and refining. In response to the exhaustion of fuelwood in Taxco, the Viceroy Mendoza issued what were possibly the first colonial ordinances limiting forest clearing in the Americas.[78] In Peru, Viceroy Toledo upbraided the miners of Charcas for having destroyed local stands of *kenua* trees (*Polylepis* spp.) without regard for their regeneration.[79] Toledo also worked to boost the supply of mechanical power needed by the stamping mills at Potosí. To do so he orchestrated the mass levy of an estimated twenty thousand indigenous Aymara laborers. It was one of the most important hydraulical projects of the early modern world. Thirty-two lakes above Potosí were dammed and their waters channeled down a single great mill-race that dropped some six hundred meters to power over 130 mills.[80] It was, writes historian Alain Craig, "the greatest single concentration of medieval mill technology anywhere in the world."[81] In neighboring valley of Tarapaya, hot-springs were harnessed to power what may have been the first steam-powered metal mill in the Atlantic world.[82] Finally, and not the least, the Spanish Crown managed the provisioning of key reagents needed for the refining process: salt, chalcopyrite, and mercury. This last was the central element for various amalgamating processes for ores whose chemical composition resisted simple smelting techniques.[83] Its strategic importance to the emerging extractive regime was such that the Crown administered the mercury mines of Huancavelica, Peru, and Almáden, Spain, directly.[84]

A well-furnished historiography amply demonstrates the depth and extent of the social costs and ecological upheavals wrought by the demands of this extractivist regime. These ranged from the dismantling of indigenous culture, political authority, and social relations to the reconfiguration of agrarian systems and landscapes, to the more intimate disruptions of bodily health and function.[85] They have been documented from one end of the colonial mining frontier to the other, from the core areas of silver production in New Spain and Peru, to the outlying districts of Parral,

Chihuahua to the gold fields of Chile. The harms wrought by this continental system of extraction would become the object of contemporary critiques. They were, like the harms they denounced, varied, expressed by indigenous chroniclers such as Guaman Poma de Ayala, Catholic moralists, and reform-minded officials. But they shared a common line of argument. They drew attention to the negative reciprocities at work at the heart of the emerging Habsburg imperial order.

Fray Domingo de Santo Tomas's view (1551) on what was taking place at the mines of Potosí has marked representations of Spanish colonial mining to the present: "Four years ago one of the mouths of Hell was discovered. Into it, every year, great numbers of people are being sacrificed by the greed of Spanish to their God."[86] The mines at Potosí "consumed more blood than ore" wrote Luis de Capoche in 1585.[87] In the early seventeenth century the mercury mines of Huancavelica—central to the larger Spanish silver mining complex—were described as churning through the population of entire regions. "Entire Provinces were finished and the Indians consumed," decried the Jesuit Salinas y Córdoba.[88] The Jesuit José de Acosta joined his pen to this effort. He had personally gone down into the mines of Potosí and the experience had deeply marked him: "perpetual night, subterranean airs, horrible and thick; the descent perilous and endless, clinging to the cliffside. To halt is a danger. To lose your footing is the end."[89] If there was to be mining, Acosta wrote, then it should be voluntary and assure the "conservation of bodies, families and communities."[90]

Against these critiques were penned the apologias for colonial mining. The extraction of gold and silver, they argued, was a political necessity. They provided the animating substances, spirit, or blood of the republic.[91] The negative reciprocities signaled by critics were inverted. The consumption of indigenous bodies and communities was excised and attention was redirected to the ways that bullion animated the Habsburg body politic. Mining was a matter of public necessity wrote the Padre Agia from Lima (1603), "the Crown's urgencies of war against the heretics and other public needs demands [it]."[92] Without the mita, argued the Viceroy Francisco de Toledo, the mines would fall. Without the mines, Spain would lose Peru. Without Peru, the fortunes of the monarchy could not be conserved.[93]

Orlando Betancor, a literary scholar, has carefully examined the arguments that were less bluntly based on Habsburg material interest, but rather grounded in providentialist readings of American natural abundance.[94] In 1571 an anonymous tract, written from the Valley of Yucay, Peru, argued that the Indies and their treasures had been gifted to the Spanish monarchs as a reward for eight centuries of war against the Muslims.[95] José Pellicer y Osau would later give the anti-Semitic variant. God had granted the Indies and "their infinite riches" to the Catholic kings for having rid their kingdoms of Jews during the great expulsion of 1492.[96] But it was the Spanish Jesuit José de Acosta who produced the most considered version of the doctrine of providential title, in the fourth book of his *Historia Natural y Moral de las Indias* (1590). Acosta's argument for Royal rights over American treasures mixed Artistotelian natural philosophy and Christian telos. It began the apparent fact that the Indies, above all other known worlds, were uniquely and profusely endowed with precious metals. Within the neo-Scholastic understanding of the natural order, these metals, like all things of the earth, were destined to serve some higher purpose. This was set by God himself, the first mover and primary cause. In Acosta's account, if the Americas were so exceptionally stocked with gold and silver it was because God wished to attract and maintain "Christian knights" so that they might civilize the barbarians and bring them into the fold of Christianity. He described it as a dowry of divine inspiration and continental proportions: "what a father does to marry his daughter well by giving her a great portion in marriage, God has also done for this land, so rough and laborsome, by giving it great riches in mines, so that by this means it might be the more sought after."[97]

The mines themselves were cast as key sites of this moral and religious conversion. As historical geographer Heidi Scott shows, it was a frame that served to justify the exploitation of indigenous labor. "They are a barbarous peoples, without any knowledge of God," wrote Pedro Camargo from the mercury mines of Huancavelica,"[who] devote themselves to nothing but idolatry and drunkenness and other despicable vices . . . Sending these people to work in the mines appears to me a service to God and to Your Majesty and in the interests of the natives themselves, for in the mines they receive religious instruction and are made to attend mass and associate with Spaniards whereby they became ladinos [Ibero-Catholic] and learn good customs."[98] The same line of argument would be extended by Spanish commentary on the mining districts of Mexico and Peru from the late sixteenth century on.[99]

· · · · · ·

In the century that separated Columbus's landing at Guanahinni and the opening of the first colonial mines at Cerro de San Pedro, a new regime of extraction was assembled. It evolved through a combination of extension—conquest, frontier expansion, and colonial administration—and reaction, through a series of administrative responses to the different crises engendered by colonial forms of mining. By the 1590s the extractivist mining of precious metals was a well-developed, consolidated, and inextricable part of the political economy of the Habsburg monarchy. It was ideologically legitimated in prevailing views of the nature and purpose of American territories as the material support of imperial designs. It was governed by a panoply of laws that worked to organize property in natural wealth and maximize the rate at which this wealth could be realized through extraction. And since this, in turn, required a constant if not increasing flow of energy, materials and labor into the mines and smelters, a core part of Habsburg colonial administration was directed at the management or conservation of subject populations, waters, woods and mines of salt, chalcopyrite, and mercury.

It was, in its own terms, and for more than one hundred and fifty years, a remarkably successful regime. The most obvious measure of this was the sheer amount of gold and silver extracted from the Americas. The outflows of bullion from pillage, and then, to a much greater extent, from colonial Iberian mining, accounted for 85 percent of the world's silver output and 66 percent of its gold output by the seventeenth centuy. A share that would increase under the Bourbon revival of mining in the eighteenth century (90 percent of global silver supply and 85 percent of gold supply).[100] In theory one-fifth of this flowed into the Royal Exchequer. And even then, since bullion serviced debt, it quickly moved into private networks of financial capital. This outflow, as well as the untraceable itineraries of private bullion remittances, are of crucial analytical importance. They show that how extractivism, as a regime articulated around the maximization of rents from nature in the Americas, drew from and was articulated to the global development of capital accumulation and circulation. The agencies involved operated at the local level of mine—and smelter-owners, at the inter-regional level of the Mexico City based merchants, and in the transatlantic operations of the trading and banking houses of the period. It was through these networks and exchanges that the great majority of American bullion circulated, both legally and illicitly, within the Habsburg jurisdictions but, also and increasingly, well beyond them: throughout northern Europe, the Islamic Mediterranean, the Indian Ocean, and, not least, to China. That is to say extractivism, as developed in the Americas under Spanish rule, was central to the global development of capitalism in nature.

Having laid out the broader views of the extractivist assemblage it is now time to turn to the Cerro de San Pedro. The following series of chapters, shorter and organized by theme, move the optic on extractivism from an assemblage of continental proportions and political economic scope to a situated view of extractivism, now seen as a political ecology of actors and relations organized around the extraction of gold and silver. It was a localized assemblage of geology, ecology and society, one anchored to a mineral body of unique provenance and composition, and one that unfolded within the biophysical settings of this region of Mexican altiplano and the social scenarios of frontier incursions into indigenous Guachchil territories and the organization of a mining and smelting complex. The local view of how extractivism took root and unfolded at Cerro de San Pedro allows a view of the historical contingencies operating within the field of social and ecological relations that enabled, and resisted, extractivism in place.

2 The Temporal, Sublime

· ·

The history of mining at Cerro de San Pedro began in early March 1592 when an unnamed miner launched his bar into the exposed ores near the top of the mountain and pried out the first chunk of material destined to be winnowed down to render the gold and silver it contained. This anonymous act set in motion the triangular relations between geology, society, and ecology that would unfold over the next four hundred and thirty years. In this sense the Cerro de San Pedro, not the mines or the community they engendered, but the mountain itself, stood as an important character in this history. It was the silent center of attention for generations of miners, mine owners, and governing elites. It was the axis around which revolved a political ecology of extractivism over the longue durée.

Silent as it might have been, the mountain of Cerro de San Pedro had its own history leading to that day in March 1592, a history that, while unfathomably deep, would strongly mark the history that followed. The history of the mountain and of the mineralized body that it came to host, determined the location of rich veins of gold and silver ore; it determined their richness, their chemical composition, and thus their treatability by different smelting and refining processes. It established the depths at which such veins disappeared and where the massive but highly disseminated bulk of the porphyry began. The mountain's physical structure and chemical composition would condition the history of mining at San Pedro: the intensity and duration of different mining regimes, the techniques they privileged, and the energy and material flow they set in motion. All of which, in turn, determined how the progressive excavation of the mountain would impact on local populations, hydrologies, and ecologies.

It is a difficult story for the human imagination to encompass because what created the Cerro de San Pedro treasure box was time, inconceivable amounts of it. Time enough to take in the exhaustion and self-consumption of a large star. Time for our own star and planet to form from this detritus. Time for tectonic plates to be set in motion and collision, for the sedimentary layering of great bands of limestone hundreds of meters thick, the cycling of magmatic cells within the mantle, the fluxing of gold, silver, and other metals up into the crust, the chemical processes of mineralization, the slow flush of mineral-bearing solutions, the growth of crystals, the weathering down of mountains. The long line of zeros stringing out after each milestone date hardly captures it. A conceptual vertigo creeps in as one contemplates these abysmal depths of time. It is the temporal, sublime.

· · · · · ·

The deepest roots of the story of the mountain of San Pedro begin 4,600,000,000 years ago in the blinding flash of a dying star. This was the supernova responsible for creating the heavier elements found in our solar system, including the silver and gold at the heart of the human story of Cerro de San Pedro. Stellar genesis began quickly thereafter, "only a few million years later," and within less than one hundred million years of swirling condensation finished producing our sun and its attendant planets.[1] From the beginning of stellar genesis on, the story that links together the events leading to the formation of Cerro de San Pedro is not so much the story of creation—all the basic material having been produced in the destruction of our ancestral star—but rather the story of concentration, of the sequence of events that gathered up the millions of atoms of silver and gold thrown out of that cosmic furnace and then packed them together into a mountain of northern Mexico.

During the formation of the solar system, our planet gathered into itself a higher proportion of the heavier elements condensing out of the larger swirl of the molecular cloud. Within the earth, gold and silver atoms amassed themselves in the core and mantle of the planet, a function of the manner in which gravity, temperature, and chemistry combined in the early formation of the earth.[2] The average concentration of gold in the mantle is roughly one part per ten million, an exceedingly low figure but one that is actually quite close to the concentrations found in contemporary open-pit gold mines (three to ten parts per ten million) and an entire order of magnitude higher than the average concentration of the metal in the earth's outer crust (one part per three billion).[3]

It is from the enriched zones of the mantle that great globules of gold and silver-laden material rose toward the earth's surface. Each traced its own particular history of ascent, histories that were conditioned by the interaction of crust and mantle. The dynamism came from the mantle where great convection cells slowly churned and launched highly mineralized bulbs upwards. The crust, for its part, presented what geologists call the "structural controls" for what came bubbling up. It provided pathways and opportunities in certain places, or blockages and resistance in others. It determined the environment into which the magmatic fluids moved: further enriching them with yet more silver and gold; mixing these with other minerals; and guiding their emplacement into different ore bodies.

The specific history of the Cerro de San Pedro is that of the intrusion of one such mineral-rich magmatic cell into the thick beds of limestone that make up the Eastern Sierra Madre. These gray-blue sheets and strata can be seen all around the valley today. One hundred and twenty-five million years ago they were still in formation beneath the warm Cretaceous waters of the Gulf of Mexico. It took some twenty-five million years of compressing and slow-cooking layer after layer of muds, marine life, and outwash to create a sedimentary layer-cake hundreds of meters in depth.[4]

The next thirty million years were spent hoisting these limestone beds out of the sea and then folding them into the various north-south ranges that roll up like a series of waves from the warm humid lowlands of Tampico and the Huasteca up to the dry central plateau of San Luis Potosí. This was the local chapter, written in the regional geology, of the Laramide orogeny, a continent-long epic of mountain building that stretched from Canada to Mexico. It was driven by the encounter between the Farallon and North American tectonic plates deep off the Pacific Coast. It was there that the North American plate, much larger and steaming west, rode over the Farallon, sliding over it and then pushing it down toward the mantle.[5] The creation of the San Pedro range was a small part of all this. The results are easily overlooked. It presents none of the glacier-sculpted grandeur of its northern counterparts, no majestically striking position or elevation, no topographical singularity or severity of aspect. Of the hundreds of ranges created by the Laramide orogeny, the Sierra de San Pedro is just not particularly particular.

For the Cerro de San Pedro, the real story came up from below. When the Farallon plate was subducted beneath North America it carried with it a thick layer of sea-floor clays and sediments that were saturated with water. This material was buried into the gold and silver-rich mantle where it began to melt. The fact that it was soaked in brine meant that it melted in particular ways with particular consequences. The entrapped and superheated water gave buoyancy to the material around it, creating a cell able to lift itself up through its denser surroundings. Water also acted like a flux that lowered the melting point of the rocks in the cell, allowing it to remain in its liquid form as it rose back up into the relatively cooler and more solid regions of the lower crust. The brine influenced the chemical complexion of the rising melt, filling it with metallic sulfide and chloride compounds. As this solution coursed up through the crust, it proved to be an excellent scavenger. It pulled in, and then chemically locked up, the tiny particles of metals it met in the column of its ascent: gold and silver, but also lead, arsenic, cadmium, zinc, iron, and mercury.

The overall image here is of a plume of superheated and highly mineralized solution billowing up from the mantle. It was anywhere between a few hundred meters to a few kilometers in diameter and it rose up for hundreds of kilometers. The rising bud of gold and silver-laden solution did not punch explosively through to the surface—this is what volcanoes do. Instead, it slowly worked its way upwards over millions of years. Slowly, but with considerable force, the kind of force needed to push through microscopic gaps of afforded by the crust, gaining a micron here, a centimeter there, but ever-gaining, inexorably rising through extraordinarily long spans of time.

The real break came sixty-four million years ago.[6] This is the moment in geological time that the rising melt was able to work its way through the crust, arriving close to the surface—less than one kilometer to go—and then nestling itself within the limestone bed of the San Pedro range. The timing was linked to the end of Laramide mountain building in this part of North America. Instead of crumpling the landscape through compression, tectonic movements paused, turned about, and began to stretch the crust. Then they stopped. Then they pushed again but now along a north-south axis that began to twist the thick limestone beds of the San Pedro range. Then they stopped again. Then they stretched themselves out. Push east. Stop. Pull west. Stop. Stretch to the north. Stop. Twist to the south. All this wrenching, stretching, and rotating opened up a multitude of faults and micro-fissures in the upper portions of the crust. These were so many opportunities for the upwelling melt to move into the limestone that formed the outer-most layer of the earth.[7]

The contact between the melt and limestone strata was explosive. The melt pushed itself into the limestone as a thick globular mass roughly half a kilometer wide by eight hundred meters in breadth. When it arrived, it was still an extremely hot, pressurized, salty, and acidic cloud of mineralized solution. The limestone was cool, permeated by fractures, and very basic. As it pervaded the limestone the melt created intense spikes of pressure inside the already weakened and fractured sedimentary rocks. Fissures were blown open, faults were pushed wide, and the horizontal breaks between sedimentary beds were jacked apart. With literally explosive force the melt cracked open a latticework of pathways allowing a further rise of hundreds of meters through the limestone. As it advanced, the solution was simultaneously cooled and neutralized by the limestone. These stepwise shifts in pressure, temperature, and acidity forced the precipitation of the metals out of the hot solution of the rising melt. The ramifying cracks and fissures of the contact zone began to fill with highly mineralized material and then congeal. These produced the veins the future miners would follow, the gold and silver-laden seams that coursed and branched through the limestone host rock. Among geologists this irradiating crown of fractured and mineralized rock is known as the aureole, a halo of gold and silver. It is the frozen afterglow of what passes for an explosion in geology—high intensity, short duration, over in the matter of centuries.

The head beneath the crown was the porphyry, a great looming mass of mineralized material that supported and fed the different vein-works. The porphyry, as we understand it now, composed the bulk of the rising melt, the portion of the plume that did not punch its way into the limestone but rather stopped hundreds of meters below the surface. There it slowly cooled itself into a large and relatively homogenous body of clays, kaolins, hornblende and other fine-grained material (plagioclase, biotite, feldspar, and quartz).[8] Like the veins of the aureole, the porphyry was packed with gold and silver as well as also other metals like lead, arsenic, zinc, and copper. Unlike the veins of the aureole, the concentration of metals in this material was substantially lower. The metals of the porphyry are present as an invisible mist of microscopic particles finely disseminated across this great subterranean body.

The explosive entry of the upwelling cell of mineralized solution into the limestones of the San Pedro range and the slower establishment of the porphyry below it, marked the first phase of the mineralization of Cerro de San Pedro. There still remained another sixty-four million years to go before that first miner dug his bar into the deposit. This provided the time needed for the forces of erosion to reduce the San Pedro range to its present-day contours and to further concentrate and enrich the body of the deposit. As the limestone capping was reduced, fresh rainwater percolated further and further down into the mountain. Water seepage provided the medium for heavier metals to filter back down and accumulate in the heart of the deposit. Water also shifted the internal chemistry of the mountain by introducing oxygen into what was an intensely sulfurous environment. It divided the mineralogy of the Cerro de San Pedro deposit in two: an upper zone of oxides, and a lower zone of sulfides. The concentrating forces of millions of years of erosion and water circulation formed the second phase of mineralization. Its slow work year-by-year work was still in progress when miners first came upon the mountain.

· · · · · ·

The geological history of the Cerro de San Pedro and the history of gold and silver mining there are intimately linked. The history of formation and composition set the stage for the subsequent of extraction and dismantlement (see figure 2). The miners of the Spanish period first encountered the tips of the different metal-rich veins that broached the surface of the Cerro de San Pedro. As other "surfacings" were encountered on the slopes and in the canyons of the neighboring hills (Cerros La Bufa, Populo, San Cristobal, Raposa, Barreno, and Blancas) miners came to see that these seven hills formed a single complex. It was a *criaderos de metales* (a nurseries of metals) centered below the Cerro de San Pedro and extending off-shoots into the adjoining hills of the valley.[9] They also found that it was exceptionally stocked with both gold and silver, an exciting and indeed geologically rare combination.[10] These outcrops were the most weathered of the mountain's ores. They were dark and crumbly and could easily be pried out with bars, broken up into chips and gravels of spall and hauled down.

Then the miners followed the ores into the mountain, excavating tunnels to delve after the mineralized fissures and faults of the aureole. These came in all sizes: from hairline brecchias (exceedingly numerous but short and constantly pinching out) to veins a hands-width or arms-span wide, to the rare *beta madres* or "mother veins" meters in diameter and running for hundreds of meters through major fault systems.[11] Occasionally these ore veins ballooned to fill cavities dozens of meters in height and breadth. To excavate them, miners built large timber scaffolds so that dozens of miners could attack the full height of the mine face. Over the first centuries of mining, a remarkable variety of textures and colors in the ores they encountered, a variety produced by location, depth, the presence of water or sulfides, and the specific chemical sequence that led to their formation. One historic inventory of the ores of San Pedro reads: "Esphatic irons; bismuthic ochres; oxides of manganese and of iron; celestine iron; cinnabaric iron; white lead, brown lead, and red lead; silvers: ashen, blackened, green or red."[12] The blackened silver was probably a silver ore mixed with galena—a lead sulfide. The red silver might be pyragyrite or fire silver, a silver sulfide that flames a deep crimson when it is broken off the mine face, but then fades as it tarnishes and oxidizes in contact with the air.[13] It was only in the nineteenth century that industrialization enabled miners to sink deep shafts

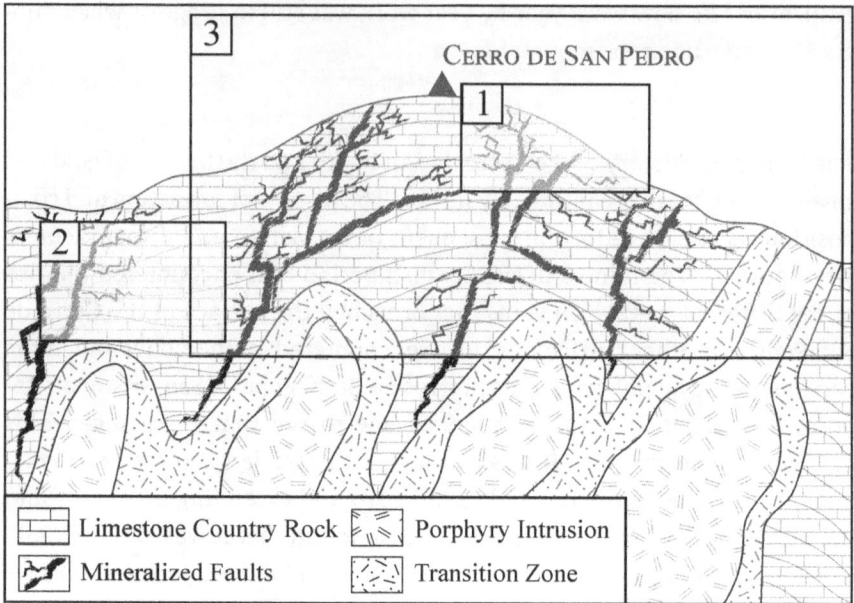

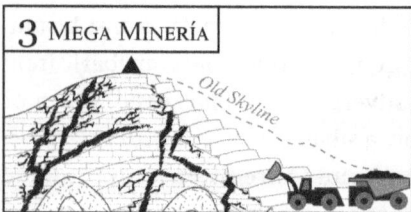

FIGURE 2 The geology and mining of Cerro de San Pedro (illustration by Geoffrey Wallace, 2021, after original by D. Studnicki-Gizbert).

into the heart of Cerro de San Pedro "nursery" where they encountered the porphyry and began to map out the outlines of its titanic bulk. Industrial tools and conveyances allowed a more wholesale approach to extraction. Entire blocks of the mountain's ore bodies were blasted out and sent out to the mills. Some parts of the porphyry were removed in this way, but in general the mining companies focused on the low-grade sulfide ores in the contact zone.

At the end of the twentieth century the porphyry was directly in the crosshairs of the last generation of miners. Airborne magnetic reflectivity surveys revealed its silhouette, looming deep underground like a hidden reef. Hundreds of drill-holes perforated its flanks at all manner of angles and the material brought back up in the core samples subjected to every available contemporary technology of measure: water was squeezed from the rock; samples were examined under electron microscopes, exposed to x-ray diffraction analysis, stable isotope analysis, radiometric dating, and infrared spectography. Beginning in the 1990s, company geologists and mineralogists worked the reconstitute the porphyry in all its aspects. Their work came to its conclusion on the eve of mining's final assault on the Cerro de San Pedro, the last campaign that would see miners blast apart the entire mountain in order to haul out its very heart. It is in 2003 that full image of the mineral body of San Pedro finally appeared into view: the bulked mass of the porphyry crowned by its halo of gold and silver-bearing veins. Core and corona. A strange subterranean replica of a star.

3 Mining in the Land of War

In general terms, the appropriation of indigenous mining zones, or the exploitation of hitherto untapped deposits, comprised the principal means of increasing the outflow of gold, silver, and the rents they accrued to the Spanish monarchy. Extension across space defined this commodity frontier. As seen in chapter 1, a number of mining frontiers existed in sixteenth-century Spanish America, each advancing behind different fronts across the human, ecological, and geographic diversity of the continent. The beginning of Spanish mining at Cerro de San Pedro in March 1592 was part of one such mining frontier, the one that moved north and west from central Mexico. For the Spanish, this entailed an important shift in context from the agrarian heartlands of the Nahua, Tarasco, Otomí peoples, to the drier *altiplano* of hunters and gatherers. The peoples of the altiplano did not mine silver or gold. Their populations lived free of the controls of state or chieftainship and thus eluded the kind of colonial labor capture witnessed in the Caribbean, central Mesoamerica, and the Andes. More important, they were long able to resist and escape subjection by Spanish arms.

All of which bore directly on the evolution of the northern Mexican mining frontier. The Spanish had to prospect to locate deposits of gold and silver. None compared to the singular bonanza that was the Andean Potosí, true, but they were more numerous and, in time, their collective output would make Mexico the world's leading silver producer. In their sixteenth century beginnings, the mines of the north were vulnerable. Like all mines, they required an agrarian hinterland to supply workers and operations with food and provisions. Unlike the Andes or central Mexico, however, these did not yet exist. Nor did the great numbers of people needed for the work of farming, stock-raising, charcoaling, or any of the other manifold labors that sustained extraction. Like an advancing army, the northward movement of this mining frontier depended on its supply lines. Along these moved the material needed for mining, the migration of central Mexican laborers, and the grains, livestock, and tools needed for planting. The military dimensions here were very real. Spanish mining in the north meant mining inside territories that were fiercely defended by the Zacatecas, Guachichils, and other nations of the Chichimeca group. It meant mining in the land of war.

This chapter gives a narrative account of the advance of the Spanish mining frontier as it moved north from central Mexico to eventually arrive, in March of 1592, at the Cerro de San Pedro. Within the lines of the narrative are a number of features that characterized this particular mining frontier. First, its progression followed a sequenced and ramifying pattern. The establishment and development of a given mining center provided the resources in capital, labor-power and supplies to prospect and open up others that, in their turn, served as the staging ground for further expansion. Second, while prospection was organized by, and capitalized upon, Spanish actors, it often recurred to local indigenous knowledge of their territories. And finally, while the consolidation of the northern Mexican mining belt required an end to the war, that end did not come through military victory. Pacification came through the subtler colonial strategies of gifting and negotiation that, in time, broke up the alliances between Guachichil groups and thus frustrated the effective defense of their territories.

· · · · · ·

Spanish precious metal mining began to extend north of central Mexico with the end of the Mixtón war. This was the two-year campaign orchestrated by the Viceroy Antonio de Mendoza against the Caxcán in 1540–41.[1] Within a few years the Spanish had opened up the gold mines of Xaltepec and silver mines at Espiritú Santo, Guachinango, Xocotlán, and Etzatlán, around Compostela, west of Guadalajara. These were relatively minor deposits, but they were important in funding further ventures to the north. Guadalajara lieutenant governor Cristóbal de Oñate, for instance, used the winnings obtained from these mines to support the work of prospection. This included the efforts of Juan de Tolosa who, while prospecting out of a camp at Nochistlán, received one day a large sample of ores delivered by an unnamed indigenous man. Following this lead brought Tolosa to a prominent mountain, La Bufa, over two hundred kilometers north of Guadalajara. There, Tolosa dug out three or four mule loads of ore samples for assaying. The results were good and within a year a full-on mining boom was in progress at the mines of Zacatecas, one of Mexico's most important, and still operating, mining districts.[2]

The interest here, however, is in the role Zacatecas played as the staging ground for the prospecting and development of mines and mining camps across the region (see map 2). From Zacatecas, expeditions of soldiers and

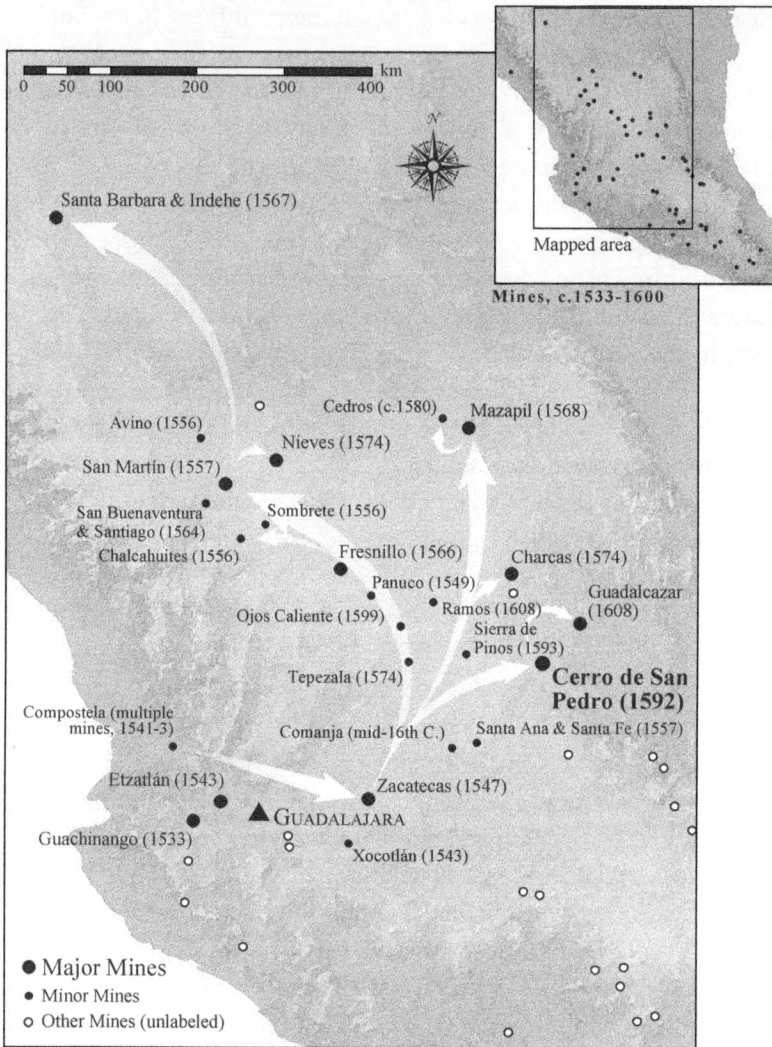

km
0 50 100   200   300   400

𝒩

Mapped area

Mines, c.1533–1600

Santa Barbara & Indehe (1567)

Avino (1556)

Cedros (c.1580)   Mazapil (1568)

Nieves (1574)

San Martín (1557)

San Buenaventura
& Santiago (1564)

Sombrete (1556)

Chalcahuites (1556)

Fresnillo (1566)   Charcas (1574)

Panuco (1549)

Ojos Caliente (1599)   Ramos (1608)   Guadalcazar (1608)

Sierra de Pinos (1593)

Tepezala (1574)   **Cerro de San Pedro (1592)**

Compostela (multiple mines, 1541-3)

Comanja (mid-16th C.)   Santa Ana & Santa Fe (1557)

Etzatlán (1543)   Zacatecas (1547)

▲ GUADALAJARA

Guachinango (1533)

Xocotlán (1543)

● Major Mines
• Minor Mines
○ Other Mines (unlabeled)

MAP 2 New Spain's northern mining frontier (© G. Wallace Cartography & GIS. Map by Geoffrey Wallace, 2021).

prospectors extended out to across the altiplano in two general directions. To the northwest they opened up mines at Fresnillo, San Martín, Sombrete, and Santa Barbara. To the northeast, the mines of Mazapil, Cedros, and Charcas.[3] Livestock estancias, agrarian fields, and charcoaling operations were opened up in the adjacent areas to provision the mines. A network of roads, the *Caminos Reales*, was organized to bind the camps together and enable the long-distance transport of supplies, lead, mercury, and bullion into and out from the mines. Three main trunk lines reached up from Guadalajara, Michoacán, and Queretaro to meet at Zacatecas. From this hub were extended secondary tracks toward other mining settlements further to the north. The extension of this colonial mining frontier, and the secondary development of commercial agriculture and transport infrastructure, would set important foundations for the geographic organization of the region that continue to define the region.[4]

For the indigenous Zacatecos, Guichichils, and other Chichimecan peoples of the area, these developments were an incitation to war. Hostilities began in late 1550 with assaults on the mule trains and wagons supplying the mines of Zacatecas and attacks on the ranches of the area. Little in the first decades of the conquest and colonization in the Americas could have prepared the Spaniards for the resistance presented by the Chichimeca. The great city-states of Mexico had fallen after a three-year campaign. But the response of the Chichimeca to Spanish invasion was of another kind altogether, venting what one seventeenth-century Mexican poet called "an elusive fury" on the Spaniards, with a rain of "hell's own arrows that did not leave a single miserable mortal alive."[5]

They waged a hunter's war against the invaders. From their highland caves and prospects they tracked the movement of Spanish militias and convoys that traveled along the flat-bottomed valleys. Raiding parties kept to the ridges and headed down the canyons during the night, launching their attacks at dawn. These were coordinated operations, with small groups of archers spread out over the terrain to multiply the lines of fire, some attacking, others hiding in wait at different ambush points to pick off fleeing enemies. And if the Spanish were able to regroup their men and launch a riposte, the Chichimecas simply melted back into the hills and forests.[6]

Sharp bowmanship was a central life skill for these hunters. Young boys were given small bows from the moment they could walk and were set upon hares, birds, and other small game.[7] Thus began a practice that would last the rest of their lives. "They sustain themselves by means of the hunt," wrote the Augustinian friar Alonso de Santa Maria, "Every day they go forth to track and chase deer, birds, and other game which they nail with their arrows; they don't even forgive the rats."[8] And they did not forgive the Spaniards. Chichimecan arrows were delivered with deadly precision through small chinks in the armor. One crack shot stapled the trigger hand of a Spanish gunner to his harquebus. Another felled a horse with a single reed arrow. "When they aimed at an eye and hit an eyebrow, they cursed it as a bad shot."[9] Even disarmed these hunter-warriors proved redoubtable opponents. Spanish veterans of the Chichimecan war ruefully acknowledged the tenacity and strength of their opponents: "Four Guachichil indians took on a hundred Mexica warriors," related one, "and although wounded they barged up to the Spanish soldiers, fighting them beard to beard, and ripped the guns and swords from their hands." Another soldier saw, "how a group of Chichimecan prisoners tore the stones from their prison's wall and waged war on the Spaniards with them all night long without surrender."[10]

For over forty years the loose and shifting confederacy of Chichimecan nations and bands held their own against the same kinds of mixed indigenous-Iberian forces that had toppled the Mexica of Tenochtitlán. The Archbishop of Mexico, Pedro Moya de Contreras, estimated that over the span of a decade (1564–74) more than ten times more Spaniards died during the Chichimeca War than during the entire suite of campaigns that composed the Conquest of Mexico.[11] Thirty years later colonial observers were still lamenting the unparalleled loss of Spanish soldiery at the hands of the Guichichil and others.[12]

The duration of this frontier war reflected the incapacity of either side to achieve a clear victory over the other. Under the Viceroy Falches, the Spanish had attempted a scorched earth campaign, a *guerra a fuego y a sangre*, to no better results than an increase in casualties. On the other hand, the Chichimecas were unable to extricate the Spanish from their positions. Mining camps were armed and able to defend themselves. Each had its complement of militias composed of Spanish and indigenous Huichol, Pamé, and Otomí troops. Miners bore arms and built *casafuertes*, strong houses of stone walls and arrow slits. This is why many Spanish mining camps and districts at the time were known as Reales de Minas, literally "Mining Forts." The strike and retreat tactics of the Chichimeca guerrilla could wreak havoc on the supplies needed for the operation of mines and smelters, and for the upkeep of their workers. In 1561, for example, raids on outlying ranches, farmlands, and charcoal-making camps forced a shutdown of operations at the mines of Zacatecas, by far the most important mining district of the region. Residents were forced to gather and eat prickly pears to survive.[13] In general, however, the Chichimecas were incapable of overwhelming and eliminating the mining camps. It did happen, but only exceptionally, and only for smaller outlying mines, isolated and deep within indigenous territory.[14] As time advanced, the Spanish mines formed a thin but very tough lattice of foreign bodies working its way into the tissue of the land.

· · · · · ·

In the late 1580s, a long generation after the beginning of the Chichimecan war, the valley of the Tangamange remained a Guachichil heartland, autonomous and free of Spanish presence. But then the logic of the Chichimecan war began to shift and the stalemate began to resolve itself in the Spaniards' favor. It was in that same decade that the first rumors were bruited in Zacatecas about the discovery of another treasure mountain to the east, in the "Gran Tunal of the Guachichils." In 1587 Miguel Caldera approached the acting lieutenant general of Zacatecas at the time, Antonio López de Zepeda, and proposes to lead a company there, at his expense, "to find the fabled treasure."[15]

Caldera played a central role in both the resolution of the Chichimecan war and the opening of Spanish mining in the Guachichil homelands of the Tangamanga. He was a complex figure, linked to both sides of the war, though ultimately a partisan of the Spanish advance. He was born of a Guachichil mother. Her name is unknown. She was captured by Spanish militia and then taken as the country wife of one the soldiers, Pedro Caldera. The younger Caldera's first years were spent in the "land of war," the indigenous hinterland beyond the Spanish towns and mining camps, before moving Zacatecas when he was a nursling. This was in 1553 when Zacatecas was still a ramshackle confusion of a mining camp in the first years of the boom. His mother tongue was Guachichil, to which he added other kindred languages of the Chichimecan language group, as well as the heavily creolized Nahuatl that was becoming the lingua franca of the indigenous population settling in the Spanish settlements.[16] His first friends were Chichimecans of different nations: Guachichil, Zacatecos, and Copuces.

Caldera followed his father's career, joining the Spanish militia at the age of twenty-one. Nine years later he made the long journey south to Mexico City where he received formal command of his first company as Capitan of the King. It assembled a typically mixed complement of soldiers: a small group of Spanish cavalrymen and gunners, with the bulk of the fighting strength coming from a contingent of Huichol warriors from Mesa Tabasco.[17] For their first sortie, Caldera led them through a retaliatory campaign against a group of Chichimecas who had raided the mines of Nahuapán. The expedition netted over a thousand captives—captains of war, warriors, women, children—who were marched in coffles to the city of Guadalajara where they were served a sentence of Spanish justice and slavery. Caldera's company raked in twenty gold pesos per head, the standard bounty of the time.[18] They also engaged in the habitual brutalities inflicted on captured indigenous combatants. Captives had their feet amputated so they could not escape, or their thumbs chopped so that they would never shoot a bow again.[19] But Spanish arms and horses could not carry war up into the cliffs and tablelands of the Chichimecan redoubts. And so when Caldera entered the mountains, he turned to his mother's side: "with his Indian friends he stripped naked and, shouldering his quiver and bow, went to dislodge the Chichimecas, to take them captive, and to punish them."[20]

But Caldera was more than an effective soldier. He was a colonizer, a man who played a central role in the foundation and early settlement of San Luis Potosí and its region. In the late 1580s, the valley of Tangamanga and its surrounding sierras was a favored place for the Guachichil people. The valley plain was speckled with marshes and wetlands. The adjoining forests of the sierras contained important deeryards. In 1590 Caldera and his troop led a convoy of over a thousand indigenous Tlaxcalan settlers to open up two agricultural colonies in the valley: Mexquitic and Tlaxcalilla.[21] This marked a new tactic in the Chichimecan war. "civilized," that is agrarian, indigenous peoples from central Mexico would be mobilized to convert the Chichimecas to the ways of settlement, cultivation, and *policia*. The recruits were mainly young couples and single men mustered up from Tlaxcala, the indigenous Nahua state slightly to the east of Mexico City that had allied with Cortes during the invasion of Tenochtitlán. At the time Tlaxcala was reeling from the devastation wrought by the epidemics: mass mortalities, social dislocation, and the dispossession of the best bottomlands by Spanish landowners. The recruits may very well have been refugees.[22] But once settled on the banks of the Rio Santiago, their role was to attract and incorporate the Guachichils, "so that they can become family and marry each other and in this way domesticate them and instruct them in the ways of political and Christian life."[23] Caldera's role in the settlement of the Tangamanga was mainly to provide logistical support. He assured the safe arrival of the Tlaxcalan settlers. He worked up the relays of emergency grain supplies from Zacatecas when the first crops of maize failed, and starvation threatened to wipe out the fledgling settlement.

The Tlaxcalan settlement on the banks of the Rio Santiago was heavy on projection but its prospects were shaky. Without the steady subsidy of food and military support it was unclear whether this seed of civilization had the energy to grow. If it did, it is because, at that same moment, work began at the Cerro de San Pedro, barely four leagues from the center of the Valley. The mines would assure the town's future. In the matter of a few years it was transformed from a plan hatched by Viceroys into a thriving boomtown of thousands.

A few months after putting his stakes into the Cerro de San Pedro, Caldera, in the company of Juan de Oñate (the son of one of Zacatecas' leading mining magnates and soon-to-be expedition leader into the Puebloan territories of New Mexico), began the work of formally founding a town. They christened it San Luis Potosí. They chose a site at the center of the valley, near the banks of the Rio Santiago. Over the next months they supervised the cord-and-eye work of laying out the grid of a city that did not yet exist—the streets and plazas, property lots, the location of churches, the buildings of government. It was the projection of *policia* in the dirt and shrubs of the Tangamanga valley, still very much Guachichil territory at that point.[24] There was something queerly demiurgic about their tracings. It was as if they know that once affairs with the Guachichils were settled, the treasure of the Cerro de San Pedro would make a town bloom from these marks in the dirt. They were and it did.

Passing from projection to reality was a challenge. In the summer of 1592 Caldera's party was small and encamped thirty kilometers away from the Cerro de San Pedro in the northwestern corner of the Tangamanga valley. Except for the two fledgling settlements of Tlaxcalan colonists, and the cords and stakes of the future city whistling in the wind, the valley was still very much Guachichil territory with numerous groups sharing its wide-open spaces, wetlands, and forests. The Tlaxcalans were barely holding on. They survived on the food rations carted in from Zacatecas, and their first meager harvests of maize. A prolonged cold snap in 1594 wiped out their crops entirely and they failed again in 1595 and 1596.[25] Brought in to spearhead the civilization of the Guachichils and to provide a bulwark for Spanish settlement, the Tlaxcalans were mainly hunched down to the desperate work of staying alive. Over three hundred Guachichils were encamped in the caves and valleys of the Sierra of San Pedro, and were not about to surrender one of their favored hunting and gathering zones to a group of Spanish miners.[26] Mining behind enemy lines pitted wealth against welfare. Juan de Zavala and his companions tried it and only just managed to return alive: "To approach the Cerro de San Pedro de Potosí was to enter into the land of war; we only went in as a company guarded by a cavalry unit and a seasoned corps of soldiers, and even then, the entire enterprise was of the highest risk and danger given the great number of Chichimecas there and their great bellicosity."[27]

Miguel Caldera knew that it would be foolhardy to try and drive the Guachichils from their homeland by force. Decades of war and personal experience had made that abundantly clear. If the Spanish were to take the treasure of San Pedro, they would have to find a way to draw the Guachichils out from the valley and into peace with the invaders. The solution was a new policy and practice that came to be known as *paz por compra* (peace by purchase). It entailed approaching individual leaders, families, and bands and offering them a regular subsidy of food and other goods in exchange for their peace. This had already been tried on an occasional basis but under the Viceroy of Mexico Luis de Velasco (1590–95) it became official Spanish policy.[28]

The deployment of peace by purchase was not wholesale work. It progressed case by case, each presenting its own mix of calculations and gambits. The trick was to identify individuals and groups that would even consider talking to the invaders and then move them from the ranks of belligerents into ranks of allies. It required a fine sense of the shifting layout of alliances and cleavages between and within Guachichil groups in order to know whom to approach, when, and with what. The negotiations themselves were extremely delicate intercultural dialogues taking place in the pressure-cooker atmosphere of a war of singular violence. It was, in sum, exactly the kind of work for Miguel Caldera. He spoke the languages, he knew the codes, he had established relations with numerous Guachichil and Chichimecans, he knew how different factions sat in relation to one another, and he had, over three decades of life and war in the region, a reputation to be reckoned with. The Viceregal authorities in Mexico City turned to Caldera to advance the cause of Spanish settlement in the north.

Caldera spent the next five years in the saddle to make peace by purchase a reality. Crossing and re-crossing the sierras and plains of the region, he covered a huge quadrant of northern Mexico bounded by Chachihuites and Zacatecas to the west, Rio Verde to the east, Charcas and Mazapil to the north and Guadalajara to the south. Throughout the center of his work and attentions remained the valley of Tangamanga and its surroundings. Caldera initiated the first parleys with dozens of Chichimecan leaders and then returned regularly to move the negotiations along. The greatest part of his efforts was dedicated to keeping the logistics running. The amount of food and goods was tremendous and assuring their timely and regular delivery by cart or by hoof over hundreds of kilometers was a considerable challenge in and of itself. From 1592 to 1598 over eighteen thousand head of cattle, twenty-eight thousand fanegas of maize and eight-four thousand yards of cloth were paid for by the Royal Treasuries at Zacatecas and Mexico City and was sent to twenty different distribution points across the Gran Chichimeca.[29] The sums involved were massive and when they were lacking, Caldera put in considerable amounts of his own personal capital to assure prompt payments to suppliers. On a thirty-day tour amongst the Guachichils of the Sierra of San Pedro, he spent eighteen thousand silver pesos that he had made off the first mines of Cerro de San Pedro to buy and disburse food and goods.[30] At a time when a good-sized house could be purchased in Mexico City for between three hundred to five hundred pesos, he had invested in a single month the equivalent of a sub-division on the peace.

While the bulk of the payments came in the form of steer, maize, and cloth, the various Chichimecan parties to the arrangement forwarded other demands, demonstrating a surprising degree of connection to what was on offer in the broader world of trade in the late sixteenth century. They specifically requested Chinese silks, Indian muslin, Rouen collars, and a complete men's suit tailored in London, axes from Biscay, and Castilian copper kettles. And then there were the tools: hoes, adzes, knives, needles, belts, pestles, and rope. Or masks, bells, trumpets, flutes, and castanets.[31] A Guachichil artist in Saltillo ordered in a full set of paints and brushes.[32] They did not, tellingly, ask for silver or gold or coins since these had no value in the nomadic economy.

The goods provided by Caldera came with conditions and at a price. The baseline was the cessation of hostilities. But as relations deepened with every mule train and oxcart delivered, so too did the terms of the peace. Guachichil leaders and their bands were required to settle in specified locales. They had to take up agriculture and the ways of colonial society. The Tlaxcalans colonists who had been struggling since their arrival in 1590, now became the principal agents of the social and cultural transformation of Chichimecans to sedentary life. As many as three hundred Guachichils were settled into rows of small, thatched huts immediately beside the Tlaxcalans. There they labored in the fields and commons of the Tlaxcalan towns and were put under the authority of their governor. They had to cut their hair and wear white cloth frocks at all times. Bodies and faces had to be kept clean of all insignia, dyes, or paints.[33]

It is unclear that all the Guachichils, or even a majority of them, accepted the new dispensations of peace by purchase. The total number of people who chose to settle among the Tlaxcalan settlers did not reach three hundred—less than a quarter of the low-end estimate of 1,300 to 1,500 Guachichils inhabiting the valley of Tangamanga and its environs in the early 1590s.[34] The settlement process was obviously fraught. In 1599 a Guachichil woman, held to be a *nagualista* by her community (that is a shape-changer who could take the form of either a coyote or a deer), stormed into the church of Tlaxcalilla, broke the crosses and ripped down the images. "Stirring up all the Indians by the force of her sorcery," a Spanish witness declared, "she attracted many Indians of the Chichimeca nation. She insisted that that they rise with her to go to the pueblo of San Luis where the Spanish have settled and kill them all. If they did not follow her, she would destroy them all."[35] Pedro de Torres, a baptized Guachichil chief stopped the rebellion before it could really get started and then settled his people back down. The Spanish seized the woman, tried her, and executed her in the very same day.

Those who refused to settle with the invaders removed themselves from a homeland that was no longer theirs. And while it is unclear where they went, it is clear that the regions to the east and north of San Luis Potosí and the mines continued to be indigenous territories. Whatever successes peace by purchase may have garnered the Spanish in the immediate region around the mines of Cerro de San Pedro, an expansive territory beyond it remained—to use the colonialist phrasing—unreduced. Chichimecan "rebellions" and raids appear in the archives of San Luis Potosí well into the seventeenth century.[36] In 1645 news of a major Chichimecan offensive coming in from the north had the belfry bells of the town ringing. The town's inhabitants were mustered up, weapons and powder were distributed, and militias formed.[37]

As for those who remained, they would soon be engulfed by the mass influx triggered by the mining boom. At their strongest these people numbered three hundred. With each passing generation, less and less Guachichils survived. The figures we have for the group settled among the Tlaxcalans of neighboring Mexquitic count ninety-eight Guachichils for 1622; forty-one for 1636. Less than fifteen years later (1650) in the town of San Luis a census taker noted, "none of the Indians here are Chichimeca."[38]

4 The Bonanza

. . . .

The moment the Guachichils accepted the entry of the Spanish into the Sierra of San Pedro, a large and quickly swelling population of workers poured into the area. By 1600, eight years after Miguel Caldera hammered the first stakes into the Cerro de San Pedro, authorities estimated that five thousand men were at work up at the mines and down in the smelters of the valley.[1] The great majority indigenous Tarascos, Nahuas, and Otomis from central Mexico.[2] In time, they were joined by a growing number of Africans, Afro-Mexicans, and mestizos. Three years later the total population of the district had risen to over six thousand men where it more or less stabilized.[3] In the 1620s it was around seven thousand strong.[4] This was the time that the traveler and Carmelite friar Antonio Vázquez de Espinosa passed through the mines of San Pedro. They hummed with activity. Dozens of mines were being worked into the slopes of the Cerro de San Pedro and adjoining hills. The valley bottom was packed with hundreds of *jacales* (miners' cottages). The caves that had once served as home and aerie of the Guachichils were under new occupation. Up and down the notch road running along the foot of the mountain, Vázquez de Espinosa counted "over fifty shops for various merchandizes and over twenty bakeries and food shops."[5]

The excitement that surrounded the opening of the mines of Cerro de San Pedro was fueled by the discovery of an unusually rich ore that held both gold and silver in its matrix. In 1593 the Alcalde Mayor of San Luis Potosí, Alonso de Oñate, reported that many of the mines produced "gold in tiles and that much of the silver contains gold."[6] A generation later Lucas Fernández de Manjón described the silver ores drawn from the mountain as "sweetened with gold"—a reference to the quartzite gold that was found finely mixed in with the silver ores, as well as the more serious candy of large nuggets (some the size of a kidney) of pure gold.[7]

The first decades of mining at Cerro de San Pedro trace a fairly typical portrait of a first strike mining bonanza: the discovery of an exceptionally rich and hitherto untapped deposit; the rapid influx of miners and other workers; the ensuing take-off in the quantities of precious metals extracted from the mines. The bonanza occupies a special place in the narratology of mining. It stands as the dramatic opening chapter of a cycle of extraction that moves from rapid growth, to apogee, and then decline. In standard accounts, much emphasis is placed on the natural abundance of the mineral deposit. It is the exceptional richness of the ores (that is, high metal to ore ratios) or the exceptional size of the mineralized body that explains the ensuing boom. Naturalizing the bonanza in this way suggests something mana-like, even providential, in the explosion of extractive work that followed discovery.

This chapter proposes another view. The bonanza at Cerro de San Pedro did not naturally spring from the mineralization it contained. It was socially produced through the construction of a system of extraction and refining that was, by 1592, well established in the mining districts of Spanish America. It was the quick assembly of that complex that accounts for the rapid increase in the tonnage of extracted ores and yield of processed metals. This was a proto-industrial system—for its mobilization of many kinds of labor and skills; for the sophistication of the metallurgical techniques; for the variety of machinery it put into motion; and, finally, for the human, energetic, and physical scales at which it operated. Unlike the large-scale manufacturing of ceramics, glass, or textiles, this proto-industrial system was dedicated to the disassembly of matter. It traced a series of stages through which ores were broken down and manipulated to separate out silver and gold and then passed these into the market.

The key element in understanding the bonanza at Cerro de San Pedro is the animating agency of capital. Pulling together a chain of extraction and refining entailed costs in fixed capital (land and machinery) and working capital (wages, provisions, inputs). Such operations it is true, could be assembled at small scales, in the form of a kind of "cottage extraction" that processed the personal winnings of individual miners in small, side-yard, furnaces. More on these later, since they played an important part in the social history of mining at Cerro de San Pedro. Here the interest is in the small number of large and highly capitalized operations that dominated the district. Owned by a clutch of some twenty individuals, they were responsible for over 70 percent of bullion output at the height of the boom. Compared to the cottage operations, they were enormous. They integrated numerous mines, mills, furnaces, as well as ancillary operations such as the collieries, farms, and ranches that supplied needed energy and materials. If the scale of the operations was so great, it was because their owners sought to increase the rate at which they processed ores. Profits, which were very high, were obtained on a per unit basis. This being the case, the logics of capital accumulation served to raise the cadence of extraction. Scale and speed were effectively the two sides of the same coin. Drawing attention to the intensity of highly capitalized mining, as measured by unit flow over time, allows a view of an extractivist mode of mining, its metabolism, and, as will be seen in the next chapter, its social and ecological consequences.

· · · · · ·

From the mine face, where gold and silver ores were broken out of their earthly matrix, to the cupels, where the near-pure metals were recovered, the extractives complex was organized into three basic steps: excavation, transport, and processing. Each stage in this sequence required its particular set of tasks, skills, tools, infrastructure, and consumable inputs. Following the physical progress of ores from mine face to furnace, and thus their transformation from ore to the metal, reveals the material organization of this proto-industrial extractives complex.

In the first months of the bonanza, extracting the silver and gold from the mountain was as straightforward as prying out the dark friable ores that could be found along different outcrops. These were the tips of the deposit's corona that had been exposed as time and weather wore down the dome of the Cerro de San Pedro. When the outcrops of ore were removed, the miners began to dig; small gopher holes at first, then trenches, and then larger pits, such as the Tajo de San Cristóbal that eventually reached dozens of meters in depth and breadth. When these, in turn, were worked out, the miners turned inward, addressing their bars, picks, and hammers to the gray-blue limestone of the Sierra of San Pedro. Miners called this *rompiendo cajas*, literally "cracking chests," breaking up the rock that ensconced the ores.[8]

Cracking out the ores of Cerro San Pedro was entirely handled through human labor and skill. Mine work was organized by cuadrillas, gangs of anywhere between four to five to a couple of dozen individuals.[9] Each cuadrilla had its own *capataz*, the foreman who coordinated the various tasks of his team—breaking out the ores, hauling, water evacuation, timbering and so on. The larger the operation, the more segmented the labor force. Historian Frédérique Langue, for instance, documents mines in Zacatecas whose workforce was divided into fourteen different *oficios* (positions). Doris Ladd inventories thirty different oficios registered in the eighteenth-century mines of Pachuca and Real del Monte.[10]

The miners of Cerro de San Pedro, like miners across the world, engaged in one of the most physically demanding forms of preindustrial labor, one often associated with corvée labor (like the repartimiento and mita), penal servitude, or enslavement. Before black powder came into the mines in the late seventeenth century, breaking out the ores from the mother rock was accomplished by hand and muscle. At the mine face miners drove long sharp-tipped bars iron into the rock and, finding its points of weakness, pried it apart. If the rock was too hard for the bars they switched to chisels and sledges to punch their way in.

Although the work of breaking up rock and ores was physically rigorous it also demanded a high degree of know-how, experience, and sensitivity. The first period of mining at Cerro de San Pedro mainly took place in an earthly body whose internal structure and composition was later described as "indeterminate, surprising, and capricious."[11] In the upper sections of the deposit—the aureole or corona—the ore was held within the shattered fissures of the limestone. Its lattice presented miners with an exceedingly complex puzzle. Veins ran along all the points of a three-dimensional compass. They stopped abruptly. They divided into dozens of tantalizingly thin hairline cracks. They broadened into large mother veins ("thick with potency") that had miners raising their hands in the air, chortling with happiness, and singing their thanks to God from the depths of the mountain.[12] Deeper still they encountered *mantos* (literally blankets), thick beds of ore that ran horizontally between two layers of limestone, and *bovedas* (vaults), cavernous galleries dozens of meters high and a hundred meters long, filled with ores, but dangerously friable and unconsolidated.[13]

To follow these veins, to find the larger cavities, to not waste effort, to avoid the various hazards of the subterranean world, all required a considerable degree of canny. Miners relied directly on their senses and experience to probe the insides of the mountain. They operated by the wavering yellow light of the miner's candle, long calumet-like brands with a lit plug of tallow at their end. They scrutinized the recently opened rock and read its signs. Visible fracture lines were extended by the mind's eye into the mass of the rock. Every ore had its color, its own way of reflecting the lick and flicker of the candlelight, its own texture. Differences in the rocks' structure produced different sounds when they were struck. The miners at Cerro de San Pedro could distinguish between ore and *esteril* (literally "sterile"), that is between the material that was worth hauling out to the surface and the waste that could safely be heaped up out of the way. And they came to know the differences between the ores they found. The Cerro de San Pedro contained at least dozen. Like any good mineralogist, miners rasped bits of ore against one another. They scratched them, licked them, smelled them.

The seventeenth-century observer Alonso Barba describes Andean miners boiling up freshly hewn ores in small pots. The resulting broth was ladled up, swished in the mouth, and then spat out on the tunnel floor.[14] This kind of discernment was important because the "temperament" of different ores determined how, and how easily, they might be processed and refined. This, as will be discussed in chapter 6, was critical for miners who sought to separate, in the depths of the mines, the ores they would keep for themselves from those they would take out for the mine owners.

As it fell to the ground the ore was shoveled into *xiquipils*—toughly woven sacks of maguey fiber—holding seventy kilograms of material or better.[15] Heaved up on the back and held there with a tumpline stretched across the forehead, *tenateros* (haulers) hoisted these immense loads up a series of shafts by muscling their way up "chicken ladders" of cut logs. And then they returned for another load, making multiple round trips in a single day. As the years advanced, and miners dug their way deeper, the ores had to be raised for greater and greater distances. By the end of the Spanish period of mining in the 1820s, the lowest tunnels at Cerro de San Pedro reached two hundred vertical meters from the surface. The limestone host rock at Cerro de San Pedro being particularly hard, miners wasted no energy on broadening or leveling their passages through the mountain. Tunnels were often shorter than the height of a man, forcing haulers to crouch low or crawl along for hundreds of meters. Miners often worked the mine face on their knees.[16] In particularly narrow or pinched tunnels, miners at Cerro de San Pedro and elsewhere, relied on the boys—the *zorritos* or "little foxes"—to get the loads through.

At each mine mouth and pit-work, the ores hauled out from the earth were gathered in tidy lines of haul sacks. There they waited until the carter or *arriero* (muleteer) arrived to load them up and deliver them to the mills and smelters for processing. They were known as *Ingenios* or *haciendas de beneficio*. In 1622 there were twenty-two such metal-processing sites around the valley and uplands. These were situated at different distances from the Cerro de San Pedro. The nearest were up-country at a small hamlet known as Monte Caldera, about six kilometers away. The remainder were found at different points of the valle de Tangamanga. There was conglomeration at the center of the valley, near the indigenous pueblo of Tlaxcalilla, and another at San Francisco de los Pozos in the south-east corner of the valley. Both were a slow twenty-kilometer journey away. The remainder were scattered across eight other points within a thirty-five-kilometer radius from the mines.

It was in the ingenio that the metals were extricated from the ores through a sequence of manipulations involving heat, water, motion, and applied chemistry. Two features characterized the ores recovered from the Cerro de San Pedro during the colonial mining cycle. The first was that they were highly weathered and came from the upper sections of the mineral deposit, the portion situated above the water table. They were mainly silver oxides of different kinds (silver sulfides being mainly located beneath the water table). The second was that these were compounded with lead.[17] The absence of sulfides and the presence of lead meant that the ores could all be processed by smelting, rather than through mercury amalgamation.[18] It was only late in the seventeenth century that mercury-based refining began to be developed in the valley of San Luis Potosí, and at first it seems to have been reserved for extracting gold from the ores.[19] This suggests that took decades of mining to reach the watered zones of the deposit. Until then, the ingenios of San Luis Potosí and environs were spared the need to secure mercury for amalgamation and the more complex processes entailed by having to refine ores through different and parallel processing lines. This made Cerro de San Pedro different from mining districts such as Zacatecas and Guanajuato in Mexico, or Potosí in the Andes, where mercury amalgamation was central to work of extracting metal from the ores.

Once through the gates of the ingenio, the mule trains and carts were directed to the main yard where the sacks of ore were dumped out on the ground. In a sulfide ore operation, they would be evaluated and then painstakingly sorted into hundreds of according to quality and temperament.[20] As a smelting operation, the first order of business was to thoroughly clean the ores to remove any excess dirt and nitrates that might impede the work of the furnaces. Ores were passed through a sequence of large tubs, five to eight in a line, until they were free of impurity. They were then ground to a small pebble-sized gravel by a large vertically set mill stone known as an *arrastre*. Its drive shaft was powered by teams of mules and horses. Almost entirely absent from the valley of San Luis Potosí were the more powerful stamp mills characteristic of other mining districts such as Pachuca, Taxco, Zacatecas, or the Andean Potosí. This was due to the lack of sufficient water flow to turn them. The exceptions were Pedro Arizmendi Gogorrón's ingenio at San Francisco, south of San Luis Potosí, where he built a large dam and reservoir to provide the required head and consistency of flow, and the ingenios built by Francisco Cardenas along the Rio de Bledos to the southwest.[21]

Once reduced like this, the graveled ores were mixed with litharge (lead oxides), and *cendrada* (ash from maguey fronds). The blend was watered, made into a paste, and then loaded with charcoal into the smelters. These were then fired to 1000° Celsius thanks to large mechanical bellows. At these temperatures, the metals rearranged themselves. Silver, gold, and whatever arsenic, zinc, copper in the ores now bound to the lead. The remainder formed a slag that was separated out and discarded. The lead and metals were cooled into bars that were then loaded into a second set of smaller furnaces. Here again, forcing blasts of air on the burning charcoal produced the extremely high temperatures needed to separate out the lead from the silver and the gold. The lead, as well as whatever other heavy metals it had picked up in the first burn of the smelter, was skimmed off as waste and left to cool as scoria. As the last traces of lead were burned off inside the furnace, the atmospheric chemistry inside the burning chamber switched and the chimney released "a beautifully colored flame."[22] The busy work of transforming ores into metals having been consummated, the furnace relaxed with a fiery sigh. What remained inside was an almost pure bar of bullion: mainly silver but, as Manjón observed, always sweetened with gold.[23]

· · · · · ·

The economic historian Sergio Serrano Hernández has carefully reviewed the records concerning seventeenth-century bullion production at Cerro de San Pedro. In addition to providing a reliable series of output data, he presents an inventory of the silver and gold registered by two hundred and thirty individuals from 1618 to 1623. It shows the remarkable degree of concentration that characterized the boom. Four individuals accounted for well over one-third of total output. The top twenty producers produced over 70 percent of the total output, a yearly average of close to thirty thousand kilograms of silver for the period. While these men minted thousands of kilograms of silver and gold every year, the average annual output of the other two hundred and ten producers was less than one hundred kilograms.[24]

The first generation of elite miners at Cerro de San Pedro and San Luis Potosí included the likes of Miguel Caldera, the architect of Guachichil pacification, as well as Pedro Arizmendi Gogorrón, and Martín Ruiz Zavala, two Basques who had been active in the mines of Zacatecas.[25] Along with about a dozen others, these men quickly took their positions within the new mining district. They were the first to register mining claims on the most promising portions of the Cerro de San Pedro. They obtained the property titles needed to build their mills and smelters. Eighteen of these were granted within a year of first strike.[26] Along with the properties, rights (*mercedes*) were acquired to a share of the local water flow under the existing Spanish regime of apportioning water for irrigation. In the valley of San Luis Potosí, water was first and foremost used to wash ores. The quantity of water needed being high, and the flows of local water courses modest, the Mexican researcher Guadelupe Salazar González indicates that these mercedes were used up by 1602.[27] Other parts of the territory were similarly appropriated to support the work of extraction. Regulations were passed that exclusively reserved blocks of forest for the production of charcoal and timber for the mine works. A three-league perimeter of land surrounding the emerging town of San Luis Potosí was set aside for the raising of mules, oxen, and horses that provided a considerable part of the motive force of the extractive complex.[28]

The ingenio occupied a strategic position within the larger extractives complex that sprang up during these years. The control of ore processing proved to be more important than the of mines. A large ingenio, with its batteries of mills, smelters, and furnaces, could handle a much greater through-put than smaller ingenios or the cottage furnaces (*fuelles*) run by indigenous smelters. Pedro Arizemendi Gogorrón built a large ingenio in the Valle de San Francisco in order to take advantage of the stronger water flow in the river there. It counted twelve smelters and a furnace for cupellation. Then he built a second ingenio at La Sauceda, with sixteen smelters and two furnaces. To these he added smaller ingenios at different points around the region (Tlaxcalilla, La Cieneguilla, Guadalcazar), and eventually a mercury amalgamation operation in the Guanajuato mining district.[29] Together, these produced close to seven thousand kilograms of silver per year, with close to half this amount coming the three ingenios of the valley of San Luis Potosí.[30]

The amount of capital required to build, or purchase, these operations was considerable. Arizmendi invested 80,000 pesos into the ingenio at Valle de San Francisco, 60,000 pesos into La Sauceda. He bought Pedro de Rojas's smaller operation for 8,000 pesos. For his part, Juan de Zavala spent 70,000 pesos to build what he proclaimed to be the greatest ingenio that had ever been built in all of New Spain and all of Peru.[31] Compare these amounts to the smaller haciendas built or sold within a fork of 500 to 1,700 pesos and one can see the gap between elite and middling operations in terms of both capital access and scale of operations.[32]

The top four operations of the San Luis Potosí district processed an annual throughput of between one hundred thousand to one hundred and fifty thousand kilograms of silver ores. This only accounts for the ores brought down from their mines at the Cerro de San Pedro. The large ingenios also processed ores from other mine owners for a per unit fee, or they rented the use of a smelter and furnace for a period of time or stipulated number of firings.[33] Unlike smaller ingenios or the fuelles, that would experience down time while furnaces were loaded or cleaned, the large operations ran unceasingly. They received ores and deposited silver on a constant basis.

At the height of the boom, the mines and ingenios of San Luis Potosí produced an average of twenty-nine thousand kilograms of silver and seven hundred kilograms of gold per year.[34] The silver output amounted to about two-thirds the yields of Zacatecas, New Spain's leading silver district during the same period. In terms of gold output, however, the Cerro de San Pedro was by far and away the Viceroyalty's leading producer.[35] In material terms this represented the extraction and processing of close to fifteen hundred tons of ore per year. And this, in turn, entailed proportionate in-flows of labor, charcoal, water and other provisions to feed the extractivist metabolism of precious mining in the bonanza.

In economic terms, all this activity unfolded within the incipient capitalist economy developing in and around the valley of San Luis Potosí and the mines of Cerro de San Pedro. Provisions, food, fodder, tallow, charcoal, pulque—all the things that were consumed by the extractive process or by the people and animals that powered it, were all bought and sold, and thus created markets of different kinds and dimensions. Labor, too, was commodified, though incompletely. Many workers received wages but since cash was tight, these were not always paid out as money but rather in kind or in the form of a share of the ores. It is important to note the various forms of compulsion that marked labor in the district: indigenous Guachichil captives brought in from the frontier war, enslaved Africans from the other side of the Atlantic, the impressment of "masterless men" through the royal laws against vagabondage. Finally, and not the least, the bullion produced by the mines and smelters was one of the leading commodities of the period. Not only for its value, but also for its reach. From the ingenios of the valley of San Luis Potosí, the bars of silver and gold won from the ores of Cerro de San Pedro were packed into heavily locked chests and hauled south to the mint of Mexico City. There they were refined one final time to separate gold from silver and to achieve the standards of purity demanded by the imperial currency of Spain. They were then taxed and stamped into the famous pieces of eight and escudos that circulated across the world by the millions. For its organization, for its role in regional economic development, and for the social relations it engendered historian John Tutino has argued that precious metal mining in New Spain formed one of the "salients of early capitalism."[36] Clearly. What is of interest here is not just whether, or even how, bullion mining in Mexico and other parts of Ibero-America produced capitalist forms of production and exchange. This has been amply documented in the historiography.[37] The issue at hand is how capitalist logics of accumulation impressed themselves into the material organization of extraction; that is: how they composed the chain of extraction and refining work, but also, and most particularly, how they motivated the increase in throughput and scale, speeding things up as a means of generating more profits but also, importantly, more rents for the Crown. From here one can begin to document and follow the new ecological relations and flows that the extractivist complex extended into local bodies, both natural and human.

· · · · · ·

At the height of the bonanza, within a generation of the first strike of 1592, the mining oligarchy of the valley entered into negotiation with the Spanish Crown to purchase a title for their city. It was finally accorded in 1655. By then the bonanza was beginning to lose its wind. The mines of Cerro de San Pedro were no longer producing so abundantly, and people were leaving the valley. So the recognition was late in coming. All the same, the Royal title they sought is worth considering for what it declared, symbolically and discursively, about the ties between precious metal extraction and the construction of the colonial order. The designated name for the new city was San Luis Potosí. This was actually a conflation of two toponyms: San Luis de la Paz, the place name for the original community of "pacified" Guachichils at the center of the Tangamanga Valley, and the Minas del Descubrimiento del Apóstol San Pedro Cerro de Potosí de la Nueva España, the full official name for the mines in the sierra. San Luis referenced Saint Louis IX of France, canonized in the late thirteenth century for his qualities as the ideal Christian monarch: the renderer of royal justice, warrior against the infidel, protector of his subjects.[38] Mexican historian Juan Carlos Ruiz Guadalajara has traced the movement of this figure, via the Franciscan

FIGURE 3 The coat of arms of San Luis Potosí (Archivo Histórico del Estado de San Luis Potosí).

order, into the Chichimecan borderlands where it served to symbolize the implantation of Christian monarchical order over the Guachichils. As for the Potosí of the mines of San Pedro, the name referenced the Cerro Rico de Potosí, then, and ever since, the world's greatest silver mines. This was clearly Caldera's intent when he named the mines in 1592. The richness of the ores, the way that they bound gold into the silver, argued for a recapitulation of the Andean precedent. This would be the Potosí of New Spain.[39]

The blazon and coat of arms that Philip IV bequeathed to the new city provided the supporting iconography (see figure 3). The field is halved into a night-sky blue and gold. In the foreground the Cerro de San Pedro drawn in its distinctive conical shape with three boca-minas on its face, each a dark portal into the underground. Atop its peak was Saint Louis *justicier*, adorned with scepter, orb, and crown. Beside him two bars of gold against the blue; two bars of silver against the gold. The visual program celebrated the links between mountain, treasure, and royal authority. Its composition, with San Luis standing jaunty atop the mountain like it was his pedestal, could not be clearer about the role the extraction of nature's treasure played in supporting the imperial order.

5 The Metabolism of Extraction

· ·

It is easy to become entranced by the magic of the ingenio, those noisy, bustling, smoking crucibles that turned ore into gold and silver. Intently we follow the steps by which these metals are drawn down from the mountain as raw ores and then became more and more entirely themselves. At first hidden in the piles of dusky ores piled in the sorting yards, they emerged in the slow stirring of the furnace slags, and then spilled out pure in a molten flash.

In this entrancement Marx saw the operations of another kind of magic: the fetishism of commodities.[1] It the kind of magic that obscures all those relations and operations, both social and ecological, that went into producing a commodity. While under its spell we only see in a bar of silver or gold the price that they might fetch on the market. But, following Marx and contemporary scholarship in environmental history, ecological economics, and allied disciplines, the fuller measure of that bar has to account for the multiplicity of relations that rendered silver and gold from the earth. The work of extraction, hauling, and refining that constituted the core of the extractive complex could not have advanced without drawing upon a broader web of human and other than human bodies. These included the people who worked and lived in the mines and smelters; the mules, oxen and horses drafted to the work of haulage or powering machinery; the fields and pastures that produced the food, fodder, and other organic inputs consumed by the extractives complex; the highland forests whose trees furnished the fuel for the smelters; and, finally, the local waters, airs, and soils that received the wastes voided by the processing of ores.

The aim here is not simply to pull into view all those things that economists enclave in the term externality. It is also to document the social and ecological configurations produced by the extractivist mode of mining that developed around at the Cerro de San Pedro. During the first forty years of the boom, the majority of bullion output from the district came from a large-scale, high-intensity, and high-throughput system of mining and refining. It depended on, and generated, large volumes of capital, and was in this sense the expression of early capitalist logics in the field of extraction. It is when this mode of mining is viewed from an ecological perspective that one can more clearly see how it forced deep and irreversible changes to local landscapes and hydrologies. The marks on human bodies were just as deep. Both at ecological and embodied scales, the extractivist mode imposed forms of exhaustion and wasting. These were the consequences of operating at this degree of intensity. The problem was one of synchronization, or rather the lack thereof, between the metabolism of the extractive complex and the ecological and bodily metabolisms that it engaged. On the one hand, the flow of material and energetic inputs that coursed into the extractivist metabolism outpaced the rates of ecological regeneration and human replenishment. This produced states of exhaustion. Wasting, on the other hand, was produced when the outflow of toxic material overwhelmed the capacity of rivers and bodies to safely process them. Pursued over decades, this mal-synchronization enabled a remarkable output of silver and gold, but it simultaneously impoverished the vitality and health of the human and other-than-human bodies of the region.

· · · · · ·

To understand the social and ecological consequences of a regime of high-intensity mining it is helpful to view the mining complex as a metabolism. This allows us to follow the flows of energy and matter moving into and out from the mines and refineries. Three steps—extraction, haulage, refining—defined the sequence through which silver and gold were drawn from the earth and separated from the ores. It is possible to describe and, to varying degrees, provide a measure of the material and energetic inputs and outputs at each step and, from there, follow how they reconfigured ecological relations.

As described in the previous chapter, the underground work of extraction was entirely accomplished by human workers. At the peak of the boom in the early seventeenth century there were many dozens of mines in operation. Arizmendi Gorgorrón declared in 1628 that they numbered three thousand, an exaggeration, surely, but one that gives the sense of the sheer quantity of excavations at work on the mountain.[2] One Spanish official estimated the total labor force at the mines of Cerro de San Pedro at around thirteen hundred workers.[3] In energetic and material terms, their labor produced the mechanical power needed for all the tasks that colonial-era tunnel mining entailed.[4] They were not, in and of themselves, a source of energy but rather more akin to an engine or what in energy studies is called a prime mover. Vaclav Smil, the Czech-Canadian environmental scientist, details the flows of energy involved in human metabolics and labor.[5] To work, people needed to be fed and kept in health, and thus they themselves required inflows of food and supplies to live and replenish. In addition to human labor, the work of extraction required organic and material inputs in the form of tools and machinery, fiber for the xiquipil haulage sacks, tallow for light, and, not least, timber. This last is difficult to quantify but the descriptions of the mining process reveal just how ubiquitous it was. Timbers served as structural supports inside the tunnels. They were used to build up large scaffolding works needed to access, and stabilize, large mine faces of the open-air *tajos* (pits) and underground grottos. They provided all the chicken ladders that allowed vertical passage up and down the mines.

The haulage of the ores from the mine to the ingenios was the second step in the chain. Here it is possible to obtain a general sense of the effort involved. Using the Venezuelan chemist Saul Guerrero's 2 percent ore grade and applying this to the itemized data on silver and gold production assembled by Serrano, renders an estimate of close to one and a half million kilograms of ore brought into the ingenios for processing on an annual basis.[6] This takes into consideration only the ores that rendered the silver and gold that was officially tithed and thus documented. An unknown percentage, possibly between a fifth to a third, of the ores from the Cerro de San Pedro were hauled out on the miners' own account and were never recorded. This stream is left out of the estimate. Transporting ores was bulk work and was undertaken by mules or oxen-drawn carts. Mules were loaded with ten arrobas (110 kilograms) of ore each. Carts could be loaded with fifty quintals of forty-six kilograms a piece, for an overall load of circa twenty-three hundred kilograms.[7] Hauling the ores of Cerro de San Pedro represented over thirteen thousand mule trips, or slightly more than six hundred and fifty ox trips a year. What the actual mix of carts to mule trains was, or how often a given animal made the day-long trip between the mines and the ingenios, is unknown. Even so, these estimates point to the hundreds of animals mobilized to the work of material transport. They too were "living engines" in that they furnished mechanical energy but needed, in turn, to be fed to do so. Materially speaking, the inputs for haulage would have also included wood for carts, axels, wheels, and shafts; tallow for grease; leather for rigs and harnesses; and metal for fasteners and tackle.

The last stage in the extractives chain was the processing work that took place in the ingenios. At the height of the boom there were twenty-two haciendas de beneficios in the region, operating one hundred smelters and furnaces. These haciendas varied in scale, with larger haciendas running between twelve and sixteen smelters, and smaller ones running one to four smelters. The number of furnaces set the processing rate of a given ingenio (between 15,000 and 18,750 kilos of ore/furnace/year) and thus serves as a rough measure with which to estimate other inputs on a pro-rata basis.[8]

As with extraction and haulage, a large proportion of the mechanical effort of moving, washing, and grinding the ores was accomplished by human and animal labor. The workforce at Jose de Briones' eight smelter operation at Monte Caldera numbered 112 people. This was a middling sized operation. Large haciendas such as those run by Arizmendi Gogorrón or Zavala, with its twelve to sixteen smelters, would have required the work of around two hundred people—a significant concentration of labor for the time. In all, an estimated fourteen hundred people worked in the ingenios of the district, close to a third of the total workforce of five thousand reported by Spanish officials.[9] The motor force needed to power the different machines of the ingenio such as its grinding wheels or bellows was provided by rotating teams of mules and horses. A mid-seventeenth-century report on the ingenios of the valley counted an average of seventy animals per operation. The ingenio of La Sauceda kept four hundred.[10]

Water and hydraulic power were also critical inputs for the ingenios. Access to water, with a preference for flowing water, was leading criteria for the emplacement of the ingenios. In the early years of the boom, water courses and springs were relatively abundant in the valley, but even so their flows were insufficient to power the heavy machinery of the ingenios. Beyond the valley, in the adjoining sierra of San Pedro and on the southwestern slopes near Bledos and San Francisco, the construction of dams and reservoirs provided the hydraulic force needed to run stamp mills.[11] In general, the primary use of water in the region was not for power but rather as a medium for washing ores and moiling the mixes of ores, litharge, and ash destined for the first firing in the smelters. For streamside operations flowing water was diverted into tubs and then back out of the haciendas through a series of canals and troughs. Those based beside ponds or springs used mules and bucket-hoists to lift the water into the aqueducts and basins of the ingenios.

In terms of material inputs, in addition to the ores, the ingenios received a constant supply of reagents for the smelting process—the litharge and the ashes—as well as the charcoal needed to produce the high temperatures needed for smelting and refining. Charcoal was produced through pyrolysis; that is by smoldering wood and other ligneous material in a nonoxygenated burn in pits or earth-covered heaps. Charcoal was lighter to transport than wood and it burned much hotter, making it a good fuel for the ingenios. Charcoal derived from mesquites and oaks burned hottest of all.[12] It was produced in *carboneras* (collieries) established across the forested highlands of the region. Some of these camps could be quite large, housing cuadrillas of workers who felled the trees, dressed the boughs and trunks, and controlled the burns. They delivered the charcoal produced in this way to the haciendas by the cartload on a constant basis. Other operations were smaller, run by families of indigenous colliers, often Tarascans, who carried their sacks on their backs or on a mule for sale at the hacienda. It took about a ton of charcoal to produce a kilogram of silver.[13] This meant that at the height of the boom the ingenios consumed something on the order of twenty-nine thousand tons of charcoal every year. It was by far the most import flow of material put in motion by the extractive complex, dwarfing all other inputs (including ores) by orders of magnitude. The demand for heat energy produced a ceaseless movement of carts, mules, and haulers moving charcoal from the forests to the smelters.

It is at the refining stage that the outputs flowing out of the extractives complex became important. Metal refining was essentially a highly labor—and energy-intensive winnowing process. Even at the relatively high ore-grades of the period, an estimated 98 percent of the ore material received by the ingenios was separated from the silver and gold and discarded. To this outflow must be added the reagents such as litharge and ash brought in to facilitate the processes of smelting and refining. Voiding took three different routes. Waste material was entrained by waters mainly in the form of the slurries and effluvia coming out of the washing tubs.[14] It was also released into the airs as emissions from the smelters and furnaces. Applying Guerrero's model to the yearly production numbers provided by Serrano, one finds that between 145,000 and 290,000 kilograms of lead were released into the atmosphere of the district on a yearly basis.[15] The amount of arsenic, zinc and other heavy metals that were present in the ores and were volatilized along with the lead is unknown. The most important outflow, by weight, were the wastes coming out of the refining process came in the form of slags and scoria. The latter was the glassy black material of shards and large gravels that congealed after the final burn in the refining furnaces. It was packed with lead and other toxic metals but since it appeared inert, and thus innocuous, it was simply dumped into large and growing heaps on the edges of the ingenios.

· · · · · ·

The ecological dimensions of a high-intensity mining boom can be assessed by following the flows into and out from the extractive metabolism (see figure 4). The amount of mass it put in motion—of ores but also of charcoal, timber, and other provisions—was significant. It represented the displacement and manipulation of thousands of tons of material, a scale of operation analogous to engineering works such as canal-building or the construction of dikes. Absent large-scale hydraulic power or fuel-driven engines, this work was accomplished by the coordinated activity of thousands of people and animals. These "living engines," in turn, needed food and other provisions to sustain their labors. In the first years of the bonanza at Cerro de San Pedro, food had to be hauled in from Zacatecas or Queretaro because the mining boom exploded in an area where precolonial agriculture was absent. The settling of Tlaxcalan colonists in the valley on the eve of the mining boom provided an important start for the development of cultivation. Then arrived thousands of migrant laborers, almost all indigenous, from Michoacán, the Bajio, and central Mexico.[16] In addition to providing the labor power for mines, smelters, and highland collieries, they dedicated themselves to the opening up of fields and pastures and the sale of the surpluses they produced. They spread out across the valley of San Luis and beyond through a network of smaller settlements dedicated to raising crops and livestock.[17]

Fields and pastures produced more than just food of course. A wide range of organic materials were produced and processed by the agrarian economy for the mines and smelters: tallow to grease the machinery and light the miners' torches; leather for belts, seals, and bladders; maguey fiber for haul bags and cordage; pulque for drink and release. Historians of colonial Latin America have long drawn attention to the way in which the expansion of silver and gold mining engendered the development of secondary markets in agrarian products.[18] From an ecological perspective, agrarian expansion, especially in the colonized territories of nonagrarian peoples, was an important form of environmental and landscape change. This was clearly the case for the valley of San Luis Potosí. The existing vegetation cover was cleared and replaced by cultivars and pasturage. Local hydrology was reconfigured. Streams were diverted to fill artificial reservoirs. Rivers were tapped to water irrigation systems that in turn fed *milpas* and *huertas* (fields and

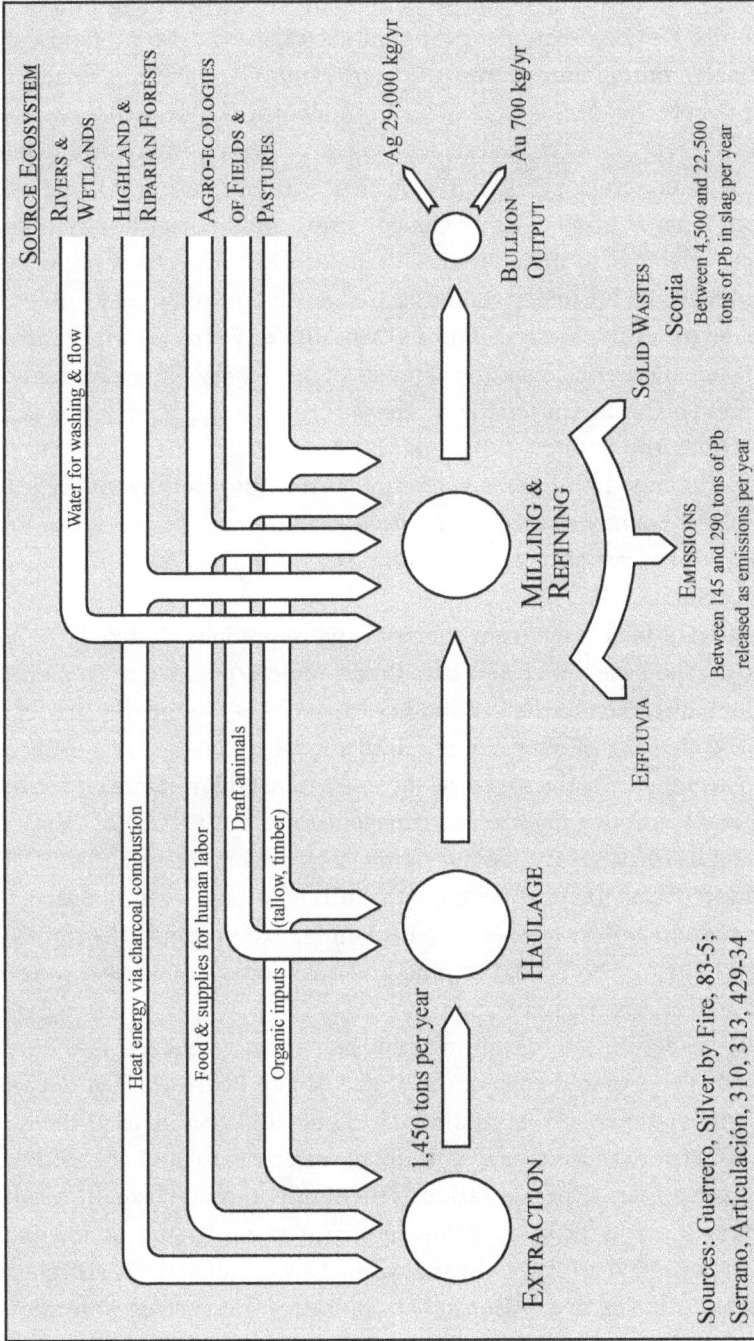

SOURCE ECOSYSTEM

RIVERS & WETLANDS

HIGHLAND & RIPARIAN FORESTS

AGRO-ECOLOGIES OF FIELDS & PASTURES

Water for washing & flow

Heat energy via charcoal combustion

Food & supplies for human labor

Draft animals

Organic inputs (tallow, timber)

1,450 tons per year

EXTRACTION

HAULAGE

MILLING & REFINING

BULLION OUTPUT

Ag 29,000 kg/yr

Au 700 kg/yr

SOLID WASTES

Scoria

Between 4,500 and 22,500 tons of Pb in slag per year

EFFLUVIA

EMISSIONS

Between 145 and 290 tons of Pb released as emissions per year

Sources: Guerrero, Silver by Fire, 83-5; Serrano, Articulación, 310, 313, 429-34

FIGURE 4 The colonial mining metabolism (illustration by Geoffrey Wallace, 2021, after original by D. Studnicki-Gizbert).

gardens). Marshes were ditched and drained.[19] Within decades, the existing ecology of the valley of the Tangamanga, once the Gran Tunal of the Guachichils, had been completely reorganized into an agrarian landscape.

Arguably, mining's most dramatic mark on the land did not come from feeding people and animals. It came from feeding the fires that were applied at different points of the extractive chain.[20] Miners lit fires at the mine face to heat and crack particularly recalcitrant sections of rock and ores. Magistral (copper-iron sulfite or chalcopyrite), an important reagent used in ore-processing, had to be prefired to make it effective. Most important, fire produced the blistering temperatures needed to smelt and refine the lead-silver ores. Colonial-era mining, whether at Cerro de San Pedro or across Spanish America, was one of the most heat-energy intensive regimes of its time. The heat required to process the ores of a single mine was easily twenty times the amount consumed by a town of five to six thousand people.[21] The amount of thermic energy needed for smelting and refining silver in New Spain was three times greater than that of the English iron-making industry, even in the eighteenth century as this last was moving into the industrial period.[22]

This kind of heat came from charcoal and to produce charcoal, the *carboneros* stripped the entire San Luis Potosí region of its trees. Quickly. In an earlier study, David Schecter and I estimated that fueling the district required the clearing of over one hundred square kilometers of forests per year, on average.[23] By the 1640s, as the extractivist, high-intensity mode of mining was beginning to taper off, travelers described a landscape entirely emptied of its trees, save "a few surviving yuccas upon the bald hills."[24] This was a transcontinental phenomenon: the hills of Minas Gerais in Brazil, the shrublands and gallery forests of central and northern Chile, the trees and quishar plants of Alto Peru, the pine and oaks forests of Mexico were all scoured to feed the smelters.

The areas deforested for mining had their surfaces opened to the work of erosion. The exposed soils of the Sierra of San Pedro washed out with every summer storm, revealing the naked blue and gray slabs of the local limestone. Without forests and soils, local watersheds could not modulate the intense pulses of precipitation. The Santiago and Tlaxcalilla rivers flooded regularly in the seventeenth century. During an August downpour in 1626 they swelled up overnight and washed away half of the parish church of Tlaxcalilla.[25] The dry season saw deeper drops in surface water flows. In the early eighteenth century the municipal alcalde of San Luis reported that the two main rivers of the valley no longer flowed all year round.

Chronic flooding, constant droughts, and "sterile" soils, he continued, were undercutting the valley's agricultural base.[26]

The extractivist metabolism not only changed local ecologies through its voracious appetite for work and heat. It also marked the land by drawing out a million and a half kilograms of ores from the earth and then dispersing this material across local waters, airs, and soils. Given the circumstances of its formation and emplacement detailed in chapter two, this subterranean matter was highly mineralized and contained high concentrations of toxic elements such as lead and arsenic. Waste material exited the ingenios in the slurries that were evacuated from the washing tubs and into the Santiago and Tlaxcalilla rivers. Not all the grimes washed away like this were toxic, to be sure. Nevertheless, heavy loading of freshwater bodies with even chemically innocuous silicates and salts will greatly disrupt their ecologies. Sedimentation raises turbidity levels and cloaks riverine plants and invertebrates with a suffocating cloak of fine particles. Wastes were also borne by the winds that lifted up the clouds of dust produced by the mechanical grinding of the ores and then dropped them across the valley. Or they were physically hauled out as score and then dumped atop the large and ever-building heaps that surrounded the ingenios. In two eighteenth-century maps of San Luis Potosí, they appear as small monticules littered around the cityscape.[27] They were dark and rough, most often shaped like sugar loaves. If the play of scale is anywhere near accurate, they towered three or four times higher than neighboring houses and occupied between a fifth to a third of a city block. These were the wastes piled up over two centuries of ore-processing. By then, however, the ingenios were no longer in operation. Their properties had been sub-divided and sold off. New houses were built within what used to be their compounds. But the growing urban metabolism of eighteenth-century San Luis could not, yet, digest the waste piles. So there they remained, looming over the houses, dark, dusty, and toxic. They were the material reminders of the wasting agency of mining.

• • • • • •

The metabolism of extraction drew upon and overloaded human bodies in ways that were analogous to its demands on the ecological bodies of the region. To increase the rate at which ores were mined and processed, and thus the rate at which profits were realized, the large mine—and smelter-owners of San Luis Potosí imposed a grueling pace and form of work. Arizmendi Gogorrón organized a quasi-carceral work regime at his ingenios. His hacienda at La Sauceda was a walled complex that combined work and life in the same space: large patios, heaps of scoria, banks of smelters and furnaces, warehouses, stables, and workshops, on the one hand; huts, dormitories, mess halls, truck gardens, and a chapel, on the other. Between them, the house of the mayordomo and the pillory.[28] All told around a hundred men, women, and children labored inside. Locked behind its walls, and under constant supervision, this group of nominally free wage workers were goaded (to death in one case), to match the relentless, twenty-four-hour pace of the smelter work. The same grueling pace was evident in Arizemendi Gogorrón's mines. They too were worked day and night. The cuadrillas were pushed to take out as much as they could, as fast as they could, without taking the time to shore up the mine works.[29]

His compere, Pedro de Salazar, another of the leading notables of the early boom, contracted a militia of armed men to go around the district and press-gang men into service. Salazar invoked Royal ordinances against vagabondage. If the men he had seized could not prove they had a master, they would serve him. Dozens of men were rounded up in this way and sent up to the Cerro de San Pedro where Salazar had them put to work excavating one of the first large adits (*socavón*) into the mountain. To make sure they kept to their labors, a thick gate of wood and iron was built at the mouth of the tunnel. Locked in, the impressed men could do nothing else but work, eat, and sleep in their underground prison.[30]

These forms of overexploitation layered themselves upon an already highly demanding form of work. Miners worked in what can only be described as extreme environmental conditions. As miners tunneled more deeply into the earth the ambient temperature rose to an almost suffocating degree. The mines at Cerro de San Pedro were known to be exceptionally hot and so miners stripped down to their breaches when working.[31] In time they would dig deep enough to enter into the water table at which point they also worked wet. This undermined their body's capacity to regulate its heat (one of the key energy expenditures of the human body at work) and subjected them, upon exit into the winds and cool air of the altiplano, to a serious thermic shock.[32] Adding the dust produced by working friable ores into these hot and damp conditions only exacerbated the risks of respiratory diseases. No direct notice of such has yet been documented for Cerro de San Pedro, but one report on the mines of Taxco in central Mexico from the period noted that the mines "abound in innumerable evils that provoke illnesses among the workers, particularly *tisi* [tuberculosis] and pneumonia that they regularly contract and frequently die from."[33] Occasionally, but unpredictably, miners encountered pockets of subterranean gases known as *bochorno*. These were accumulations of carbon monoxide and hydrogen sulfide that made candles sputter and breathing difficult. At higher concentrations, it could fell miners instantaneously.[34]

But the real sites of passage between the noxious substances of the underworld and workers' bodies were the ingenios of the valley of San Luis Potosí. The grinding mills produced a great deal of transportable and inhalable dust. Smelting vaporized lead and other heavy metals into the air, delivering acute and occasionally lethal doses into those who worked around the furnaces. A century later, Francisco de Gamboa provided a particularly vivid account of the dismal conditions that characterized this kind of labor:

> Here the unfortunate smelter [workers] suffer much . . . for the furnace is hot in the extreme and the crowbar is heavy . . . The smoke and vapor from the slag, which are quenched by pouring water upon them, and which are consequently carried down to the feet of the workmen, are poisonous; and as they drink water incessantly to relieve their exhaustion, they lose the use of their hands and feet, and become bloated. They are subject also to violent pains in the stomach. . . . this evil is inevitable and severe in its attacks on those employed in this laborious business, so important to the public.[35]

The atmosphere of the ingenios must have been thick with fumes. Smelting and refining ran day and night without cease. In the early seventeenth century, the chimneys, as Guerrero observes, were only slightly higher than a person. They were thus lower than the walls that enclosed the haciendas.[36] This impeded the dispersion of the fumes, fogging the enclave with lead-laced vapors that were constantly inhaled by everyone inside, whether they were at work or at rest.

· · · · · ·

Workers in the mines and smelters of Cerro de San Pedro and San Luis Potosí, found their bodies enmeshed into the extractivist metabolism. By providing labor, they gave of their energies. The rhythms and discipline of work sapped them of health and vitality. By working in the particularly dangerous and noxious environments of the mine and the ingenio, they absorbed its harms. These exchanges limn out the basic relations that existed at the interface between labor and extraction. In an extractivist scenario, the greatest part of the economic and social benefits generated from these relations was appropriated by the monarchy (through rents) and a small group of capitalist actors (through profits). This represented, in sum, the political ecology of negative reciprocity, in which health and vitality were drawn from workers without a counter-balancing flow of compensation, benefit, or comfort.

The negative reciprocities produced by the operation of the extractivist metabolism also operated at the larger scales of the regional ecology. The stripping of forests for fuel reduced the ecological vitality of the region. In the span of fifty years the highlands passed from a forested landscape—with all that comprised in terms of species diversity, ecological complexity, and net primary production—to a much sparer landscape of gravels, exposed bedrock, and the hardier suite of xeric plants (cacti and brush) that were able to hold on. As for the wetlands and rivers of the valley, for decades they had served as the dumping ground for the mass discharge of toxic slurries. With the progressive removal of tree cover, the valley's waters received less and less inflow from the highlands, and slowly but inexorably dried up. In 1635, Antonio de Espinosa wrote that the situation had reached the point that haciendas had to cease operations for lack of water to wash their ores.[37] After some forty years of high-intensity mining, and its attendant exhaustions and wastings, the land was spent and could no longer give.

6 The Miners' Share

In his sixteenth-century polemic, the Dominican Fray Domingo de Santo To-mas compared the mines of the Andean Potosí to a devouring maw. To produce silver, the King ordered the muster of thousands to work in the mines. But that work, for its dangers, for the over-exertions it demanded, and for the noxious tolls that it imposed, consumed bodies. Santo Tomas's image stands as a master metaphor for the negative reciprocities operating at the heat of an extractivist mode of mining.

In New Spain, the question of what mining brought to, or took from, the people and the land was a matter of some debate among elites. They, too, recurred to images of the body to capture the relations between extraction and society. José Fabry, an early eighteenth-century promoter, argued the case of positive reciprocity. In his view, mining was a vitalizing heart that produced a "superabundance of capital." It set in motion systolic and dia-stolic movements that irrigated and thus nourished the body politic. "Min-ing spills from its coffers the vital humour that is money and sends it through the innumerable conduits of the Body Politic, from the summits of the Great Aristocracy to the lowest of plebes."[1] The opposing view was offered by Domínguez de la Fuente in his report on the mines of Guanajuato. Like Santo Tomás he argued that the extraction of silver drew upon the vitality of min-ers. "They work so that others might rest. They die so that others might live." "They spend their time buried in the caverns of the earth, only join-ing the world so a few moments in the evenings and during feast days—a pitiful state of life that can be mistaken for death."[2] Another observer of the eighteenth-century mining boom in Guanajuato was Francisco Antonio de Mourelle. He drew attention to the workers who "had their lives con-sumed in the deep holes of the Valenciana." His account went on to detail how the work deformed miners' bodies and the general resentment that this produced: "people sigh for the unhappy victim[s] and detest the name of the mine-owner despite his great wealth, and abhor the wealth that comes from that kind of gyre."[3]

Absent here are the voices of those at the heart of the matter. We do not hear what the miners or residents themselves might have thought about the negative reciprocities wrought by mining. They did not passively resign themselves to a situation of overexploitation. In the specific case of Cerro de San Pedro there are simply too many cases of conflict, resistance, and outright insurrection in the archives to accept a theory of quiescence. Contestation, often violently exercised, was a common feature throughout the social histories of the mining districts of Spanish America.[4] What remains to be explored, however, are the views and practices that animated subaltern politics in and around the mines, and the place that ecological relations played therein. The fact that, beyond brief interventions in judicial proceedings, the direct testimonies of workers were rarely recorded makes this a delicate task. Still, enough evidence exists—of the actions, practices and, occasionally, opinions—to reconstruct an operating moral ecology in mining districts such as Cerro de San Pedro. In opposition to the negative reciprocities of extractivism, wherein local life was consumed and wasted in the name of the removal and appropriation of commodity wealth, this moral ecology sought a basic measure of justice in the relations between people and the mines. If the former surrendered energy, health, and vitality to the mines, then the mines should in turn give miners the means to sustain themselves.

· · · · · ·

At Cerro de San Pedro, and throughout Spanish America, three claims were made to the silver and gold drawn from the underground. Two were formalized while the third was customary and indeed illicit. The first was that of Spanish Crown which asserted radical title to the metals of the earth as part of its *realengo*. The second was that of the owners of a given mine. They claimed the riches extracted from the earth as their personal property. Their claims were formally demarcated in the *registro de minas*, the Crown's grant of exclusive extraction rights of the metals contained beneath a circumscribed polygon.[5] The third claim was the enacted customary right of miners to independently recover a portion of a mine's riches. Ubiquitous across the mining district of the Americas, in New Spain this was known as the *partido*, which I liberally translate as the miner's share. This claim was never recognized in Spanish law, and for centuries it was denounced as theft by mine owners and officials.

The earliest documented records of the partido in Mexico date to the very first years of the Spanish mining boom in Zacatecas in the 1540s. Unlike the districts of central Mexico where labor could be secured through the repartimiento, the northern mines were developed in areas where the local peoples were fewer in numbers, highly mobile, and either at war against, or in retreat from, the Spanish. Northern mine owners thus had to attract their work force through something other rather than compulsion. The problem they encountered was that while the mines rendered large amounts of silver, the districts was paradoxically poor in cash. The dearth of cash was a feature of other mining districts, including Cerro de San Pedro, and it made the payment of wages difficult. To resolve this obstacle, Spanish mine owners in Zacatecas instituted a kind of share-mining arrangement. Every miner was responsible for hauling out a fixed number of sacks of ore per week. This was the *tequio*, from the Nahuatl *tequitl* for "share" or "cut," a term also used to describe tribute or group work.[6] In exchange for this quota, the owners provided mine workers with lodging and food rations, and they cautioned the *pepena*. The pepena was any ore that miners recovered beyond what was stipulated under the terms of the tequio.[7] In time, this came to be known as the partido or share system and it became common practice across the mining districts of New Spain, and indeed the continent.

The roots of the partido system began at the mine face. This was, recall, the territory of the cuadrillas. Mine owners and officials rarely, if ever, ventured down into the tunnels and shafts of the mines. Which veins to follow, how to excavate the ores and mother rock, where to build supports, leave pillars, cut drainage, who would do what: all of these matters were handled within the cuadrilla. The subterranean spaces of the mines were where Carter Goodrich, in his classic account of Virginia coal miners, sourced a "miner's freedom." The labyrinthine lay out of mines, the dangers that they enclosed, the organization of work into small and independent teams "gave miners entire freedom of action and untrammeled privilege of working when and as they please."[8]

This kind of autonomy was critical for the miners' share. In enabled the cuadrillas, rather than officials or mine owners, to handle the sorting and selection of ores that came off the mine face. Miners gauged which ores should be given up to the mine owner as the tequio and which parts should be kept as their own as pepena. Mine owners, or rather their stewards, only came into possession of the ores of "their" mine when they emerged by the sacksful at the mine-mouth. In the Pachuca and Real del Monte mining district, elites forced a mixing of the sacks to combine the ores destined for the owners and those taken as the miners' partido. This was known as the *revolutura*. It was a complex game played between miner and mine owner that pitted skills of observation, sleight of hand, and verbal wits.[9] At stake: who would walk away with the richest share.

The mines of Cerro de San Pedro were particularly enticing for miners. High value, low-volume gold-laden ores, or pure plates and rounded globules of gold could be buried into the sacks counted as pepena, leaving the meaner ores for the owner. In one remarkable episode, the miners of Cerro de San Pedro entirely dispossessed the owner of his mine. This took place in 1628 when the miners at the El Rosario mine encountered a massive envelope of soft gold-bearing ores that came to be known as the Bolsón de Oro, or the Golden Purse. As news of the find circulated, over a hundred miners representing dozens of cuadrillas piled in to work it out. They came from other mines, and they worked the deposit during the night. They were working beyond the formal terms of any agreement with the owner of the mine José Briones. Technically, they were claim jumping. Briones of course complained to the Alcalde Mayor, Alonso Nieto Dolores, but was ignored. The Alcalde simply refused to interfere with the cuadrilla fearing that any such meddling was sure to be met with a stout response from the miners. His acquiescence underscores the de facto power of the miners' claims. A handful of xiquipil sacks seized at the beginning of this episode gives a sense of the stakes involved. They were found to hold seventy-thousand pesos' worth of gold. This compares to the building costs of Juan de Zavala's ingenio, one of the largest in the valley of San Luis Potosí at the time.[10]

While the appropriation of the Bolsón de Oro was exceptional, the recovery of gold-silver ores of the Cerro de San Pedro as pepena was a long-lasting and day-to-day part of local subsistence. It survived well into the twentieth century, not only in Cerro de San Pedro, but also in districts such as Parral and Batopilas in Mexico and Potosí in Bolivia.[11] The partido underwrote an entire household economy. Instead of being smelted in the large Spanish haciendas of the valley, the ores brought home by workers were roasted in small household smelters known as *fuelles*. These were roughly the size of a bread oven and could only process small batches of ores per charge. In exchange for the ores, the artisanal refiners gave bread, maize, pulque to the miners. This was exchange was known as the *rescate* or bartering of the ores. For miners and their families, the pepena was a form of raw currency.[12]

Between the control of the mine-face, the appropriation of the highest-value ores, and the existence of cottage-scaled smelting, there emerged distinct circuits of extraction, processing, and value-formation. They coexisted with the large-scale, capital-intensive, circuits controlled by the top mine—and smelter-owners of the district. One hesitates to call these non-capitalist circuits, since they did generate and circulate modest amounts of capital from the work of silver and gold extraction. All the same, they operated at much smaller scales and at lower rhythms. They were embedded within local exchanges and relations. In sum, they worked according to other logics than the extractivist sector and, in this sense, were positioned in tension with it.

In 1599, the Viceroy of New Spain Gaspar de Zúñiga y Acevedo set out to break the economy based on the miners' share. He passed a decree prohibiting the rescate of ores and banning the artisanal smelters across New Spain.[13] In Cerro de San Pedro and San Luis Potosí, the decree was immediately protested as a prejudice to the people.[14] A year after the ban was passed, a local official reported that many indigenous miners and their families had moved out of the district. Without the pepena, he explained, food and upkeep had become too expensive.[15] New mines were opening up in the Sierra de Pinos, a two-day cart journey to the west, as well as to the north. They had gone to try their luck there.

Overall, only a minority left the valley and mines of San Luis Potosí. Subsequent calls to enforce the ban show that the household mining economy of rescate and fuelles continued.[16] In 1616, the cartel of large-scale capitalist mine—and smelter-owners, including the likes of Pedro Arizmendi Gogorrón, Juan de Zavala, and Pedro de Salazar, succeeded in passing a resolution that prohibited the smelting of any metal outside their ingenios. The resolution aimed at cutting out their many small-scale competitors.[17] In 1683 their successors tried to control the mines and the workforce. They called for the expulsion of all independent miners from the Cerro de San Pedro. Only those who could prove that they worked for a mine owner or lived on the grounds of his operations were authorized to remain.[18] But again, these decrees should be taken as the traces of a losing struggle rather than a measure of elite control. Fuelles were inventoried and registered in the documents well into the eighteenth century. So, too, were cases of rescate.[19]

By the seventeenth century the miners' share was ubiquitous across all the mining districts of the Americas. In the gold fields of Panama, the Pacific coast and the central valleys of Tierra Firme, Afro-Creole slaves transformed the weekly Sunday rest into the day where they could mine the placers on their account.[20] Similar customs prevailed in the gold fields of Minas Gerais in Brazil where African and Creole slaves mined in conditions of relative autonomy away from the immediate supervision of mine owners. They were held to providing a certain daily quota to the masters, but the remainder was theirs to keep and, as in Tierra Firme, Sundays were fully at their disposal.[21] For them, the miners' share was a critical means to purchasing their freedom. In the Andean mining districts of Potosí and Oruro the practice was known as the *kajcheo* where it took the form of weekend raids in which cuadrillas forced their way back into the tunnels to collect the richest ores stocked there over the workweek.[22] Accounts of the Potosí kajcheo describe a mobbing-like movement as miners massed together to barge their way into the mines. Economic historian Enrique Tandeter provides an estimate—based on one of the few extant runs of solid documentation—of the total amount of silver obtained in this way. He finds some four thousand miners taking out a full third of the mine's production in 1759.[23]

Historians of Latin American mining such as William French and Monica Navarro analyzed the partido in economic and social terms, either as a substitute for, of supplement to, cash-based wages, or as part of a moral economy, that is as a customary claim to a subsistence right.[24] Both dimensions were clearly present in the claim and practice of the miners' share. What deserves to be emphasized, however, is that such claims took place within a field of socioecological relations; that is within relations between laboring bodies and the conditions and ambiences of work of the extractive complex. Claiming the best ores, and defending that claim, counterbalanced the exactions and dangers that this work imposed. Against the negative reciprocity of the mine's maw, the moral ecology that surrounded the partido emphasized the positive reciprocities of what subterranean riches could give to those who, in their labors, gave of their bodies and vitality.

$$\cdot \ \cdot \ \cdot \ \cdot \ \cdot \ \cdot$$

In the mid-eighteenth century, a series of linked insurgencies swept through the central mining districts of New Spain. They took place in the midst of a general renewal of the Viceroyalty's precious mining sector and, indeed, were in important ways provoked by the measures deployed by the Bourbon monarchy. Over the course of two years, 1766 to 1767, one after the other, mining districts were shut down by mine and smelter-workers and their allies: Real del Monte, Pachuca, Cerro de San Pedro—San Luis Potosí, Guadalcazar, Guanajuato, Bolaños, and Zacatecas. They provoked a serious enough disturbance to temporarily stop the year-on-year increases in Mexican bullion production. Defamed by elites in every possible way, these uprisings expressed in political speech and action the moral ecology for subsistence, bodily health and renewal, autonomy, and justice.

The troubles began in Real del Monte and Pachuca, some ninety kilometers to the north-east of Mexico City. It was there, in July 1766, that miners and hacienda workers abandoned the mines and works. It has been called the first workers' strike in Mexico.[25] Akin to eighteenth-century sailors "striking their sails" to halt maritime labor, the collective action of these workers served to interrupt the flow of extraction and processing at the heart of the extractivist metabolism. Judging from their actions, however, the aim was not the destruction of the mines or their wholesale appropriation, but rather restoration.

At the heart of the matter was the partido. The leading silver magnate of the district, the Conde de Regla, considered it the "rock of scandal and the apple of discord of the infamous plebe."[26] In concert with Bourbon officials, the Conde had abolished the partido. Workers responded by dropping tools and filling the streets. They quickly numbered in the thousands. They attacked the jail and other sites of royal authority. When the Crown's representative Francisco de Gamboa arrived to settle the scandal, the crowds welcomed him and clamored for royal justice. A commission presented their petition. The terms are suggestive for the relations it framed between the bodily exhaustions of extractive labor and the compensations provided by the partido. "For many years now," it read, "we have worked in the mines enduring hunger and fatigues because the only relief the miners draw from the mines is the partido that they take as their share . . . it is what feeds and fortifies them for the labours and atmospheres of the mine." "Without the partido," they continued, "we have become consumed, annihilated and finished."[27]

Rather than an object of discord, miners presented the partido as the keystone of their laboring and bodily relations with the mines. It was a form of restitution, materializing what nature gave to the miners in exchange for their health and vitality.[28] It was what made hard work in toxic environments bearable. It kept them alive. For the Conde, the partido represented an unknown, but still unacceptable, portion of what he considered as his treasure. He refused to concede. And so the miners waited him out. With work suspended, no one manned the mule teams and whims that drained the mines, and the waters began to fill the tunnels. After eight months, the Conde admitted defeat. If this continued much longer, he would lose the mines for good. In March 1767 Gamboa restituted the partido, and the miners and workers returned to work.[29]

For the Visitor General José de Gálvez, only just arrived in New Spain, it was imperative that matters remained there. He was particularly concerned with the situation at Cerro de San Pedro. By the 1760s the mountain had been all but abandoned by the local elite. Mining continued but under the control of small crews of miners. They worked on their own account in mines that were formally held by others. The ores were processed in the fuelles and were neither minted nor tithed. In Gálvez's view, the popular control of the gold and silver of San Luis Potosí was an aberration, an inversion of the proper social and political order. Worse: the miners of Cerro de San Pedro were reported to be particularly insolent with their betters, and the district a hearth of rebelliousness. With work and order at Real del Monte finally restored, Gálvez decided to rein in the miners of Cerro de San Pedro and put an end to growth of what he called this "monstrous polity" marked by the "deformity, dissonance, and the bad order of its parts."[30] An edict was sent calling on the people of Cerro de San Pedro to find a master or face impressment for vagrancy. As a follow-up, a second edict banned the bearing of arms save by royal privilege.[31]

For the people of Cerro de San Pedro, the orders to disarm and surrender themselves to the Spanish masters of San Luis was pure provocation. On the day that the edicts were posted, a crowd formed and attacked the lieutenant mayor of Cerro de San Pedro. Then, on May 27, hundreds of workers from the mines and smelters of Cerro de San Pedro, Real del Pozos, and the environing sierras descended on the city of San Luis Potosí. They were known as the *serranos* and they dressed in red sarapes. In the main plazas of the city they were joined by hundreds more: the people of the city's indigenous barrios, the herders and farmers of its surrounding ranchos, and a delegation of one hundred indigenous Pamé fletchers from Armadillo (forty kilometers away) who had arrived to serve the cause. Here was, in the flesh, the monstrous body of Gálvez's phantasies. "The streets filled with *zorillos*, the houses with plotters and the fields of scandals."[32] They were headed by the Governors of the city's indigenous barrios and two miners of Cerro de San Pedro, Juan Antonio Osorio, and José Patricio Alanís, also known as "el cojo—the cripple." They first addressed themselves to the Alcalde Mayor of the city but finding no satisfaction proceeded to stone and then invade the Cabildo. Over the next hours the crowd broke down the gates of the jail, smashed the public pillory and gallows, raided the powder magazine and the Royal tobacco storehouse, and then ransacked the Spanish merchant houses for their cloths and goods.[33] "They then spent the rest of the day in the stalls of the city offering drinks to whoever passed by."[34]

It would take two months for Gálvez to arrive in San Luis Potosí and began the work of suppressing the insurgency. His journey north from Mexico

City was delayed by a regional rebellion of the Tarasco indigenous communities of Michoacan and miners' uprisings in Guanajuato and San Luis de la Paz. In that time, an interregnum was established, a temporary alternate political order. It was a republic that federated the miners of Cerro de San Pedro with the people of the Indigenous and Creole barrios of San Luis Potosí. It was a short-lived assertion of popular sovereignty over local territory: the mines, the fields, the city.

Over the next months a number of petitions and plans were tendered by the insurgents. "We are," they wrote, "the true masters and miners of this Real [de Minas] because those in the city who call themselves miners neither work the mines nor enter them." They were poor, but they did what they could to work the mines. They had sold their women's dresses to buy the candles needed for the tunnels and "now these people want to take what's ours."[35] The insurgents signed their documents as the real *señores* of the mines. Together with the people of San Luis Potosí, they drafted a new constitution for the valley. It would be ruled by an elected triarchy composed of the three kings of Biblical inspiration: one for the Indios, one for the Negros, and one for the Españoles.[36] Writing ten years after the events, the Mexican *letrado* José Granados y Gálvez imagined a mountainside castle for one of these kings, *el cojo* José Patricio Alanís. It was established in one of the mining tunnels of Cerro de San Pedro. There he held court "like another Pluto," complete with crown and scepter, dispensing titles, orders, and dignities. On the gates of this underground palace was inscribed, "New King, New Law."[37] Granados y Gálvez's sketch was political grotesque, it described the monstrous inversion so feared by Gálvez. It turned the royal coat of arms of the city on its head. Instead of the emblem of Royal Justice ruling atop the Cerro de San Pedro, the plebe had claimed dominion over its treasures and ruled, as it were, from below.

When Gálvez heard the news of what was happening in San Luis and its environs, his fury and indignation were absolute. The plebe had "dared crown themselves as little kings."[38] He would expunge them and he would restore justice and authority where it ought to be—atop, not inside, the Cerro de San Pedro. Gálvez eventually arrived with his militias in late July. He blockaded the valley and sent calvary units into to city and up to the mines. Mass arrests brought over four hundred individuals to trial—women and men; miners, peasants, and workers; indigenous, Afro-Mexican, and otherwise. Two hundred and fifty-seven were sentenced on August 7: to public whippings and other bodily mutilations; to penal servitude in the presidio of Havana; to hanging. The leaders were hanged and then decapitated. Their houses were torn down, their families exiled, and their lands salted. Their cadavers were spiked on the emptied lots to hang there until "time consumed them."[39]

At the end of that long and grim day, Gálvez appeared on one of the balconies of the plaza. "In view of the innumerable concourse of corpses, still warm and hanging from the gallows" he delivered a long oration on Royal justice and the punishment of "*aquellos infelices.*"[40] He then delivered the new royal orders. The miners of Cerro de San Pedro were dispossessed en bloc. Only "masters" could legally hold title to a mine now. All miners had to have a master. The partido was abolished, so too were the small-hearth smelters. And as a final incitement to local elites, the mining district of San Luis Potosí was exempted from taxes.[41]

In those same years, miners' insurgencies erupted in other mining districts of Mexico: Guanajuato, Bolaños, in the neighboring Real of Guadalcazar.[42] In the case of the last, tumult sprung from a mass game of La Chueca, a no-holds barred form of Tarascan field hockey, that had been improvised by the miners of the district to provoke the authorities.[43] In Guanajauto thousands of miners descended to occupy the city, playing guitars.[44] The defense of the miners' share was at play in other miners' rebellions of the eighteenth-century Americas: Oruro in 1781, Tumaco and Barbacoas on the Pacific coast in 1791, as well as a long string of revolts that threw the Brazilian gold fields into turmoil from 1704 to 1756.[45] Across these different territories and contexts the common demand was the miners' claim to their share of the subterranean bounty of the Americas. It materialized the exchange between the work and vitality of the miners and the wealth of the earth.

7 The First Death of Cerro de San Pedro

In Spanish America the mining boom, that period of frenetic activity triggered by the opening of untapped deposits, was known as a bonanza. The word came from the world of the sea. It originally described fair-weather sailing, those times when the winds and currents aligned themselves with the ship and sent it cutting smartly across the water. In the mining districts the bonanza came to describe those periods when the ores were both easy to recover and generous with the silver and gold they rendered. Or it marked the beginning of a particularly rich and potent vein. A bonanza meant windfall mining, a situation when nature's underground bounty was abundant, even prelapsarian. The counter-season was the borrasca, the sea tempest. It described the times when a good vein began to pinch out, when the amount of metal recovered from the ores dropped, when the water began to pool up in the tunnels, or when an entire district fell into abandonment.[1]

Cerro de San Pedro's first bonanza began in the early 1590s. For the next forty years, the capacity to maintain high rates of ore extraction and refining resulted in correspondingly high outputs of silver and gold. In the 1630s, at the crest of the bonanza, the records of the Caja de Fundición show that they supplied a remarkable 96 percent of all the gold legally minted in New Spain (at a time when Mexico was the most important gold producer in the Americas). In terms of silver production, the mining district of San Luis Potosí ranked second only to Zacatecas.[2] And then the decline set in. Decade after decade, for the next hundred years, production dropped. The troughs of the borrasca came in 1712 (for silver) and 1745 (for gold), years when Cerro de San Pedro only rendered 4 percent of the silver and 15 percent of the gold that it had produced during the boom.[3]

The century-long decline of mining at Cerro de San Pedro was situated within the general levelling off of New Spain's economic growth. In the case of the mining sector, after the explosive growth of the sixteenth and early seventeenth century, the secular trend in precious metal output flattened out. The reasons for the slow down are complex because the pursuit of precious metal mining was linked to a range of external factors that included population dynamics, the agrarian economy, global bullion trading, and climate change. But there were also dynamics that were internal to mining itself, or more precisely, to the kind of high-intensity mining that had characterized the boom. By focusing attention on these allows a view of the internal contradictions that existed within the extractivist mode of mining. At the Cerro de San Pedro, the borrasca was in important ways produced by the bonanza. As miners advanced into the mountain, the material and chemical challenges involved in extracting and refining ores rose. In metabolic terms, this forced an increase in energetic and material inputs simply to maintain output and profits. Without profits, elites withdrew their capital and shuttered their mines and smelters. This was the pattern observed in Cerro de San Pedro.

In the early years of the bonanza the work of extracting ores from the Cerro de San Pedro mainly consisted of digging out soft dark ores in open-air trenches and pits. While not easy work by any stretch, these were boon days compared to the operations that followed. Tunneling through the veins that radiated through the upper layers of limestone required the collective effort of one or more cuadrillas. As time went on, and miners bore deeper and deeper into the mountain, increasing amounts of labor and energy had to be expended on managing waters, hauling ores to the surface, and shoring up the tunnels and bodegas. All of which had a cost. It presented mine owners with an unhappy choice. If they maintained the rate of extraction, the profit margins on every quintal of ore brought to the surface declined. On the other hand, if they did not make the necessary investments in labor and dead works, they had to accept a gradual slowing down of, or even the outright end in, the outflow of ores. For larger capitalistic operators, those who integrated multiple mines to large smelters such as Martín Ruiz Zavala, Miguel de Maldonado, or Pedro Arizmendi Gogorrón, slowing down was not an option.[4]

By the 1690s, the elite mine owners of San Luis Potosí came to the position that continuing in this way was no longer tenable. To maintain profitability, large-scale investments had to be made. These works were essentially infrastructure that aimed to reach what were thought to be rich untapped zones of the Cerro de San Pedro deposit. A plan was developed to drive an adit into the heart of the mountain from its side. This would provide access to the *criadero* or nursery from which all the ore veins were thought to originate. Its width and steady incline would also drop the costs of hauling ores and provide outflow for waters. It was a massive undertaking for the period, one that would need the coordination of many cuadrillas, and it exceeded the capital resources of individual mine owners. A cartel was formed. The Crown extended a sizeable loan. In 1694, the excavation of the *Socavón del Rey*, the King's Adit, began but work was abandoned long before reaching its destination.[5] In the same years, another syndicate was formed to mine out what remained of the large pocket of soft ores on the northern slopes of the mountain. These had been previously worked by small teams of miners in trenches, but their structure was dangerously friable and constantly caving in. A large-scale and coordinated operation was proposed, a *tajo abierto*, an open pit that would muster hundreds of workers and excavate the top of the mountain wholesale. The indigenous alcaldes of the barrios of the city were instructed to provide the manpower. The Crown furnished a forty thousand-peso loan. But here too the project was abandoned within a few years, leaving a large crumbling and canyon-like gash in its wake.[6]

In parallel with the growing difficulties besetting the work of extraction, the smelters in San Luis found that they were able to extract less and less metal from the ores that they did receive. This diminishment was expressed as a *falta de ley*, that is a drop in ore grade, and it was seen as a key cause of the decline of the district by observers in the late seventeenth and early eighteenth centuries.[7] As the chemist and historian Saul Guerrero has clarified, however, what was happening was not so much a drop in the average amount of gold or silver held within a given unit of ore, but rather shifts in their chemical composition. These shifts in ore chemistry were related to the downward travel of miners deeper and deeper into the mountain. The ores encountered during the early years of the bonanza were silver chlorides formed by the geological weathering by air and water. They were compounded with lead. These chemical attributes allowed them to be smelted without need for mercury and the more complex processing sequence that amalgamation required.[8] As the mines descended, miners dropped below the upper strata of weathered rock and into the water table. There they increasingly encountered silver ores that were composed with sulfides. Gamboa described these as "rebellious ores."[9] They required higher and higher temperatures to smelt, and increasingly complex sequences of dressing and chemical amendment to spring the metals they contained.[10] The majority of these sulfide ores never entirely surrendered their treasure. Consequently, a significant fraction of the silver and gold they contained was discarded into the waste piles or washed out into the rivers.[11] In his account of the mines at Real del Monte, Humboldt observed that an enormous amount of silver was lost in this way, carried away in the slurry and slimes that were thrown in the river.[12]

To the difficulties building up around ore extraction and refining was added the rising challenges of securing charcoal and timber. This, too, was a problem of the system's own making. As seen above, high-intensity mining drove region-wide deforestation across the sierras around Cerro de San Pedro and San Luis Potosí. Some twenty years after first strike, the smelters were supplying themselves with charcoal produced in forests over a hundred kilometers away.[13] The expanding radius of deforestation raised the energy costs of smelters at the same time as the shifting chemistry of ores drove energy demand upwards.[14] Deforestation also cut into the supply of the sizeable timbers miners needed for shorings, ladders, and other structural work. By the eighteenth century the miners at Cerro de San Pedro complained that these were impossible to find at any price.[15]

The problem here was not the cutting of forests for charcoal and timber as such. In principle, adequate woodland stewardship could maintain the wood supplies needed for pre-industrial mining and smelting operations. Royal decrees and Viceregal instructions legislated in this sense.[16] In the case of naval-grade timber, Spanish forest administration was able to assure supplies for hundreds of years.[17] The contemporaneous early modern British iron industry showed operators as agents of an incipient form of sustainable yield forestry. They were not foresters, but they were interested in maintaining local fuel supplies.[18] In Cerro de San Pedro, however, a more extractivist logic took hold under which constant increases in demand outstripped rates of forest regeneration and exhausted a key biological foundation of the industry. Overexploitation of forests, and the negative check to extraction that it caused, was also a problem in other mining districts such as Guanajuato and Real de Catorce.[19]

By the early eighteenth-century Cerro de San Pedro and San Luis Potosí found themselves in the depth of the borrasca. It was a season of abandonment. The ingenio built by Miguel de Maldonado was a great ruin, its walls crumbling, its roofing plundered of its tiles, and its banks of furnaces quieted, its chimneys a brooding line of sentinels. The same silence reigned over the smelters of Juan de Valle, Antonio de Espinosa and others.[20] An enterprising group of former workers was caught dismantling the chimneys of an ingenio so as to better glean out the rich residue of silver and lead condensed on their inner walls.[21] Up in the Cerro de San Pedro mine owners tried to protect their mines from "ore robbers" who not only stole but also dangerously weakened the mines by "pillaring," that is removing the rich ores that remained in the structural columns.[22] But these were half-hearted efforts. By then the elites in San Luis Potosí had lost their appetites for sinking more money into the mountain. On tour through the valley in the 1740s, José Villaseñor y Sánchez wrote that mining, "is much diminished: for the lack of grade in the ores, and the lack of spirit amongst the owners who are unwilling to risk their capital to meet the costs of working the mines."[23]

Since the collapse of the projects for the King's Adit and the open pit at the turn of the century, the Potosino elite had been selling off their mining properties to former workers at a rebate. And what the miners did not purchase, they simply occupied. By the end of the 1740s historian Mónica Pérez Navarro shows that the Cerro de San Pedro was under the de facto control of the miners.[24] From taking a share of the ores under the moral ecology of the partido, the miners now held the mountain in full.

The occupation of the Cerro de San Pedro by cottage miners in the wake of the boom operated by a large-scale, extractivist mode of extraction shows that terms such as abandonment or exhaustion need to be considered with care. There was indeed an abandonment here, the abandonment of that cartel of highly capitalized mine and smelter owners who, during the boom, were responsible for over 70 percent of bullion output. Their participation lasted for about fifty years, roughly two generations, before the pinch on profits began to be felt and they began to scale back. Small scale miners and cottage smelters, on the other hand, operated both during and after the boom. By the time José de Galvez arrived in 1767, this sector had been drawing their subsistence from the Cerro de San Pedro for over one hundred and seventy years. The longer duration evidenced by cottage mining rested on the smaller scales and lower intensities at which they operated. They could be more selective in the ores that they removed. This allowed for good rates of metal output even at the lower processing speeds of a small fuelle. Their smaller scale of operation meant that although charcoal was dear, lower quantities of fuel were needed—a cartload could keep a fuelle in operation for weeks. Capital costs, both fixed and mobile, were lower and more easily covered. Labor came from familial networks. Played out over time, these small-scale, low-intensity operations assured the sustenance of the household economy. And, of course, there was always the prospect of those micro-bonanzas that Cerro de San Pedro was reputed for, those pockets of sands that hid stones and platelets of pure gold.

The notion of exhaustion, as Guerrero's observations show, needs to be revisited. The deposit of minerals held within the Cerro de San Pedro still had a great deal more to give, provided the appropriate techniques of extraction and processing. More broadly, too, the ecological and bodily exhaustions incurred by the extractivist mode of mining disappeared as the boom tapered off. As fuel demand dropped and time advanced, some measure of afforestation took place. How quickly, at what scales, and with what ecological consequences has not been researched for the highlands around San Luis Potosí. What is known is that on the eve of the next extractivist boom in the 1890s the Sierra de Álvarez was once again cloaked in timber. Post-mining afforestation has been documented for the Real de Catorce mining district and it shows not only a return of forests but also of increased surface water flows in previously desiccated areas.[25]

As for laboring bodies, it is true that the hazards associated with underground mining continued to exist, even for small crews of gambusino miners. Pillaring—that is the working out of seams of high-grade ores contained in the columns that were left to support the tunnels—was an extremely risky endeavor. The exertions, the encounters with pools of bochorno gas, the chance of a cave-in, all these confronted gambusino miners. By the same token, the lead emissions from a yard-side fuelle were no less noxious to human health, even if doses and exposure times were lower given slower processing rates. In the event, for the families of miners and smelters the risks and tolls appear to have been acceptable. They were, after all, at the Cerro de San Pedro because they had purchased a claim or were squatting one. That is, they were not there under compulsion. They did not labor behind walls under the supervision of others but for themselves. They had found their sustenance in the ruins of the extractivist bonanza.

This degree of popular autonomy and control over what was considered to be the Crown's patrimony was unacceptable for José de Gálvez. The severity of his repression may have pushed the district into a true bust. The prohibition of "masterless" miners, the mass prosecutions, the executions, and the salting of household plots presented a grim scenario for those who might have thought of returning to work the mines. And, coming into the 1770s, there were other booms underway, notably in nearby Real de Catorce, that would have presented much more promising prospects. Even so, this abandonment, too, was temporary. In the early nineteenth century, following the passing of the Spanish monarchy and now under republican rule, the gambusinos had returned to the Cerro de San Pedro and resumed operations.

Part II   **Extractivism Revived**

. . . .

8 Extractivism Revived (1740s–1940s)

· ·

Mining would boom again at Cerro de San Pedro in the early 1890s, three hundred years after the first strike bonanza of 1592. In the long borrasca between the two, however, mining as such never entirely ceased. It was continued by small scale system of gambusino mining and cottage refining that was quietly embedded in the local rural economy. The boom of the late nineteenth century thus marked a return of a particular kind of mining, a system that operated at large scales and high intensities, and was, once again, articulated with capital of the highest levels.

This revival is interesting for what it reveals about the relations between extractivism and exhaustion. On the one hand, the extractivist logic of intensification, of ramping up the rhythms of extraction and refining, of constantly increasing the outflow of silver and gold, inevitably sped the travel toward states of exhaustion. This trend was, as seen, more complicated than the simple drawing down of a set stock of mineralized ores. It also involved the lowering of average metal to ore ratios, as well as the growing material challenges posed by mining at further depths and smelting ores of increasingly intractable chemistry. These trends drove up costs, reduced profit margins, and eventually led to the withdrawal of large investors. On the other hand, however, if these challenges could be resolved then supposedly exhausted deposits could once again provide the material base for new cycles of capitalist extraction. The route there consisted in further intensifying the extractive metabolism; that is through stepwise increases in the inflow of energy, labor, and other inputs needed to raise throughput of ores and thus return to profitability.

The efforts required to quicken the metabolism of mining were not only material in nature. Considerable investments of capital were required. The state had to renew its support through laws, administration, and outright subsidies. Finally, a renewal of vision was needed to return the work of extraction to its position as a leading political project, one that was not only possible but in fact predestined by nature and providence. All of which to say that the revival of an extractivist mode of mining required a broad "re-assemblage" of extractivism as a political economy.

This chapter draws back from the specific account of Cerro de San Pedro to take in the broader view of how extractivism was revived at the scale of Mexico as a whole. It sets the broader context for the revival of extractivist mining at Cerro de San Pedro and San Luis Potosí from the 1890s and forward. Between the late seventeenth century decline of mining in Mexico and the late nineteenth century there were, in fact, two important cycles of extractivist revival. The first took place in the eighteenth century under the administration of the new Bourbon dynasty. Cerro de San Pedro did not participate in this expansionary phase for the reasons suggested in previous chapters. This marked the one occasion when it fell out of step of the broader shifts that made up the long history of mining in Mexico. The Bourbon project to revive bullion production was remarkably successful. It not only pulled the mining sector out of its slump; it pushed the volume of output well past the records achieved during the first bonanzas of the late sixteenth and early seventh centuries, making Mexico the world's most important supplier of silver in the eighteenth century.

In 1810, the boom was abruptly cut short by the insurrection launched by Miguel Hidalgo and the political strife of the first decades of independence in Mexico. Although much emphasis has been placed on the physical damage to the infrastructure of mining and refining by insurgents, the real impact came through the interruption of the capital flows that underwrote and animated the work of mining. Absent this capital, the metabolism of high-intensity mining could not continue or recover. From 1810 to the 1880s the sector lapsed into a deep recessionary trough in which only gambusino mining survived.

The second cycle of revival came under the administration of Porfirio Díaz, begun in 1876/77. His presidency opened Mexico's mining sector to foreign capital, the key to transforming the scales and rhythms of precious metal mining. The great majority of this capital came from the United States, and with it an industrial mode of bullion extraction. The industrialization of mining brought previously worked (and reworked) mines back into operation and then ramped up their output to unprecedented levels. The industrial mining bonanza, which would last until the Second World War, was the product of greatly increased flows of capital, the powerful new energetics of fossil fuels, and the new technologies of industrial extraction and processing. In strictly commodity terms, industrial extraction in Mexico achieved outputs of silver and gold far beyond anything achieved under the Bourbons, or the previous frontier mining booms of early conquest and colonization.

Almost all of it came from deposits that had been considered mined out, exhausted, and abandoned.

......

The Bourbon return to bonanza began with the idea that the mines of Mexico were far from spent. Renewing with older cornucopian traditions, observers, promoters, miners argued that the greatest portion of their riches remained to be taken. In the case of Cerro de San Pedro, mine owners reported that the greatest prosperity remained to be discovered since everyone affirmed that the mountain continued to guard large belts of virgin gold. Rumors also ran of a plug of solid gold, half a yard thick, lost when a tunnel caved in.[1] As for Villaseñor, visiting in the early 1740s, he clearly thought that the Potosino mining elite was wrong to retract their capital from the mines at Cerro de San Pedro. "Experience shows," he wrote, "that it is in the further depths that the greatest riches are to be found."[2] The Mexican arbitrista José Antonio Fabry argued that all the mines of Mexico, whether abandoned or in borrasca, and irrespective of the quality or grade of their ores, were capable of doubling production within five years.[3] The problem, he wrote, was the lack of profitability. If the Crown were to assure a plentiful and cheap supply of mercury, margins would go up, spirits and the mines would all return to operation.

Reformist officials and ministers of the Bourbon Crown addressed themselves to the challenge of lifting the borrasca. To assess the overall state of the sector, the Crown systematically surveyed the mining districts from Oaxaca to northern Chihuahua. Through questionnaire, inspection, and tabulation it compiled information on geographical context, climate, production rates, the kinds of ores and their grades, the depth and structure of mines, as well as the inputs consumed by mining and refining: materials, energy, and labor. Surveys were run periodically between the 1740s and 1770s, until they were systematized in bi-annual reports submitted by local mining officials to the Tribunal de Minería in Mexico City. They made the domain of Mexico's extractive system legible as a whole and in its constitutive parts. They provided the knowledge foundation for state intervention.[4]

Materially speaking Bourbon reforms in the mining sector aimed at increasing the flow of material and energetic inputs into the mining districts. One key input was mercury. Fabry's argument rested on the fact that mercury increased the chemical range of refinable ores. By 1743, amalgamation was long-used and widespread in Mexico. The problem was supply. The

Crown intervened directly to increase the supply of mercury recovered at the mines of Almadén, Spain, Mexico's main purveyor. It mainly did so by seeing to the recruitment and health of the miners. Spanish historian Rafael Dobado documents how the Crown created an early modern "micro welfare state" at Almadén, complete with hospitals, the first system of workers' pensions, subsidized food, and housing, as—not least—a blanket exemption from conscription for anyone choosing to mine out the quicksilver.[5] By the 1790s Almadén was furnishing Mexico with sixteen thousand quintals per year, a four-fold increase since the 1740s. The Crown heavily subsidized the price of mercury, and halved it, twice, over the course of the eighteenth century.[6]

Also increased was the supply of black powder to the mines. This explosive assisted miners who had hitherto only broken out ores through the bodily work of bar and hammer. It was packed into paper cartridges and sealed with pitch or wax.[7] These *cohetes* (rockets) greatly dropped the time it took to excavate a given unit of material. It allowed miners to penetrate deeper into the earth, finding new vein systems and ore beds. The deepest, and most famous, was the Valenciana mine, the record holder of the pre-industrial world at over five hundred meters (630 *varas*).[8] Explosives allowed adits to progress more quickly, setting new records for the length of these excavations, and resolving the problem of waters for a growing number of mines.[9] Like mercury, the price of powder was also cut. Other forms of support came through fiscal incentives that aimed to encourage capital investments. Entire mining districts became tax-free zones, relieved from paying the alcabalá and other taxes.

The Crown legislated new use and access rights to forests for charcoal and materials, to pasturage for work animals, to clean and abundant water to power machines and wash the ores. The claims of mine owners and refiners to these ecological resources were given precedence over common and private property—they could not be denied. It is here, interestingly, that measures aimed at conserving water quality and forests were also legislated for the first time in the mining districts.[10] Given the central importance of labor in the extraction and smelting of precious metals, the Bourbon Crown revived earlier laws against vagrancy that allowed for the conscription "masterless" men.[11] Men were literally "lassoed," that is press-ganged into the mines.[12] Wages were cut by 40 percent.[13] The partido was abolished.[14]

The Bourbon program of reviving the mines of New Spain proved to be remarkably successful. José Fabry's projection that they would double their production was in fact exceeded by a healthy margin.[15] Most of this was the result of accelerating the rhythm of extraction in the old stalwart mining districts such as Guanajuato, Zacatecas, and Pachuca. The Bourbon bonanza not only resolved the borrasca; it made of New Spain the world's most important source of silver, accounting for over half the world's supply by the 1790s.[16] The revival operated under the Bourbons was primarily driven by the quickening of known and previously worked mining districts. Of the thirteen leading mining districts identified by Alexander von Humboldt at the peak of the boom, only four were discovered in the eighteenth century.[17] One of these, Bolaños, had already gone through a cycle of boom and abandonment following mass flooding of the mines in the 1760s.[18] Measured by their registered output, the contribution of the four new green-field districts was far less (3,432 tons or 12.8 percent of total) than that of the nine existing brown-field districts (26,651 tons or 87.2 percent of total).[19]

Surveying the scene at the turn of the nineteenth century, von Humboldt opined that this was only the beginning. There were many more mines to be stoked back into life. He had inventoried and mapped out over three hundred. The geological evidence promised many other deposits to be found across Sierras. "It is almost unnecessary to agitate the question if the produce of the silver mines of Mexico has attained its maximum," he wrote. "Europeans have yet scarcely begun to enjoy the inexhaustible fund of wealth contained in the New World."[20]

· · · · · ·

Von Humboldt could not know that some five years after writing on the inexhaustibility of Mexican mining, the entire sector would come to a sudden and long-lasting crash. Humboldt's report on silver mining in New Spain was mainly concerned with the scale of what he saw. Five thousand workers in the ingenios and mines of Guanajuato. Over fourteen thousand mules turning the grinding mills. The deepest shafts known to him. The overwhelming number of ores that could be worked for their gold and silver.[21]

The collapse of the mines of Mexico began in Guanajuato during the fall of 1810. It was there that the rural insurgency led by Hidalgo met with local mine and smelter workers to win the first battle of Mexican independence.[22] After taking Guanajuato, the insurgents moved on to lay waste to Zacatecas, and then did the same at Real de Catorce to the north of San Luis Potosí. Another arm of the insurgency destroyed the mines of Real del Monte and Pachuca. Former workers targeted and sabotaged mines in order to shut down this cornerstone of elite power.[23] As Fausto de Elhuyar would subsequently note, in a matter of months the most important mines of the Viceroyalty were destroyed.[24] The battles and general strife of central Mexico disrupted the vital supply lines to the frontier mining districts of the north. Apache, Comanche, and the other indigenous nations took advantage of this retreat to recover control of their territories, shutting down the movements to and from the mines, and raiding them directly. Dozens of camps and districts across Chihuahua, Sonora and Coahuila were abandoned.[25]

At a more structural level, the key consequence of the Independence struggles was the interruption of capital flows. This was, according to the British observer Henry Ward, "the real evil of the Revolution. It was not the destruction of the material of the mines, however severe the loss . . . but the want of confidence and the constant risk to which capitals were exposed . . . that led to the dissolution of a system which had required three centuries to bring to a state of perfection."[26] Without capital, mine owners could not pay for the upkeep of equipment and infrastructure, nor could they provision themselves in the fuel and other materials needed for their operations.[27] Paying workers became very difficult, a situation that encouraged the return in force of the partido as well as the creation of local currencies and scrips in the mining districts.[28] For lack of energy, labor, and materials activity in the mines of Mexico slowed down dramatically or ceased entirely. With no one to rebuild and man the pumps, the subterranean waters quietly reclaimed the deeper workings of the mines. As they did the costs of drawing away the waters rose, and the prospects for recovery receded. The shock of Independence had set in motion a feedback loop between miners and the underworld environment that moved toward abatement and cessation.[29]

A dozen years after Hidalgo's revolutionary insurgency, British "mining captains" were sent to Mexico to evaluate the state of the new republic's mineral reserves. It was part of a wider early-nineteenth century prospecting effort undertaken by British capital to identify and assess the natural wealth of Spanish America.[30] Captain Garby, commissioned by the Anglo-Mexican Company, made his way to Guanajuato's Valenciana mine. This had been, without contest, the greatest silver mine of the Bourbon mining boom, the mine that had so completely riveted Humboldt's attentions during his visit in 1803. In 1823, the Valenciana was perfectly stilled. Its massive central shaft—broad enough to be fitted with stairs instead of ladders—was filled close to the brim with water, dark and quiet. Once the single-most important producer of silver in New Spain, the mine had effectively become a very large, very deep, and stoutly built well.[31] Of Guanajuato, Garby's compatriot Joel Roberts Poinsett, wrote, "nothing can be more ruinous and gloomy" and that no amount of work would ever drain the Valenciana.[32]

The ruins of the Valenciana set a scene that was repeated clear across early nineteenth-century Mexico. Cut short in full bonanza by the struggle for independence, the mining districts were littered with the broken remains of pumps and whims, their great ingenios and mills now silent and surrendered to the elements. Where thousands of miners had once worked, only small squads persisted in venturing back underground. Abandonment had made the dangerous work riskier still. Fires set by insurgents destroyed or greatly weakened the timber shorings. Tunnels were unstable or collapsed. Passages were blocked. The underground waters reoccupied the tunnels and works. At Real del Monte, the great Morán adit had collapsed, plugged the drainage, and raised the waters throughout the mines.[33]

The Mexican mining elite's view of the state of the mines in 1820s was deepened by their memories of having seen them at their historical peak. In 1803, the year of Humboldt's visit to Guanajuato, Lucas Alamán was an eleven-year-old growing up in the greatest mining city in the Americas. Every day the young Alamán was brought to the family's haciendas of El Patrocinio and into the mines of Cata. He watched the azogueros finessing the amalgam, he followed the miners down the shafts and through the tunnels. What was for Humboldt a marvel was, for Alamán, daily life during Guanajuato's turn of the century bonanza.

Seven years later, he witnessed firsthand the days that broke the boom. By then Alamán was eighteen and had taken up the charge of his family and its mining operations. It proved a bitter coming of age. "The Revolution of the priest Hidalgo" as he called it, took lives from his household and imprisoned others. By December 1810 he escaped with his people to the relative safety of Mexico City—still the capital of the Viceroyalty. He would not return to Guanajuato for another fourteen years. During this time he educated himself in drawing, European languages, the natural sciences, most particularly the new sciences of mining: chemistry, mineralogy, physics, geology, engineering.[34]

Alamán is best known as the leading light of early nineteenth-century Mexican conservatism. But as the intellectual historian Charles Hale showed, this portrait needs to be nuanced, particularly in what concerns his interventions in the realm of applied economics.[35] In Mexico, Alamán was a leading proponent of what would become a liberal ideology of extractivism. In this he was joined by the likes of José Joaquín de Eguía and Fausto de Elhuyar, who had also known the mines of New Spain at their peak. Both were professors and scientists at Mexico's Colegio de Minería. The latter had served as Humboldt's personal guide in the mines.[36]

Alamán, Eguía, and Elhuyar believed that the crisis engulfing the Mexican mines was not a problem of exhaustion but rather application. Below the waters there was an inexhaustible supply of ores. This superabundance was in Mexico's nature. The Creator had blessed it with gold and silver, the "most distinguished of his gifts."[37] Continuing in this vein, Alamán explained that since in Mexico, and the Americas more broadly, "the operations of Nature are almost universally on her grandest scale . . . [its] metalliferous productions are proportionally extended."[38]

What was needed was a means of resuscitating the mines, wrote Eguía "so that its moribund body can return to life."[39] The animating substance was capital. But if Mexico was blessed with an abundance of natural wealth, it wanted for capital. So little of the great fortunes of Spanish period having survived the disruptions of Independence, new infusions would have to come from abroad. This, in turn, required an updating of Mexico's colonial political economy. Whereas under the Spanish monarchy the mineral wealth of the Americas was seen as something to be kept within the boundaries of the imperial body politic, the silver and gold of Mexico were now imagined as part of the international system of free trade and investment promoted by early liberalism. In a providentialist gloss on the liberal theory of comparative advantage, Elhuyar argued that the Creator had signaled specific products for each country from which to derive its fortunes. Mexico drew gold and silver.[40] It would once again provide the world economy with bullion. In the view of the British mining promoter Sir William Rawson, it was simply "philanthropic . . . to bring into activity [the New World's] energies, and to elicit her natural resources as early as possible." Mexican silver and gold would buy European manufactures. Both would contribute according to its "natural" advantage. From these exchanges would come civic, intellectual, and even moral improvement.[41]

Domestically, it was hoped that the revival of mining would allow the new Mexican state to emerge from the crisis of Independence and rebuild its national economy. Beyond the revenues and profits obtained from the gold and silver extracted from the mines, Eguía argued that mining would once again provide the "principal gyre" of Mexico's economy. Reviving the mining districts would reanimate agriculture, manufacturing, and consumption.[42] Alamán echoed the argument. Mining was Mexico's "true fountain of wealth" not only for what it produced, but for what it consumed. "Its use of a multitude of bodies, machines, and animals, for both extraction and refining, gives rise to an immense amount of consumption, the equivalent of a considerable amount of food exports."[43]

Thus the extraction of gold and silver was reinscribed at the heart of the political economy. Longstanding ideas of inexhaustible subterranean wealth, its providential origins and purpose, were now entwined with the new ideals of nineteenth century liberalism: private enterprise, free trade, and comparative advantage. Bound together, this bundle of ideas formed a liberal ideology of extractivism. It would inform elite action and policy from the aftermath of the Independence to the Mexican Revolution.

In the 1820s, Alamán and others traveled to Europe to find the capital needed to revive the mines of Mexico. The second Conde de Regla, head of what had been the richest family in the Viceroyalty and owner of the mines at Pachuca and Real del Monte, headed for England.[44] Alamán ended up in Paris where Humboldt introduced him to Mr. Andriel, "an adventurer"— that is a capitalist. Together, Alamán and Andriel formed the United Mexican Company, one of the first private companies to invest international capital into the postcolonial Mexican mining sector. In 1823 the pair sailed for London. The city at the time was flush with capital and ramping up toward one of the greatest speculative booms of its history. Much of the money was poured into joint-stock companies investing overseas and into bonding schemes organized to fund the insurgents, and now new governments, of Spanish America.[45] It took Alamán and Andriel less than a year to raise a million pesos. The following year they had six. This was over twenty times the value of La Cata, Alamán's family mine.[46] But United Mexican had larger ambitions than the revival of a single mine. By 1828 it was invested in over forty mines across nine different mining districts from Chihuahua, Zacatecas, and Guanajuato to Oaxaca and Temacaltepec.[47] United Mexican was one of seven mining associations operating in Mexico. It was all part of the larger speculative boom on Latin American state bonds and mining. Over £24 million was invested in the mines of Latin America, of this Mexico accounted for 20 percent of the total.[48]

The project of directing British capital into the revival of the mines of Mexico proved to be very short-lived. By 1825 the greatest part of the promised capital disappeared when the London Bubble burst. Investors who survived the crash ran headlong into resistance in Mexico, both ecological and social. The extent of subterranean flooding was beyond the energies or resources of the British companies. Their early steam pumps could not yet keep up with the inflow of Mexico's subterranean waters. The problem was not with the pumps themselves, but in the lack of fuel to run them. Humboldt had warned of this energetic bottleneck decades earlier.[49] In Guanajuato, the Bourbon mining boom had burned through the surrounding forests—there was no longer enough wood or charcoal to power up the necessary steam head need to raise water from the bottom of the Valenciana.[50] Cave-ins and gases further slowed the British advance. Strikes, riots, and the partido—which in Guanajuato apparently represented half of all silver—and the ongoing militarized conflicts inside Mexico did the rest.[51] By 1850 there was only one survivor—the Pachuca and Real del Monte Company—but it was done. After sinking £7 million in investments, it was still deep in water and just as deeply in debt.[52]

Twenty years later, in 1870, the restored Mexican government took stock of the state of the nation's mines. The results were desultory but unsurprising. Annual gold production for the entire country was still at a trickle, at less than twenty tons per year.[53] (For comparison, this was the equivalent to the annual output of the mines at Cerro de San Pedro alone.)[54] Silver output in 1873 was a bit less than half of what it was in 1809, on the eve of Independence.[55] Officials recorded over nine hundred mines across the territory of the republic but the great majority were classified as inoperative or abandoned. Even the powerhouse districts of Guanajuato, Zacatecas, and Pachuca that had driven Mexican silver mining since its beginnings in the mid-sixteenth century, were described as "virtually abandoned."[56]

The government of Porfirio Díaz put into action the program of renewal that Lucas Alamán and others had been arguing for since the 1820s. Its guiding concept was "fomento"—the nineteenth-century keyword and precursor for state-directed economic development and modernization. As historian Casey Lurtz observes, the central object of fomento was territory: arable lands, forests, and, of central importance, the subterranean territories of mineral, and in time, fossil fuel deposits. Increasing the production of natural commodities was to be the foundation of future prosperity. Fomento, in Mexico, released natural prosperity through policy and institutional practice.[57]

Under the long presidency of Porfirio Díaz and its caretaker government of Manuel González (1880–84), mining was positioned as the well-spring of national economic development. The program was most clearly laid out in Santiago Ramírez's book *Riqueza Minera de Mexico* (1884) commissioned by the minister of Fomento, Carlos Pacheco. Ramírez's text recapitulated pre-existing arguments made by early nineteenth century liberals and colonial-era proponents before them. Mineral wealth defined Mexico's nature and thus mining was naturalized as its national vocation. The nation was bound "through intimate relations and precise laws" to the riches of its subterranean nature. These were signaled in the special qualities of its topography, climate. "The more thoroughly one observes its physiognomy, the more clearly and more precisely the mineral character [of Mexico] reveals itself."[58] "All of Mexico is a mine, and thus, across its entire extension, the influence of mining is generalized."[59] Mining's need for labor created society and political community. Its need for machinery, foodstuffs, materials, services, trade, transport, "creates an extraordinary movement whose individual manifestations are infinite in number."[60] Citing Eguía, Ramírez resumed: "mining is the principal gyre of these lands."[61] "If it receives the impulse that it so urgently needs . . . it will exert its vivifying and beneficial influence. Production will rise, industries will be revived, the sources of work will multiply and Mexico will become great by mining."[62]

The history of mining forced a certain updating of older tropes. Gone, for instance, were the simple cornucopias of the past. Mining would realize the destiny of the polis, not by harvesting the providential windfall of natural treasures as the like of José de Acosta had posited, but through contention and effort. "The development of the Nation's resources and faculties is developed through its constant struggle against the environment . . . through that unceasing work it gives birth to, feeds, conserves, and vivifies its improvement and prosperity."[63] That struggle, in the world of mining, was the work required to face the growing difficulties of extraction. This was a product of history, of the many centuries of working and reworking the same mines. In Ramírez's text, however, the resulting drop in ore grades was transformed into new forms of inexhaustibility. To the extent that ores of lower grades would be worked through great application, it actually multiplied the overall number of treatable ores. "Mining production can in this way proceed indefinitely . . . because the poorer the ores recovered from the mines, the more abundant they are, proportionally." Moreover, the greater efforts required to work the lower-grade ores, "represents so many more workers employed in the mines, in the smelters, in the countryside, in transports."[64] Deepening exhaustion thus led to the economic increase of the nation. Cast in this way, Ramírez argued that mining was "an inexhaustible wellspring of work, life, and consumption."[65]

The government of Porfirio Díaz would extend all manner of support to the foment of the mining sector. It came in the form of personal deals and customized concessions, the work of professionalized mining inspectors, and the dispatch of armed forces to quell worker militancy. However, the Porfiriato's first enabling act, its first move to revivify the industry, came through juridical amendment.

The Mexican constitution was reformed to move mining from state to federal jurisdiction. A new mining code was then passed in 1884 by presidential decree; that is without congressional debate.[66] It began with a declaration of the Mexico's radical domain over the mines (a carry-over from the Spanish period). Operators acquired titles that gave them use rights to a given polygon of land. These could be held indefinitely as long as a modest rent was paid for each title. It allowed well-capitalized operations to hold large swaths of Mexico's mineral dominion as a reserve and thus provide the long-term security of tenure needed to secure large capital investments. The exploitation of mines was deemed to be of public utility. This justified rights of preeminent domain over waters and land. Holders of mining titles could expropriate existing owners and access common resources such as forests and waters. The 1884 Code also included some regulation in favor of workers' health and claims—ventilation and maintenance to assure the security of workers, free hospital care for sick and injured workers, the exemption of miners from regular military service. It even sanctioned the partido, guaranteeing the free sale of ores taken out by the miners.[67]

This first code was judged insufficiently liberal, and it was replaced with the new Mining Law of 1892. The American mining promoter Richard Chism wrote that this was "one of the most illustrious monuments to the executive ability of General Díaz and his cabinet. [One that] started it on a career of prosperity that is entirely without precedent."[68] In London it was received as the "best of all extant laws" and held as an example to the rest of the world.[69] It was a much pared-down version of the 1884 code. It made no mention of the nation's fundamental title over the wealth of the subsoil. The state would no longer regulate working conditions, health, or the payments made to workers. Owners and managers were given entire freedom to run their operations as they saw fit. Mining titles were inalienable, there was no limit to how many titles a company could hold, and they were held in perpetuity. The standard easements such as tax breaks and subsidiary rights to water and expropriation were also provided.[70] No mention was made of the partido.

Manuel Fernández Leal, one of the secretaries of the ministry of Fomento, commented that the guiding aims of the codes of 1884 and 1892 were "facility of acquisition, liberty of exploitation, and security of retention."[71] They provided the juridical assurance needed to attract capital investments from abroad. The timing could not have been more opportune. The establishment of Mexico's new liberal mining codes coincided with a world-wide boom in natural resource investments led by markets in London and New

York, along with secondary hubs in Montreal, Johannesburg, Sydney, and San Francisco. In the mining sector, British capital was mainly directed at South America and its imperial territories in Africa and Asia.[72] For Mexico, foreign investments overwhelmingly came from the United States. In 1888, US investments in mining totaled three million dollars. In the 1890s it climbed to $55 million. At the outbreak of the Revolution of 1911 it reached close to a quarter billion dollars.[73] US capital represented close to 80 percent of the investments in the Mexican mining sector overall, at a time when Mexican ownership was at less than 3 percent.[74] And finally, unlike what was for Lucas Alamán a disappointingly short rain-burst of British capital, American investments flowed almost continuously (with only the Revolution imposing temporary hiatuses) until the 1940s—a run of over fifty years.

· · · · · ·

The revival of extractivist mining under the Porfiriato rested on increasing the rates of extraction and processing. This was accomplished by transferring the industrial hard rock mining system that was, by then, well-assembled in the mining districts of the US west. Professionalized geologists and engineers drew up very precise, large-scale, and even three-dimensional, maps of ore bodies.[75] They produced detailed flow sheets that charted out the movement of ores from the mine face deep underground to the final pour of molten silver and gold. Each station along the way was carefully blueprinted, assembled, and calibrated. Designing for flow allowed for bottle necks to be identified and eliminated, and for the extractive and refining system to operate without interruption, with as little waste as possible, at the highest rates possible.[76]

New sources of coal and hydroelectricity generated motive power that was orders of magnitude greater that what even the largest teams of horses or mules could produce. The most powerful high pressure steam engine ever assembled in the nineteenth-century United States was first used in the Comstock mines of Nevada.[77] At 300 horsepower of output it represented the equivalent effort of between 375 and 650 draft animals.[78] In time, gasoline engines, especially the Crossley engines, replaced steam engines as the favored power generator, mainly because they were more efficient in producing the DC current needed to power pumps, hoists, compressors and milling engines.[79]

The Georgetown mines of Colorado saw the arrival of the first mechanical rock drill powered by compressed air and capable of delivering three hundred blows a minute. (A crack double-jacking team working with chisel and hammer could manage about twenty.) Giant Powder—that is dynamite—also arrived on the scene to crumple out tons of rocks at every blast.[80] Miners no longer followed veins but were tasked with the excavation of entire blocks of low-grade ores, some layered hundreds of feet thick. They dug up from below—a process known as block caving—building story after story of thick timber works rising up into the ever-growing cavity. Tunnels were leveled out and widened to admit rails and wheeled carts. These allowed a single miner to move close to a ton of ore and send it up to the surface. Bracing his body against the cart, he slowly set its wheels into motion and then, picking up speed, guided it to the shafts where mechanical whims whisked loads up hundreds, even thousands of feet in a matter of minutes.[81]

Until the turn of the twentieth century, the chemistry of ore refining continued to work through lead (through smelting) or amalgamation (with mercury). Through the Washoe process, though, mercury amalgamation was greatly sped up, with large pans processing up to seven hundred kilograms of ore every five hours.[82] In 1905, Mexico's first cyanidation mill was established at Guanajuato.[83] The cyanidation process further increased the rate at which slurries, or "slimes," of finely milled ores could be treated. A smaller plant, such as at the San José de Gracia mine in Sinaloa, could work fifteen hundred tons of ore per year. This was the same rate as all twenty-two ingenios working at the peak of the seventeenth-century boom at Cerro de San Pedro. Eventually cyanide plants were able to treat twenty-seven thousand tons per year, and then two hundred and thirty thousand tons-year.[84] Cyanide, furthermore, was a much more potent amalgamator and this greatly improved the rates of metal recovery. Smelters likewise grew with industrialization. The furnaces and roasting ovens were scaled up to the point that they could process three to four hundred thousand tons of ore or better every year.

Tailings left after centuries of mining were suddenly worth quite a bit more than waste. Since they did not have to be excavated, they provided a cheap source of ores to work on.[85] At Pachuca, Dahlgren described the operations of the American Pachuca River Concentrating Company processing between fifty and seventy thousand tons of tailings per year.[86] These had sedimented in the riverbed through the effluvia and voiding of centuries of refining. It was, he wrote, contracting "men well-versed in the art of picking up a river and carrying it off a proper distance" to fully access these river-bottom deposits.[87] Robert S. Towne similarly started by feeding his plant the tailings that had built up over the last three hundred years at the Cerro de San Pedro. The Guggenheims bought the mines at El Magistral, near

San Lorenzo, Chihuahua, with no intention to reopen the mines but simply to work the estimated ninety thousand tons of its accumulated mining waste.[88] Formations and entire sections of mineral bodies that generations of earlier miners had left in the ground because they were judged too lean were now targets of industrial acquisition and extraction.[89] The great silver deposits of the colonial period such as Guanajuato, Pachuca and Parral were first choices but, on the booming nineteenth-century market in Mexican mining titles, any Spanish-era mine, any *Antigua* was considered a plum.[90]

· · · · · ·

The late nineteenth-century quickening of the mines of Mexico marked a remarkable return from abandonment and exhaustion. By the turn of the twentieth century, registered gold production had increased more than seven-fold from the levels recorded in 1872. This boom, fomented by the Porfiriato's liberal regime of resource and realized by American industrialization, was the third in a series of mining cycles to pass through Mexico. Taking stock of this last cycle in light of its predecessors brings to light certain key trends in the long history of extractivist mining.

The first is the transformation of the commodity frontier. Under the Habsburgs all increases in bullion output came either from the appropriation of existing indigenous gold mines, such as those of the lower Rio Balsas watershed, or through the discovery and development of untapped silver deposits. The pattern here is a classic one in which increases in commodity output proceed through what geographer David Harvey termed appropriation by dispossession.[91] This logic was still at work during the Bourbon revival of the eighteenth century, as the northern movement of the mining frontier renewed itself to open new districts in sierras and arid plains of Sonora and Chihuahua.[92] Nevertheless, what is notable about the Bourbon bonanza was the fact that the majority of increased bullion output did not come from opening new mines but rather by increasing the productivity of existing mining districts. Cerro de San Pedro, for essentially political reasons, was not party to this trend. In Zacatecas, Guanajuato, Pachuca, Real del Monte, and Taxco, to name only the most important, the renewed influxes of capital and the various forms of royal support allowed extraction to scale up, speed up, and produce more silver than ever. In the third bonanza, the industrial boom, the secular trend toward brown-field mining was confirmed. In 1901, in the midst of the new bonanza, a team of Mexican mining engineers produced a comprehensive inventory and map of the mines of the Republic. The majority had been mined for the first time under the Habsburgs. Another fraction was opened up during the Bourbon revival of the eighteenth century. Only a small number were new, green-field, operations. That is to say that the late nineteenth-century mining boom ushered in by government of Porfirio Díaz was almost entirely based on the return to activity of districts that had passed through at least one cycle of mining and abandonment. The majority had been through two.[93]

The second observation concerns the shifts in the cadence and duration of extraction, that is, its temporality. Districts that were in decline, or considered to be abandoned outright, could be brought back to profitability by increasing the rhythm of extraction and refining. This was a basic, well-established, and by now thrice-repeated, formula. Hastening the rhythm of the extractive system allowed a much greater volume of ores to be worked for the silver and gold they carried. This, in turn, generated the revenues needed to offset the costs associated with meeting the challenge of deepening underground operations, flooding, heat, gases, refractory ores and dropping ore grades. These included the costs of energy (coal, fuel, electricity), machinery and infrastructure, consumable materials, and, of course, labor—in the mines, in the refineries, and in the offices. It all amounted to a serious amount of capital but, as Alamán and others had seen, capital was what was required to sufficiently speed up the cadence of the extractive complex to return it to profitability. The investments were large, but given the scale of the venture, so too were the returns. The trade-off of intensification was a contraction in the duration of activity. Acceleration, scaling up and intensification could resolve exhaustion, but it simultaneously hastened the rate at which operations moved through a deposit and thus hurried the return to that moment when available technologies, materials and energetics could no longer continue to profitably extract metals. The diminishing duration of each extractivist cycle in Mexican precious metal mining up until the twentieth century illustrates this trend: the Habsburg cycle lasted well over 110 years (1540s–1650s); the Bourbon bonanza lasted seventy years (1740s–1810); the industrial one around fifty years (1890s–1940s).

9 The Industrial Bonanza

· ·

The industrial bonanza in Cerro de San Pedro and San Luis Potosí was composed of three parts. The first was material construction of Mexico's first industrial mining and refining complex. The basic elements of this package were transferred from the US West beginning in 1890 by the American operator Robert S. Towne. At Cerro de San Pedro, the industrialization of mining entailed the mechanization of extraction and haulage. In San Luis Potosí, an entirely new refinery known as the Hacienda de Morales was built on the outskirts of the city. It combined mill works, banks of smelters and furnaces, and a cyanidation plant.

The second key feature of the industrial bonanza was its capacity to greatly increase the extractive metabolism. The fact that the Cerro de San Pedro was considered spent and all but abandoned hardly deterred Towne. By the 1890s, industrial hard rock mining was a proven system. Once established, it would produce a new surge of ores from the three-hundred-year-old mines of Cerro de San Pedro. The Morales plant was capable of bulk processing over a thousand tons of ore a day. At that rate, there was profit to be made, not only from the low-grade ores recovered from Cerro de San Pedro but also from over dozen districts down the length of the republic.

Finally, the industrial boom was a social phenomenon marked by rapid population growth and the return to a kind of urban life that had not existed at Cerro de San Pedro since the seventeenth century. Just as it had in the colonial period, this was mainly driven by the arrival of hundreds of workers. Urbanization and the building of communities happened very quickly. At first people, sheltered in camps and caves until, in time, entire barrios of small cottages were built. These communities were of a kind with many other urban enclaves that emerged around the reawakening districts of the late-nineteenth century Mexican mining boom. They contained many of the features normally associated with life in larger cities: electric lighting, crowds, an animated public life, and, in time, cars and buses.

The working-class barrios of mining districts such as Morales and Cerro de San Pedro were, however, different from other urban centers of the

Porfiriato. They were inescapably shaped—materially, metabolically, and socially—by the high-intensity extractive complex organized by Towne. Barrios were established atop, or immediately adjacent to, the mines and smelters. The largest single segment of the working population labored inside them. There they experienced the high-intensity metabolism of industrial mining directly in their bodies. Everyone lived within the peculiar environments produced by mass waste voiding and rapid and unsupported urbanization. These can only be described as difficult environments, places where basic amenities such as water access and sewage were limited, where parasites and illness were endemic, and where airs, waters, and soils were laced with the toxic output of industrial extraction. By tracing the material, metabolic, and social dimensions that composed the industrial bonanza, this chapter lays the groundwork for a discussion of their social and ecological consequences, the main subject of subsequent chapters.

• • • • • •

The man who played a central role in both the reanimation of the mines of Cerro de San Pedro and the creation of Mexico's first industrial-era mining complex in the valley of San Luis Potosí was Robert Safford Towne. At the age of twenty-two he graduated from Ohio State as a B.Sc. Mining Engineer. At the time, in 1880, this was an entirely new profession.[1] He immediately found work with the Consolidated Kansas City Smelting and Refining Company working out of Argentine, Kansas.[2] The "Con K" processed silver and gold ores coming down from Colorado by train.

After two years in Kansas, news of a rich silver bonanza in Mexico turned Towne's attentions southward. In 1882, he traveled to the Sierra Mojada in the desert quarters of Coahuila state. This was the location of a bona fide first-strike bonanza (one of the last in Mexico), the mines just been accidentally discovered by a Mexican militia captain out patrolling for smugglers. By the time Towne arrived, over three thousand miners were already working their way into a steep thousand-foot mountain face. There, a multitude of small black bocaminas showed where miners had dug into the cliffs and rock-bands. Getting up to these aeries and getting ores down to the plain edged the miners of the Sierra Mojada into the terrain of the mountaineer. To speed things up Towne developed a set of aerial tramways. These airborne ore-freighting systems looked like nothing else than a ski-slope gondola (and may very well have been its technological forbearer). Large ore-filled buckets sailed through the air, each cradling hundreds of kilograms of material, passing from pylon to pylon down from the mine head to the yard below.[3] Once he received the ores, Towne's job was to get them to Argentine, Kansas—over thirteen hundred kilometers away. In the first years, he hauled the ores out by mule trains to Escalon station where they were loaded onto northbound trains.

In 1887 Towne convinced Con K to halve the distance between Coahuila and Kansas by building a new smelter at El Paso. They contracted him to do it. Even as the machinery was being assembled in Texas, Towne was simultaneously overseeing the construction of a branch line connecting the Sierra Mojada mines to the recently completed Central Mexican railroad. From his experience in Kansas, he well knew the importance of rail-links for the new kind of mining. Mining districts connected by rail could haul in heavy machinery and coal and ship out ores, thousands of tons at a time, cheaply. Large, centrally located, smelters could process the ores of many mines and thereby realize important economies of scale. In linking the mines of the Sierra Mojada by rail, Towne opened a new flow of Mexican ores into the United States. Cheaper, easier to process on account of their admixture with lead, and more profitable because of their high silver content, the influx of Sierra Mojada ores had Colorado mine operators lobbying Washington to put a block on Mexican silver. What they got was the McKinley tariff of 1890, a protectionist levy that raised the price on Mexican ores in a bid to support the mine and smelter operators of the US West.[4]

The McKinley tariff was not an embargo per se but it motivated Towne, and almost immediately afterward the Guggenheims, to build smelters on the Mexican side of the border where ores could be processed duty free. Towne headed for San Luis Potosí.[5] At the time, the city was already a key node of Mexico's emerging rail network, situated at the juncture of two trunk lines to the United States (El Paso and Eagle Pass), an eastern line to the future oil town of Tampico, and a southern line to Mexico City and Oaxaca. Towne made his proposal directly to the Mexican presidency and in 1890 was signing for the country's first industrial smelter concession. Signing for the Republic was Carlos Pacheco, President Porfirio Díaz's one-armed right-hand man and minister of the all-powerful Ministerio de Fomento.[6] Under its terms, Towne was responsible for building a six-tower smelter, with all its attendant mills, furnaces and buildings. The metals processed would be taxed by the federal government. In exchange, Towne obtained the land for the smelter, an operating license, water rights, as well as concessions for new railroads and huge tracts of timberland in the nearby

Sierra de Álvarez. Towne would, in parallel, buy the key mining concessions at Cerro de San Pedro. He bought the largest tracts: La Victoria, Barreno, Begonia, and with these began working his furnaces.

At the time, Cerro de San Pedro was the preserve of cottage miners. Individually and in families, each applied themselves to one of a few dozen claims in and around the Cerro, slowly probing the ore veins and hauling out the richest material. Coaxing them on were the regular encounters with pockets of high-grade gold bearing material. In 1883, for instance, a lucky party of miners working the Santo Domingo mine took out a single piece weighing an astounding two hundred ounces of gold (just over five kilograms).[7] Their labors kept eight small smelters running and producing an estimated 115 kilograms of silver per month. In both design and scale, these charcoal-fired furnaces differed little from the ingenios and fuelles of the colonial period.[8]

Towne's operations completely transformed this scene. Engines and generators were brought up. Warehouses, offices, new mine heads were built. Old tunnels were cleaned out, broadened, and leveled to run tracks and carts. New shafts and adits were driven into the mountain. The mine works were outfitted with all the standard pieces of an industrial mine: electric cables, pneumatics, power whims, pumps and, in time, electric lighting and mechanized ventilation to air-cool the lowest levels of the mines.[9]

Industrialization allowed miners to push their way into the heart of the Cerro de San Pedro. They passed through the latticed ore veins of the limestone capping rock. The new shafts doubled the depths reached in the eighteenth century. They were now over six hundred meters into the earth, deeper than the Valenciana shaft at Guanajuato, the deepest mine that Humboldt had ever seen. The miners set their drills and picks into the large bodies of softer ores, even, for the first time the giant mass of the porphyry that formed the heart and core of the deposit. As Santiago Ramírez had correctly observed, these deep-laid ores may have been low in gold or silver content, but this was more than offset by their massive abundance.

The ores of Towne's mines at Cerro de San Pedro were loaded on to a narrow-gauge train known as the *Piojito* ("Little Flea"). It rolled straight past the small cottage smelters and headed down the valley to San Luis Potosí where, about five kilometers west of the city, Towne built his smelter, the Hacienda de Morales. It was outfitted with ten large blast furnaces and seven roasting furnaces. When the new metallurgy of cyanidation arrived in Mexico in the first decade of the twentieth century, Towne immediately built the El Carmen cyanide plant. Combined, cyanidation, high-temperature smelting, and high-speed mechanized milling dramatically increased the scope of what were considered profitable ores and the amounts that could be processed. The Morales plant would eventually treat four hundred thousand tons of ore a year.[10] It represented a through-put three hundred times greater than the largest colonial-era ingenio.

But Towne had more than San Pedro in mind when he built his smelter. San Luis's multiple rail links connected him to mines producing the length of Mexico. At its peak, the Morales plant processed ores from Chihuahua and Coahuila to the north and as far away as Oaxaca in the south. Thirteen mining districts representing dozens of mines constantly fed the plant with ores.[11] (See map 3.) Towne was simply reproducing in Mexico what he had learned as a young mining engineer working for Consolidated Kansas. Rails allowed smelters to scale up and centralize control over an entire interregional complex of mines.

The Guggenheims immediately followed Towne's lead. They, too, obtained concessions to establish industrial-scale foundries at Monterrey and Aguascalientes. The latter was the largest smelter in North America.[12] At the beginning of the twentieth century, Daniel Guggenheim took control of the ASARCO smelting consortium and developed a network of low-grade mines organized around a string of centrally located smelters on both sides of the US-Mexican border. Towne died in 1916, in the midst of the first, violent, phase of the Mexican Revolution. His company, the Compañía Metalúrgica Mexicana, did not last for long after his death. The challenges of keeping the flow of supplies, ores and metals moving in times of civil strife, drove the company into debt. ASARCO bought out the CMM in 1923.[13] With this acquisition, ASARCO, became the single largest integrated extractive complex in Mexico. It enjoyed quasi-monopolistic presence and power in the mining sector until the 1960s when, under the policy of Mexicanization, it was bought by Mexican capitalists (as ASARCO, then IMMSA in 1974, and then since 1978 Grupo Mexico under control of Larrea family).

· · · · · ·

One of the basic metrics of the extractivist metabolism is the quantity of ores extracted and processed over a given unit of time. Prior to industrialization, a leading preindustrial refining complex in Mexico was the Hacienda de Regla in the state of Hidalgo. It processed the ores from Real del Monte, a long-running and always important mining district. Still in operation in the 1870s, Regla represents, both in scale and through-put, the culmination of an Ancien Regime extractives complex. It was generously

SOURCE DISTRICTS FOR FUEL AND ORE SHIPPED TO THE MORALES PLANT, SAN LUÍS POTOSÍ

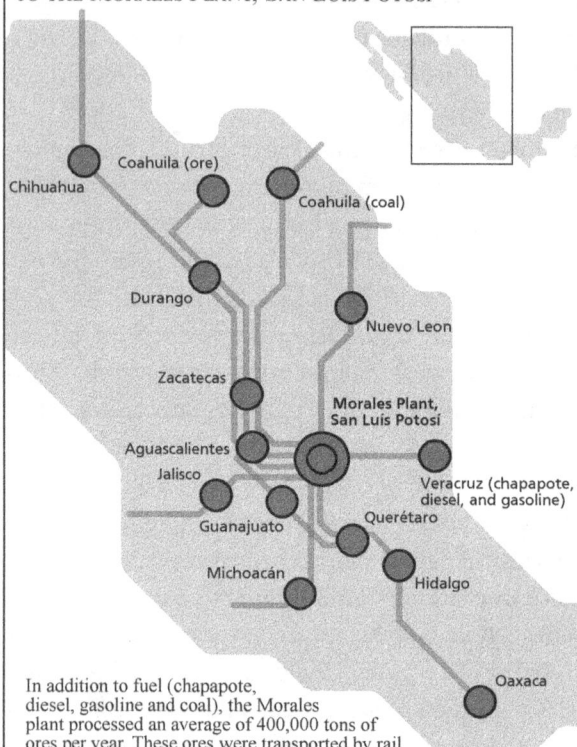

Chihuahua

Coahuila (ore)

Coahuila (coal)

Durango

Nuevo Leon

Zacatecas

Morales Plant, San Luís Potosí

Aguascalientes

Jalisco

Veracruz (chapapote, diesel, and gasoline)

Guanajuato

Querétaro

Michoacán

Hidalgo

Oaxaca

In addition to fuel (chapapote, diesel, gasoline and coal), the Morales plant processed an average of 400,000 tons of ores per year. These ores were transported by rail from thirteen different mining districts across Mexico.

THE MORALES PLANT AND THE BARRIO DE MORALES

Every day the Morales plant released an estimated 1 ton of heavy metals (As, Pb, Zn) as emissions and produced over 1,000 tons of solid wastes.

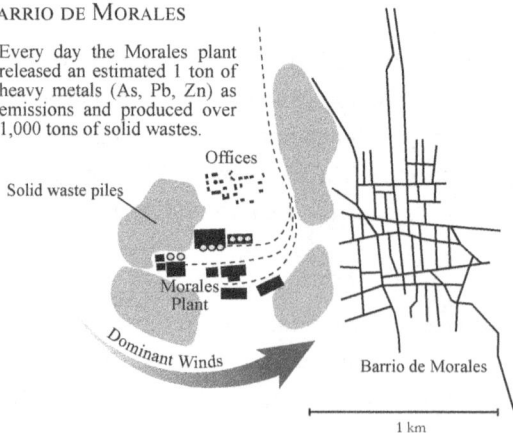

Offices

Solid waste piles

Morales Plant

Dominant Winds

Barrio de Morales

1 km

MAP 3 Ore processing at the Hacienda de Morales (illustration by Geoffrey Wallace, 2021, after original by D. Studnicki-Gizbert).

powered by a diversion of the Amajac River whose waters were channeled through a set of sluices and reservoirs into an elaborate system of water wheels, axels, and pinions. They produced the motive force for banks of large wooden stamp mills processing an estimate thirty tons of ore per day.[14] This was a great deal more than the 0.5 to 0.7 tons that even the largest ingenios of the valley of San Luis Potosí could muster in the seventeenth century. But Regla was in turn dwarfed by what could be achieved under the industrial system. In the first year of operation, the Morales plant processed eight hundred tons of ore per day.[15] In 1911 it was up over a thousand.[16] Admittedly Morales was one of a small group of top-tier producers at the time, but even among the far more numerous middle—or small-tier processing plants average processing rates at the time ranged in the range of one hundred to two hundred and fifty tons per day. Even the smallest refineries operating remotely in the mountains of Chihuahua or Sonora could render around fifty tons per day.[17]

The jump in processing rates ushered in by industrialization required a corresponding increase in the scales of the energetic and material flows moving through the late-nineteenth and early-twentieth Mexican precious metal complex. The hydraulic power on display at the Hacienda de Regla represented a top-end energy system for its time. As Andreas Malm has shown for early nineteenth-century Britain, waterwheels were capable of outproducing early coal-based steam engines.[18] The impact of industrialization on water power was less its replacement by coal burning, as eventually transpired in Britain, but rather its transformation into hydroelectricity. Rivers were dammed and their kinetic power poured into new generators and electrical grids. In Mexico, the boom in industrial mining and the development of hydroelectricity were in fact closely connected. For operators, electrical cables enabled a much more practical and efficient means of distributing power through the works. It illuminated the tunnels, and it powered the machines—pumps, ventilators, whims, mills. It was, furthermore, cheaper than coal-fired alternatives, dropping the energy bill to a quarter. Thus, where feasible, mining companies built their own hydroelectric plants.[19] By the 1920s 40 percent of all privately generated electricity was produced by mining operations.[20] They were joined by power companies from Canada and the United States that invested in the development of large dams capable of generating tens of thousands of kilowatts of power. Their principal clients were mining companies, who by the early twentieth century were consuming a full third of Mexico's electrical output.[21]

San Luis Potosí, however, lacked the hydrological conditions necessary to power a small colonial-era water wheel, let alone the large power outputs required by Towne's operations. Instead, Towne had to rely on the combustion of different kinds of fuels. During his first years in San Luis Potosí, Towne fed his furnaces and engines with charcoal—exactly as had been done for hundreds of years. The contract he signed included a very large timber entitlement in the neighboring Sierra de Álvarez.[22] He used it extensively, building another narrow-gauge line into the mountains to fetch the wood. The account books for Towne's operations show that the majority was oak, now returned after the great deforestation of the early seventeenth century.[23] Charcoal consumption was an intermediary source of energy in the transition to fossil fuels. Even in the short-term, local forests could not sustain the steep jump in energy consumption required by the new generation of machinery and furnaces. Back in the 1820s, the British mining expert Poinsett discovered the energetic limits of charcoal. He calculated the wood stands around Guanajuato would be insufficient for generating the amount of steam head needed to drain and keep the Valenciana mine dry. That was for a single pumping operation for one mineshaft.[24] The problem was two-fold: charcoal-fired steam engines were highly inefficient, and charcoal simply lacked the energy density (kilojoules per kilogram) of coal.

Thus, within a few years of having begun operation in 1892, Towne was already casting around for alternatives to forest-based charcoal. He began by burning *chapopote*, a mineralized tar found in various deposits along the Gulf Coast, that he had transported up to San Luis Potosí by rail in car loads of leather bladders.[25] When coal became available, however, Towne immediately made the switch.[26] Mexican coal deposits had been identified in the eighteenth and early-nineteenth centuries, but it was only in the early 1880s that the Ministry of Fomento devoted serious attention and resources to their development as a new energy base for the country.[27] Officials in the ministry noted that the "destruction and neglect of forests" demanded an alternative source of fuel.[28] High quality coal was found widely distributed across Mexico but the most important and largest fields were found in the northeastern state of Coahuila. It was there that the first large-scale coal mining and coke producing operation began in 1899. Within a few years the Mexican Coal and Coke Company was producing close to fifty thousand tons of coal and coke every month. This production, along with that of its proliferating number of competitors (two dozen by 1911) was primarily destined for consumption by the country's mining and refining operations and was delivered there through the national railway network.[29] Between 1891 and the onset of the Revolution, Mexican coal production rose seven-fold from two hundred thousand to 1.4 million tons a year.[30] Industrial smelters were the main consumers of this energy, and, indeed, they also recurred to imported coal to satisfy their needs.[31] Diesel fuel would come on line in the 1920s as a product of the oil boom in Tampico. Under ASARCO, the mines Cerro de San Pedro burned diesel to run electric generators, air compressors for pneumatics, whims, and a new fleet of trucks that now undertook the work of hauling ores down to the Morales plant.

As during the colonial period, the main output of the industrial extractives metabolism was not silver or gold but rather mining waste. The Morales plant would process up to four hundred thousand tons of ores a year. By then, the average ore grade for silver was no longer measured in percentages as under the Spanish but rather parts per thousand. The rest was volatized into the atmosphere during the smelting process, or physically discarded in the form of gangue, tailings, and slag. One of the reasons why Towne chose the Morales tract to establish his refinery was for its ample flatlands that "afforded an unlimited dumping ground for slag."[32] He filled them high. For some years, he realized some profit from some of the leaden slag by passing it through a Comet crusher and selling it as bedding to the Mexican Central Railway.[33] Most of the waste, however, stayed right where it was dropped.

· · · · · ·

For all its new sources of capital, new technologies, and new forms of power, industrial mining still had a great need for labor. Every station of the extractive chain, from the mine face to the final pouring out of gold and silver bars depended on human skill and work. Wages were the leading cost of every ton of ore extracted from the earth.[34] Even in the smelters—with their high rates of energy consumption—labor was only surpassed by the energy costs.[35]

To work the old mines at Cerro de San Pedro, Towne hired six hundred workers.[36] At Morales, the workforce numbered thirteen hundred.[37] Across Mexico, the estimated population of miners and millworkers ranged between eighty to one hundred and thirty-five thousand people, depending on the year.[38] These people were mainly adult men, though a significant percentage were boys (between 3.5 and 5 percent) and women (1 percent).[39] They constituted an important constituency within the emerging industrial proletariat of Mexico. On the eve of the Revolution one in seven wage-earners in Mexico was a miner or a smelter worker.[40]

Back in the 1860s, before the presidency of Porfirio Díaz, the population of Cerro de San Pedro had dwindled to the point that it lost its legal standing as a municipality. Its territory and people were divided between its neighbors: the municipalities of Monte Caldera and Cuesta de Campa. Insofar as the Mexican state was concerned, Cerro de San Pedro had ceased to exist.[41] The arrival of six hundred new miners in the 1890s brought it back from the dead. They came from the ranchos of the region. Others same from further out, from the states of Zacatecas, Michoacan, Guanajuato, Durango, and Jalisco.[42] The miners were either soon joined by their families or founded them. Shops opened up. Tradespeople moved in. By 1900 Cerro de San Pedro was once again its own town and listed in the gazetteers. The census taken in that year counted 633 inhabitants. By the next census in 1910 it had more than tripled to 2,202 inhabitants. After the Revolution it reached close to five thousand people, a figure that most residents remember today and one that brought Cerro de San Pedro's population back to the numbers reached during the first bonanza of the 1590s–1620s.[43]

Five thousand inhabitants did not a city make, but in people's memories Cerro de San Pedro felt like one. Marcos Rangel who was still a youth in the last decades of the ASARCO period remembered: "When you look at it now it's easy to forget that this was a town of more than five thousand people. We had electricity, water fountains, mail, cars and buses. There were schools, more than one, with good books, rulers, everything."[44] The town was big enough to be divided into different barrios or neighborhoods. Heading down the valley, near the mine-head and power station, was the *colonia Americana*, a small suburb of tidy white adobe houses built to house the company's managers, engineers, mechanics, and doctors.[45] In one of the side-gullies, behind the Church of San Nicolás, a small red-light district emerged—*la Cocinera* ("the Kitchen")—with its pulquerias, gambling houses, and weekend brothels. There was a baseball league whose teams were composed of workers from different sections of the mining works, crowded religious processions, Carnaval, and other fiestas of the Catholic and national calendar.[46]

Down in the valley, the Morales smelter provided another pole of demographic growth. Hundreds of workers were needed to build the smelter, then well over a thousand to run it. Unlike Cerro de San Pedro, which was re-born as it were, Morales was established on pasture lands outside the city proper. A tramway was built to facilitate the commute for workers who continued to live in San Luis. A growing number chose to establish themselves on site. They lived in the rows of company housing built on the side of the slag piles and sorting yards or in the larger barrios that grew upon the informally occupied lands of the former Morales ranch. On the eve of the Mexican Revolution, over four thousand people lived in, and immediately around, the plant, creating an industrial enclave on the periphery of the city of San Luis Potosí (population ca. seventy thousand at the time).[47] As in Cerro de San Pedro, this was a divided community between its barrios of Mexican workers and their families clustered around the two churches, and its colonia of foreign managers, perched on the hillock overlooking the plant and the barrios, surrounded by a wall.

· · · · · ·

The revival of an extractivist mode of mining at Cerro de San Pedro, after over two hundred years of small scale gambusino mining, was driven by the influx of capital and technologies from the US mining sector. At first brush, the new array of techniques appeared to be transmuting exhaustion into abundance but what they aimed at, and what they effectively accomplished, was a restoration of profitability. By increasing the scale and cadence of throughput by a factor of thirty or better, the new complex organized by Towne was able to achieve considerable profits from these ostensible abandoned mines. The surge of both mechanical and heat energy provided by coal, diesel, and electricity was critical to this shift. So too the reassembly of large labor forces. Together they enabled the bulk extraction and processing of low-grade ores. They also produced, as a necessary consequence, a commensurate rise in mining waste.

The interest in showing the intensification of extractivist metabolism is not only to show the shifting, historically contingent, links between capital and resource exhaustion. It is also to focus on the biophysical relations that linked the operations of industrialized metal mining to the bodies, both ecological and human, upon whom this work depended. As Brett Walker's study of the Ashio copper mines of Japan shows, the relations between the metabolism of industrial mining, ecologies, and human bodies, are complex and manifold.[48] This complexity meant that the stepwise increases in metabolism represented by the late nineteenth-century industrialization of mining did not translate to nature and society in a neat and linear way. Intensification entailed important shifts in scale and speed, true, but also qualitative changes in kind and in organization. This, in turn, bore directly on the composition and consequences of the new ecology of industrial mining.

10   The New Ecologies of Industrial Mining

During the two-century interval between the colonial and industrial bonanzas, the extractivist mode of mining was no longer the principal driver of environmental change at Cerro de San Pedro and in the valley of San Luis Potosí. Small-scale, gambusino, mining continued, of course, but its scales and intensities were quite modest and its mark on lands and bodies comparatively light. Still, this was very much a mining landscape. The legacies of the Spanish bonanza continued to exert their influence. The most important, arguably, was the dearth of soils. These, unlike the hardier mesquites and oak, had not been able to reestablish themselves. The result was that while the ecologies surrounding the mines were mainly conditioned by an agrarian smallholder economy, this economy was limited to what could be pursued in a post-extractivist landscape of diminished fertility: the raising of goats and sheep, a small amount of charcoal-making and mining, and, here and there, where conditions allowed, the cultivation of agave and its processing into pulque. This is to say that while the abatement of high-intensity mining offered a reprieve and space for other ecological logics to develop, its deeper legacies endured and impeded the possibilities of return to prior conditions.

When Towne established his smelter at Morales and took possession of the mines at Cerro de San Pedro, he set in motion a new cycle of extraction-driven environmental change that transformed the landscape once more. The metabolism heuristic, with its inflows and outflows of matter and energy, again provides a useful way of understanding the links between the renewal of mining and the environmental changes it entailed. As seen in the last chapter, the industrial mining metabolism churned at significantly faster rates, and operated at much greater scales, than the Spanish bonanza. This stepwise increase in energy and material flows resulted in a commensurate surge in the transport of highly mineralized subterranean material up to the surface. Almost all of it was treated as waste and was heavily stocked in metals such as arsenic and lead, both toxic for human bodies. These metals infiltrated and contaminated entire sectors of the city, most especially the barrio of Morales. This neighborhood had sprung up quickly as people arrived to work in the smelters but, in the rush, the basic needs of clean water and sanitation were neglected. This produced the ecological conditions for hookworm and other diseases to thrive. The speed and intensity of the industrial mining metabolism also determined working environments inside the mines and the smelters. There, new risks and harms associated with mechanization were layered upon those of previous centuries. Finally, the new ecologies associated with the industrialization of Cerro de San Pedro and San Luis Potosí were no longer entirely local in their scope. Some part of the environmental effects that it produced were to be felt in the coalfields of Coahuila and the oilfields of the Gulf Coast.

· · · · · ·

The suite of power sources needed for industrial mining and refining was composed of hydroelectricity, coal, and diesel. They were the industrial analogues to the hydraulics, charcoal, as well as human and animal-power of the Ancien Regime. Beyond the important difference in the amount of energy they furnished to the work of mining, the industrial period substantially shifted the geography of supply. Colonial haciendas tapped local forest stands for their embedded heat energy. Industrial mines and smelters, on the other hand, were connected to national power and transportation grids that allowed them to access energy stocks in coalfields, oil patches, and water reservoirs hundreds of kilometers away. This web of provisioning lines extended the ecological footprint of a given operation, fragmented it across different source regions, and mixed it with the power requirements of other operations and consumers. The new geographical configurations make it difficult to establish clear one-to-one links between the industrial revival of a single mining district and the local environmental changes that it contributed to. The links were there, though. The flooding of lands and communities by dams; the Gulf coast oil boom and all its attendant consequences on local peoples and ecologies; the environmental impacts of coal mining in the Coahuila belt : the industrial mining boom had its part to play in each of those stories. Some of which, such as Myrna Santiago's treatment of oil fields of Veracruz or Germán Vergara's work on the shift to the coal economy, have already found their historian, while others await.[1]

That said, the impacts of high-intensity, high-throughput mining were also felt locally. Deeply. For the people of San Luis Potosí and Cerro de San Pedro, the most important came from the outflows, rather than inflows, of the extractivist metabolism, through the mass voiding of wastes that laced local airs, soils, and waters. On an average day, the Morales smelter blew well over a ton of lead, zinc and arsenic out its stacks.[2] Metal precipitation produced an invisible rain of metals and compounds that dropped on local communities. The greatest part of this material fell upon the working-class barrios of Morales. The remainder was borne up and over the city by the prevailing winds and then dropped out over the farms and ranchos of the valley's east side (see map 3).[3]

The other source point for heavy metal contamination were the heaps of tailings and other solid mining waste. At Morales these were piled in ziggurats tens of meters in height and hundreds of square meters in extent. Because tailings were laced with heavy metals—lead, arsenic, mercury—and were fiercely acidic (with pHs of 2 or lower), life was unable to reestablish itself upon these mounds. Their texture ranged from the finest flour-like silt to gravels to clumps of leaden slag that look like fragments of a meteorite, all congealed and bubbled. Their palette was composed of dusty grays, dunnish yellows, greens, and a hard metallic black. Known in Mexico as *escombros* ("debris") or *estéril* (literally "sterile") these monumental heaps survive to the present as the dusty brooding monuments of industrial ore processing. They define the landscapes of mining districts across Mexico and indeed across the Americas. Thousands have been inventoried from Chile to Canada.[4] In Canada we call them orphan mines, but really they are orphaned waste.

These silent and hulking piles appeared inert, like the muted background of life. In reality, they were chemically active and constantly infiltrating the living ecologies that surrounded them. Winds moved fine particles of toxic material off the heaps of the Morales smelter in great clouds, pushing them directly into the neighboring barrios, with the Fraccionamiento de Morales, again, positioned first to receive them. Once inside the neighborhoods, the metallic dust spread across the surfaces of daily life: into people's houses, into their soils. Another portion of the tailings was leached out by rains and carried down into the aquifer that lay beneath the valley of San Luis Potosí.

Rates of heavy metal contamination in the local population became, with time, very high. Mining waste harms people by entering into their bodies. Tracking whether, and to what extent, this is taking place requires the combined work of environmental chemistry and public health science. This a field in which Mexican researchers excel. A number of studies have been conducted specifically on San Luis Potosí. What they find is that the two most important elements moving from mining waste to bodies—as measured by prevalence and volume—were lead and arsenic. Ubiquitous in-house dusts, garden soils, and drinking water, these metals were, and continue to be, absorbed or ingested by residents. Samples of people's hair, blood, skin, and bones register levels of arsenic and lead contamination at between five and eight times the acceptable national norms. Children are the most effected, then women.

What has been documented for the valley of San Luis Potosí is part of a national pattern. Over thirty such studies have been conducted across thirteen mining districts in Mexico. The accompanying map and table identify the study locations, the principal metals of concern, and the relevant studies (see map 4). There are variations in the chemical composition of the waste and the environmental contexts that condition their dispersion and thus availability for uptake by human bodies. All the same, taken as a whole, their findings consistently show how mining waste created toxic living spaces and the chronic wasting of human health.

The harms produced by persistent heavy metal contamination come from the movement of metals into the human body. They entered as an extremely fine dust, fine enough to pass through a cell's membrane and into its mitochondria.[5] The damage they provoked was a function of their accumulation. Small amounts of certain metals, such as arsenic, zinc, and manganese, are vital to human health. In higher concentrations, however, they disrupt cellular reproduction and metabolic function. This leads to the wasting of muscle and bone tissue (arsenic, lead, cadmium), or to massive inflammatory responses and skin lesions (arsenic, lead, mercury). Most heavy metals produce microscopic lesions that interrupt the normal signaling functions of the nervous system. This produces the tremors, the palsies, and the damage to different parts of the brain (arsenic, mercury, lead, manganese).[6]

All of this has been documented in the Mexican studies. It should be recalled that these studies have been conducted over the last fifteen years or so. What they capture are the cumulative effects of three generations of heavy metal contamination. Among the surviving photographs of Cerro de San Pedro and Morales one finds the typical snapshots of community life—the assembled baseball team grinning for the camera, a Sunday outing—but their backdrop was not a field, park, or plaza but rather large dumps of dusty tailings.[7] The photographs date from the 1940s. It is their grandchildren who are being tested today.

· · · · · ·

The industrial mining complex's massive loading of heavy metals and other mineralized discharges into the local ecology wrought harm through the interruption of normal biological processes. Its chronic effects were based on

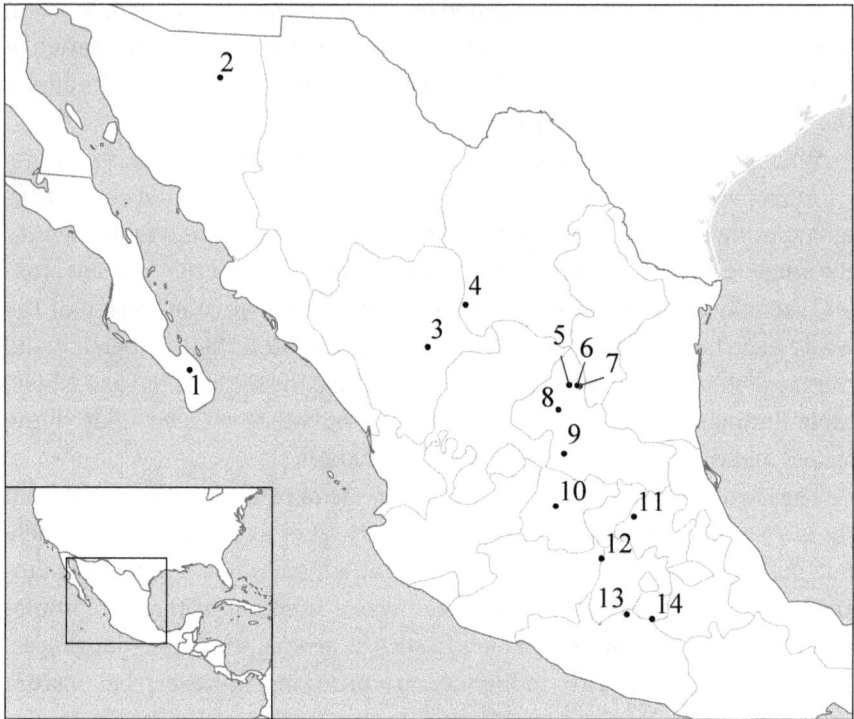

KEY	CONTAMINANTS	REFERENCES
1: San Antonio, Baja California	As	5, 27
2: Nacozari, Sonora	Cu, Mn, Zn	8
3: San Jose de Avino, Durango	Cd, Cu, Pb, Zn	28
4: Torreon, Coahuila	Pb	11, 33
5: Real de Catorce, San Luis Potosí	As, Pb	21
6: Villa de La Paz, San Luis Potosí	As, Cu, Pb, Zn	6, 14, 16, 22, 23, 32
7: Matehuala, San Luis Potosí	As	19
8: Charcas, San Luis Potosí	As, Cd, Pb, Zn	25
9: Morales, San Luis Potosí	As, Cd, Pb	7, 15, 29, 30, 31
10: Guanajuato, Guanajuato	Hg, Pb, Zn	1, 4, 24
11: Zimapan, Hidalgo	As, Cd, Pb, Zn	22, 23, 32
12: Tlapujahua & El Oro, Michoacan	Pb, Zn	3
13: Taxco, Guerrero	As, Cd, Pb, Zn	2, 12, 18, 26
14: Huautla, Morelos	As, Cd, Mn, Pb	9

MAP 4 Industrial mining contamination study sites in Mexico
(© G. Wallace Cartography & GIS. Map by Geoffrey Wallace, 2021).

References

1. Ramos-Arroyo, Prol-Ldesema, and Siebe-Grabach, "Características geológicas"
2. Méndes-Ramírez and Hernández, "Distribución de Fe, Zn, Pb, Cu, Cd, y As"
3. Corona Chávez et al., "The Impact of Mining"
4. Mendoza-Amézquita et al., "Potencial lixiviación de elementos"
5. Espino Ortega, "Afectación de suelos y sedimentos"
6. Gamiño-Gutiérrez et al., "Arsenic and Lead Contamination"
7. Calderón et al., "Exposure to Arsenic and Lead and Neuropsychological Development"
8. Meza-Figueroa et al., "The Impact of Unconfined Mine Tailings"
9. Mussali-Galante et al., "Evidence of Population Genetic Defects"
10. García, Armienta, and Cruz, "Distribution and Fate of Arsenic"
11. Del Razo et al., "Arsenic Levels in Cooked Food"
12. Armienta et al., "Geochemistry of Metals from Mine Tailings"
13. Armienta et al., "Environmental Behavior of Arsenic in a Mining Zone"
14. Monroy, Carrizales, and Diaz-Barriga, "Arsenic and Heavy Metal Pollution"
15. Carrizales et al., "Exposure to Arsenic and Lead of Children"
16. Castro-Larragoitia et al., "Arsenic Mobilization in Aquatic Sediments"
17. Armienta et al., "Origin and Fate of Arsenic"
18. Espinosa and Armienta, "Mobility and Fractionation of Fe, Pb and Zn"
19. Jasso-Pineda et al., "An Integrated Health Risk Assessment Approach"
20. Ceniceros et al., "Geochemical Distribution of Arsenic, Cadmium, Lead and Zinc"
21. Chipres, Castro-Larragoitia, and Monroy, "Exploratory and Spatial Data Analyses"
22. Jasso-Pineda et al., "DNA Damage and Decreased DNA Repair"
23. Monroy-Fernandez et al., "Arsenic and Lead Contamination in Urban Soils"
24. Ramos-Gómez et al., "Movilidad de Metales en Jales"
25. Romero and Gutiérrez Ruiz, "Estudio Comparativo de la Peligrosidad de Jales"
26. Hernández et al., "Dispersión Hídrica de Arsénico"
27. Corral-Bermúdez, Rivera-Quintero, and Sánchez-Ortiz, "Percepciones y Realidades de la Contaminación"
28. Diaz-Barriga et al., "Arsenic and Cadmium Exposure in Children"
29. Romero et al., "Solid-Phase Control on Lead Bioaccessibility"
30. Rosado et al., "Arsenic Exposure and Cognitive Performance"
31. Gamiño-Gutiérrez et al., "Arsenic and Lead Contamination in Urban Soils"
32. Vargas et al., "Lead Exposure in Children Living in a Smelter Community"

this cumulative loading of nonorganic material into people's bodies. The same complex also produced an ambient ecology that was particularly convivial to the flourishing of life forms—viruses, bacteria, parasites—whose propagation incurred serious consequences for health of workers and residents. Typhoid and smallpox outbreaks erupted regularly and rapidly spread from one district to another.[8] Measured by persistence and ubiquity, however, the greatest stalker of human bodies in the mining districts was the hookworm.

The symptoms of hookworm infestation—the loss of blood, chronic anemia, exhaustion—were known as the *mal de minero,* the miner's disease.

Like tropical plantation zones or the Panama Canal works, whose mix of climate, congregated human hosts, and ample mosquito-breeding sites assembled the conditions for the catastrophic spread of yellow fever and malaria, industrial mining districts produced the ecological setting that hookworm needed to multiply to epidemic proportions.[9] Hookworms are small nematodes that migrate into the small intestine where, thanks to their eponymous hook-like teeth, they latch on to the intestinal wall of humans to draw blood.[10] When they are deeply ensconced in the human body, as well as during their time outside the body as eggs and larvae, hookworms are basically invisible. The effects of an infestation are not. A single hookworm steals a small thread of blood from its host; a congregation of hookworms provokes serious forms of anemia. On his tour of the mines of Pachuca, the Mexican physician Angel de la Garza Brito crossed some children who appeared healthy from afar but in closer conversation showed themselves to be extremely pale and utterly exhausted from the simple exertion of walking home. He followed them and interviewed their mother. She related that the entire family had fallen prey to the "mal de los mineros"—the miners' sickness. It had completely consumed their energies ever since her husband had begun to work at the mines.[11] Mexican newspapers called it the germ of laziness, a sickness that "enervates men's strength and makes him averse to work."[12] Those who suffered "were white as wax, or a clayish yellow . . . with anxious faces, often with an idiotic expression . . . Many sufferers have a vacant look, as if they were absent."[13] When "miner's anemia" arrived at the Morales plant in San Luis Potosí in 1913, it halved the workforce, laying low six hundred workers in a matter of months.[14] By the 1900s and 1910s it was estimated that half of the workers in Pachuca were infected.[15] Other early foyers of infection included mines in Baja California Sur and Sonora—also key destinations for migrating mine workers.[16]

The passage of hookworms into people's bodies is often imperceptible. Apart from a small rash of dermatitis at the zone of entry and perhaps a dry cough provoked by the worm's traverse of the esophagus, the human host is none the wiser.[17] It is only when the population of hookworm increases that anemia and associated complications set in. For this to happen, hookworm habitats need to be multiplied and the corridors between them kept open so that a growing number of worms can make their migration. This was precisely the ecological matrix that mining districts of the late nineteenth and early twentieth century mining boom offered the hook-worm. Common deposition zones included the abandoned side tunnels of the mines, the privies located corners of the patios behind the cottage, or the small pits and gullies that provided distance and discretion. In the barrios, hookworm eggs and larvae were dispersed when human feces were mixed with straw and swept up off the patio floor. In Cerro de San Pedro, human and animal wastes were pushed into roughly dug culverts that lined the cobbled streets. There they remained to quietly mellow and decompose until the annual summer rains flushed them down into the valley of San Luis Potosí. The eggs and larvae were slowly moved by periodic rains or the strong winds that kicked up the dust. Workers themselves were another medium of dispersion, as they moved the grimes and dirts that accumulated on their bodies during their work-shifts to their houses and neighborhoods. The hookworms' passage into the body was further eased by the daily press of bare or sandaled feet on the packed earths of households, rooming houses, and patios. The propinquity of people in the barrios agglomerated the human hosts for the worms and multiplied their breeding and birthing grounds.

The boom, with its large-scale and wide-spread movement of workers, accelerated the hookworm's colonization of the mining and smelting districts.[18] "There are grave reasons to fear that all the states are infested," Ricardardo Manuell, a Mexican public health official, warned in 1906, "bearing in mind, on the one hand, the active movement among our wage-earners, and on the other, the favorable conditions for the preservation of the larvae of the uncinaria [hookworm] . . . The numberless mines which are scattered through our territory are more than enough to present inexhaustible breeding places for the larvae."[19] In 1913 the Instituto Medico Nacional confirmed his fears, releasing the results of the country's first epidemiological survey of hookworm, with an accompanying map.[20] The hookworm, by then, was present in all the mining regions of Mexico.

In the absence of a reliable vermifuge, health authorities understood that their best chances against the multiplication of hookworms was to dismantle the ecology that sustained them—a line of defense whose successes had been demonstrated in controlling insect-borne diseases in the Panama Canal zone and other parts of the tropics. Thus, under the reformist postrevolutionary state, Mexico's Instituto Medico Nacional prescribed quarantine for infected miners, the disinfection of mines, the building of latrines, furnishing showers and lockers for mine and smelter workers, and assuring a supply of clean water to assure the cleaning of hands and bodies.[21] Similarly Garza Brito advocated the systematic cleaning of mine tunnel floors and walls with a lime chloride, requiring miners to void in pans emptied in prescribed latrines, banning the transport of food or drink into the mines, and providing showers, change rooms, and laundry services.[22] Many of these recommendations would become central demands of miners' unions.

· · · · · ·

Mine and smelter workers of the late nineteenth and early twentieth centuries were enmeshed in an industrial metabolism that, in speeding up processing rates, greatly raised the chronic harms and acute dangers inherent in working environments. Meshed with machines, forced to keep up with its rhythms and absorb its atmospheres, workers' bodies became, as they had in the colonial period, the central nexus of the social and ecological relations of extractive mining.

Mines and smelters had always been dangerous places and the taxing environments and serious threats from the past remained. At Cerro de San Pedro, ASARCO pushed the Juarez shaft down past the six hundred-meter level. During his brief tour in 1926, the inspector Juan Martínez noted that the temperatures were exceedingly high and work arduous.[23] The miners of the Cerro de San Pedro toiled in this heat every day, handling heavy equipment, cracking out the loosened ore, loading and pushing the bogie carts, all along the length of their eight-hour shift. Some miners worked entirely naked.[24] The different noxious gases that invisibly stalked miners in the preindustrial period did not dissipate with industrialization. Miners continued to keep a wary eye on their candles and lamps. They were watching for the telltale sputter of the flame that signaled the lack of oxygen.[25] Rockfall remained a threat. Not only the catastrophe of a collapsing tunnel, but also to more quotidian, though no less lethal, release of loose material. Safety helmets only began to appear after the Revolution. Until then a miner "could be felled by a piece of rock no larger than a baseball."[26]

Coming into the twentieth century, however, the intensification of the extractive process layered in new dangers and harms for workers. With time, tools and systems were brought into the mines to insulate workers from the harms wrought by this hazardous techno-nature. Ensuring their systematic application was one of the key gains made by miners' unions over the course of the twentieth century. But from 1890, through the Revolution, and to the Second World War, that is across the arc of the industrial bonanza, the tolls excised on miners' bodies deepened, either directly through injury or indirectly through the new toxicities that machinery laced into working environments.

Before electricity illuminated the mines, miners worked by candlelight. They would bring a pocketful with them, sticking one on their caps and placing the others on ledges and rock caps. They worked in a world of flickering shadows. In the preindustrial period this was simply one more of the ambient bothers that miners had to deal with. Electric lights arrived late in the industrialization process, after the Mexican Revolution, and even then, only in the larger, better capitalized operations. This meant that, miners worked for decades with heavy steel machinery that was powered up and moving at speed in that same world of shadows. There were hundreds of ways that improperly gauged or unexpected contact with machinery could maim or kill. These were so numerous that in the occupational health statistics they were simply piled into the category "machinery."[27] Work in the cyanide mills or smelters above ground might have been better lit, but the machinery was just as lethal. There the *Mexican Mining Journal* noted "high trestles without railings, exposed gears everywhere, no protection against heavy-duty belts liable to fly off at any moment and whip someone into eternity. These surround [the worker] at every turn, and if he is slow in getting around, he gets fired."[28]

Industrialization laced new toxic compounds into the underground airs. Running combustion engines for pumps and other equipment released carbon monoxide which, being heavy, pooled up in the lower depressions of the mines. Don Tomás recalled how, after sixteen years as a machine-operator he quit. It was on account of the exhaust. "I got out," he said, "because of the smoke. There was a lot of smoke, a lot. When they began to mechanize the mine there was a lot of machinery and all that smoke from the machines used to blow out my nose like the tailpipe of a car."[29] The discharge of dynamite further saturated the tunnels with noxious compounds of nitrogen and carbon.[30] Their smell pervaded the mines. Miners who were green passed out from the gas; the veterans acclimatized.[31] But even the latter were liable to receive a stiff dose: "Twice I vomited blood from my mouth, the second time more than the first, I threw up like three liters and after that I couldn't walk."[32]

The atmosphere in the smelters and foundry-works was not necessarily better. The furnaces produced copious amounts of noxious gases—vaporized lead, hydrogen sulfides, coal and tar volatiles, and concentrations of carbon dioxide. Inadequate ventilation rendered the atmosphere in the smelters intolerable and in fact noxious to human health. Company managers and industry spokesmen disputed this because the science did not support it, but workers knew it as a visceral fact.[33] Accidental mishandlings of the blast furnaces released clouds of dense metallic vapors that laid mill workers out cold. At Towne's Morales plant outside of San Luis Potosí, workers incapacitated by such vapors were taken out on stretchers on an almost daily basis.[34]

The former miners of Cerro de San Pedro remember how the work in the tunnels transfigured them. It painted them with a deep red mineral dust. As they walked home at the end of the shift they appeared as a tired phalanx of Carnival Devils—"red as the Devil," Don Marcos Rangel recalled, "with only the whites of our eyes peeping out"—startling the townsfolk into laughter.[35] There were many source points for dust in the mines, mills, and smelters; anywhere, in fact, that the mineralized rock was broken up and pulverized: the blasting of the mine face, the prying down of loosened material from the walls and roofs of the tunnels, the crash of ores down the chutes, their reduction to powder at the stamp-, rod-, and ball-mills. But the leading source of fine particles was surely the pneumatic drills. As they hammered their way into the mine face, a thick plume of dust blew out from the hole. It then dispersed and thickened into a dense fog. Within minutes everything was enshrouded. For the next hours—until the drill had finished its three-meter travel—miners navigated through the mist. The dust was particularly bad when miners worked upwards, tunneling out a rise or an overhead stop. That's when they received the plume directly in their faces.[36]

The dust made breathing toilsome and, mixed with sweat and oil, it made for a pasty mess that caked itself on to skin and clothes. But on balance, when compared to death from rock-fall, electric shock, dynamite explosion, gas poisoning, the dust was, like the heat or the exertions of mine work, something to be endured. Besides, there was no arguing with the fact that the drills greatly accelerated the work.

Over time, however, dust was the agent of the most serious of consequences. The first was heavy metal contamination, since the source material producing the dusts was laced with lead and arsenic. Inhalation was one vector of incorporation. In the early twentieth century, the atmospheres of the mines were said to provoke *saturnismo*—a condition that combined psychological depression and loss of body control.[37] The second was silicosis, the disease provoked by the intrusion of microscopic silicate flecks in the lungs. Compared with lead or arsenic, silicates are chemically speaking innocuous for the human body. If silicates are dangerous, it was because they are diamond sharp. As they cut into the inner surfaces of the lungs, they trigger an inflammatory response that produced a spreading mat of fibers to contain the crystal and protect the surrounding tissue. The problem was that this response did not switch off. It produced a "progressive massive fibrosis" which, in time, came to fill and ultimately choke off the lungs.[38] This dense black mat was what early twentieth century X-rays revealed when they imaged miners' chests. The actual amount of silica needed to trigger silicosis was very small: one to three grams inhaled over time. Continued exposure accelerated its development, but by the time one felt its effects it was too late to stop it. To this day there is no effective cure or therapeutic for silicosis. As the miners themselves recognized, "the troubles really come out over time. By the time they get you, there's nothing you can do."[39]

· · · · · ·

The arrival of high-intensity industrialized mining produced an exceptionally exacting set of ecologies for workers and residents of Morales and Cerro de San Pedro. The sheer amount and density of heavy metals present in the environment was unprecedented. In the two late eighteenth-century maps discussed in chapter 5, the cartographers painted piles of scoria and other wastes, here and there, into the cityscape of San Luis Potosí. Even if they were brought together in a single pile, theses tailings would be dwarfed by the massive mounds that covered close to three hundred hectares of what used to be the pastures of Morales. Colonial piles could be a bit taller than a small house. In Morales they formed a multi-storied structure. They were embedded into the local environment, and until the tailings are finally buried or otherwise isolated, the dusts that blow off them will continue to pervade local bodies for generations. For those who worked in the smelters or the mines, high-powered and mechanized extraction took a constant and chronic toll on their bodies. Their homes and neighborhoods were particularly vulnerable to epidemics. Taken together these were the basic features that characterized the social-ecological contexts that came to surround the people of Cerro de San Pedro and San Luis Potosí. They came to frame the political response of workers and residents to the embodied injustice of industrial-era extractivism.

11 De Profundis

As mining revived in the nineteenth century so too did the pulse of worker activism. Beginning in the 1870s strikes and riots burst out in the mining districts with regularity. Rodney Anderson records twenty-five strikes for the period leading up to the Revolution (1872–1909) though this is by no means an exhaustive catalog.[1] They were ubiquitous across the mining belt, from Sonora to Oaxaca, in small remote mining operations in the Sierra Madre of Chihuahua or Sinaloa and in the large mining centers such as Pachuca and Guanajuato. In the state of San Luis Potosí, the first strikes took place to the north, in the mining districts of Charcas and Mathuala (1884) and Real de Catorce (1886, 1891, 1898, and 1900). The first worker action in San Luis proper took place at the Morales plant on July 31, 1903. This was more of a protest than a strike as such. Workers interviewed by the local press declared that their original aim had been to call attention to their demands rather than a confrontation with the company. Had they decided on the latter, they would have arrived armed, and the results would have been much the bloodier.[2] As it was, they were met by police, a unit of the mounted gendarmerie, and another unit of *rurales*, all dispatched by the governor to extinguish the protest.

The newspaper report stated that the workers sought an end to the swindles operated by the company store, and to achieve an improvement in wages, both regular demands of Mexican mine and smelter workers during this period. That said, the Mexican social historian Moisés Gámez documents a deeper vein of complaints that began a decade earlier over the toxic atmospheres of the plant. The first complaints about the noxious vapors vented by the furnaces were recorded in the local *El Estandarte* newspaper in 1892. The company quickly denied that there was a problem. "From the furnaces," it stated, "neither gases, nor smoke, nor vapours are emitted." But then, in the single month of May 1893, thirty workers were taken out on stretchers, laid low because, "the lack of measures taken by the company to remedy the risks." A decade later, the situation had still not improved. The official gazette of the state of San Luis Potosí registered an unceasing string of accidents at the plant.[3] Six months later workers gathered to voice their protest and were met by the local constabulary. Upon request of company management, a unit of armed rurales was permanently stationed on site at the plant.[4] Work proceeded, as did its tolls.

A well-developed social history documents the militancy of Mexican mine and smelter workers leading up to and then past the Mexican Revolution of 1910–20. It emphasizes the racialized inequalities of labor in the districts, how Mexican miners and smelter workers challenged unequal wage scales in which they made half or less than American or other foreign workers for the same job; how they militated against the more day-to-day inequities of life in which Mexicans were relegated to the status of second-class citizens in their own country.[5]

The ecological field of relations that defined work and life in the districts was arguably just as central to the working-class politics of the industrial period.[6] This field encompassed the zone of interaction between the physical environment and the body; between the biological changes wrought in the body and the sensory, subjective experience of those changes (the dark register of different pains, the debilitation of function, the slow realization of decline of lung capacity); and between somatic experience and the realm of political consciousness, expression, and action.[7]

The militancy of mine and smelter workers came *de profundis*, from the depths: from the temporal depth of centuries of miners' struggles for justice; from the physical depths of the tunnels; and from the interior zones of experience. Seen through the lens of moral ecology, it emerged out from a notably unequal matrix of social and ecological relations. The Mexican workers and their families at Morales and Cerro de San Pedro took on the brunt of what high-intensity extraction demanded in terms of health and vitality. In exchange, they received the next-to-lowest wages for mine and smelter work in Mexico.[8] Immediately beside them were the colonias of American and other foreign managers, whose inhabitants were protected and cushioned from these tolls as a matter of privilege. When those same foreigners denied the harms, as they did in 1893, and then refused to better conditions, but instead brought in the constabulary, or fired the maimed, as they did for years, militancy became something more than protest. It became a socially organized effort to achieve justice and redress.

· · · · · ·

Plumbing the sources of mine and smelter activism requires attention to what Rob Nixon termed the slow violence of chronic environmental harms.[9] It means looking past the catastrophic drama of the cave-in or gas explosion which, in Cerro de San Pedro and other districts, was signaled by siren call and whistle blast.[10] Instead, it draws our attention to the dour peal of the funeral bell that marked an almost daily time signature in the districts. At the Nacozari copper mines, Ralph Ingersoll noted the "continuous processions, on the road to the graveyard, of funerals, in which the little blue coffins of babies predominated."[11] This regular, more muted, even intimate, rhythm captures, better than the collective drama and tragedy of the emergency siren, the beat of extractivism's toll.

Occupational health statistics only began to be systematically recorded for Mexico's mining sector after the Revolution. They reveal that same chronic tempo of harm. For a six-month period between January and June of 1926, inspector Juan Martínez recorded 186 accidents in the mines of Cerro de San Pedro.[12] That is a touch over an accident per day. At the Morales plant in San Luis Potosí the average was about the same: 392 accidents for the calendar year 1924.[13] National statistics show the very short odds dealt to miners and smelter workers. In 1925, on an estimated population of 120,000 workers, over 27,000 had suffered an accident, that is, a touch more than one in five.[14] The numbers for 1929 to 1931 were worse. The Secretaria de Trabajo calculated between 6 and 7.23 accidents per 1,000 person shifts.[15] For an operation of the size of Cerro de San Pedro this translated into an average of 2.5 accidents per day. These statistics do not record the exactions of chronic silicosis, a disease whose existence and etiology remained a matter of debate well into the 1930s.[16] Public health statistics were also unable to fix a precise bead on the extent, frequency, or severity of epidemics, hookworm infections, heavy metal contamination, or any of the other diseases that affected the entire communities of working-class Mexicans. Nor did they tally the innumerable minor injuries that were not considered worth a visit to the clinic and thus went unrecorded.

These harms were first and foremost registered in the body. Americans who related their time in the mines of Mexico noted the long scars, lesions and sores that etched the bodies of the miners.[17] For mine and smelter workers the wasting agency of their work environments was felt in the loss of breath, the lack of strength and energy, the loss of motor control. These were all intimate corporeal experiences. Which is also to say that these were either invisible or escaped easy observation by outsiders. And if they were the product of the acceleration of the extractive regime, such harms were chronic rather than acute in nature. They had long latency periods before they were expressed. The deterioration that they provoked was cumulative in nature, progressing over months or years or even—as in the case of the congenital diseases provoked by heavy metal contamination—across generations.

Elites, whether American engineers and managers or Mexican state officials, did not deny the suffering of workers, nor the fact of their debilitation. They felt that Mexican workers were particularly susceptible to disease—"the resistance of the people was virtually nil," wrote Ingersoll, "the simplest disorder snuffed out their lives like candles in the breeze."[18] They believed Mexican miners to be exceptionally weak. It was written that they could not handle the overhead work of pushing up a vertical raise; that it took two Mexicans to control the same mechanical drill that an American worker could manage by himself; that they did not have the strength to drive in a blasting hole by single-jacking—that is to wielding a sledge and drill in each hand to pound in one by the other.[19]

The source and reason of the various afflictions, they felt, was to be found in the comportment and character of workers and their families. "In matters of hygiene," declared the US engineer Allen Rogers, Mexicans workers were "absolutely ignorant, of course, and in mining camps it is one of the hardest problems to keep the place clean."[20] Even Ralph Ingersoll, among one of the more fair-minded and sympathetic North American observers, believed that Mexicans lived in squalor and misery because they were accustomed to it and that they didn't take baths because they weren't "particularly interested in experimental adventures of that sort."[21]

These libels were of a kind with others current at the time. Mexican mine workers were described as irresponsible, chronically drunk, and lazy.[22] They were racialized as part of the "national character" and temperament. "The population," intoned Walter Weyl in his report to the US Department of Labor, "is marked by an almost constitutional indolence" that he attributed to the heat and humidity of the lowlands, to the lack of severe winters which excused Mexicans from the work of "improving houses, clothing and diet" and, in the highlands, to the negative effects of altitude on "the muscular power and the nervous energy of the population."[23]

Discourses that naturalized Mexican squalor, weakness, and indolence were not only libelous and racist, they served to veil the social and ecological causes of filth, disease, and impoverished constitutions. That weakness might be produced by hookworm infestations was not considered. Nor were the considerable efforts that working Mexicans, especially women, took to keep bodies and living spaces clean. There was no acknowledgement of the difficulties involved in securing ample supplies of clean water, difficulties that were exacerbated by the pollution from the mines and smelters, or by the fact that company managers controlled the water supply.

To be fair, there existed a reformist streak within the broader class of foreign mine owners and professionals. Like other improvers in other industrializing settings, they believed that the districts could be spaces for betterment. But even here, the racialist lens skewed the view. Instead of targeting the social and environmental causes of ill-health and disease—a direction that would inevitably lead to uncomfortable questions about the material organization of the extractive process and the social relations that it depended on—the reformers focused on the Mexican character. What they did not publicly admit was the possibility that these sufferings might be produced by the material exigencies of industrial mining (the speed and techniques needed to achieve high throughput processing) and the dwelling conditions produced by rapid boom-town type growth. Instead, they planned moral uplift and transformation. They hoped to turn Mexicans into idealized Americans: clean, healthy, sober, disciplined.[24] The project was the object of some debate among the managers and engineers of the mining sector. Against the reformers were those who believed that such a transformation would take years, generations, or never.[25] For the skeptics, the intractability of the Mexican character meant that investments in clean housing, clean water, showers, health care were so much money lost.

Harry Franck, who actually spent a season sweating it out with a cuadrilla in the mines of early twentieth-century Guanajuato, provides closer insight into the miners' frame of consciousness. His account is not an emic account, but it reads as one well-attuned to the play of power and experience within the mine. Franck knew that the miners of Mexico were exceptionally strong and untiring, men who could easily hold their own beside the quarrymen, cotton-pickers, and rail-road workers of the United States that he had previously worked with. On the other hand, he railed against the engineers and bosses who could not bear the extreme conditions of the mines and collapsed from the smoky and dust-ladened air ("Brummel . . . fell down at once with dizziness and went to bed . . .").[26]

Franck knew better than most what lay beneath the quiet stoicism of those who toiled in the mines. Others interpreted this as fatalism or as the racialized reflection of their innately less sensitive, more brutish, constitution, ("the hospital doctors asserted that the peon has not more than one fourth the physical sensitiveness of a civilized person."). Franck felt that miners kept their true thoughts hidden. Their cant, filled with its

Nahualismos and other indigenous loan words, mixing whistles, words and signals, sheltered their conversations and allowed "discussion of the alleged characteristics of their superiors in their very presence without being understood." Their "manner was little short of obsequious outwardly, yet one had the feeling that in crowds they were capable of making trouble, and those who had fallen upon 'gringos' in the region had despatched [*sic*] their victims thoroughly." And: "As they saw it, these foreigners had made them go down into their own earth and dig out its treasures, paid them little for their labors . . . Like the workingmen of England, they were only 'getting some of their own back.'"[27]

Recovering how Mexican mine and smelter workers of the late nineteenth and early twentieth century saw things is hampered by the lack of sources written in their hand. Oral histories provide an alternative. A former worker, Antonio Herrera Cervantes, emphasized the hard bodily labor that campesinos undertook once inside the Morales plant: "those *rancheros*, they were the ones who put up with the worst of the toil, it was pure cart work, everything done by hand."[28] Their exertions were compounded by their bodies' exposure to the harsh conditions of the plant. They worked barefoot and were forever injuring themselves in the new and unforgiving world of heavy steel and jagged stone.[29] The finely milled ore dust filled the air, constantly churned up by the valley's breezes, "we swallowed it constantly, that and the smoke," that is, the thick leaden fumes of the furnaces.[30]

Up at the mines of Cerro de San Pedro the atmosphere was hardly better. There too the dusts slowly wrecked the miners. A man interviewed for René Medina's oral history recalled: "There were many *cascados* in those days. The *cascados* were those who walked around all day, coughing and coughing and coughing. Those were the ones who were already wrecked . . . because they already looked like corpses. They couldn't handle much walking because it drowned them, that is they were in rough shape. They were the people who had been hit hard by the mines. The ones who stuck at the work, died. They were really affected, young that is. They died. They didn't last long."[31]

The term used here, *cascado*, is an important term in the popular etiology of environmental health in industrial mine work. It referred to the wasting of the lungs. Don Luis, a former miner in Zimapán, explained: "All my *compitas* were finished like that, the ones from La Encarnación, del Mezquite, de La Hortiga, all of them. Their lungs were wrecked. I've suffered too, not for being an idiot, not because I was a drunk. My lungs are screwed. They must be like a husk (*como un cascarón*) because I worked my fair share."[32]

Thus cascado quite literally meant "to be rendered a husk," or "to be husked." It meant becoming a wasted shell of a person.[33] It was the depressingly evocative term for silicosis. Recall that silicosis was a chronic affliction that took time to manifest (chapter 10). As it advanced, it destroyed lungs. It cut breathing to a laborious gasp, to the point that people felt like they were drowning. Deprived of oxygen their bodies withered. The afflicted began to look like walking cadavers, that is, husks.

The response of workers to the intimate and embodied exactions they experienced was tightly controlled. For reasons of reputation and self-regard it was important to present a posture of strength and fortitude. Managers and fellow workers were clear about the exceptional toils that awaited those who took up mine and smelter labor. Herrera Cervantes remembers how he and other country-folk were greeted at the gates of the Morales plant: "Here we want people with strong arms and backs," warned the foreman, "people who can work, who can move. So: if you're up for it, c'mon in. If you can't, get on home."[34] Don Gabriel, up at Cerro de San Pedro, echoes: "For them [his uncles], it was endure, endure, endure. You didn't go to the infirmary. Only if they carried you in."[35] Historian René Medina corroborates, noting that the records of the *Hopital Juarez* show that significantly fewer nonurgent entries for male miners as compared with women or other occupational classes.[36] In the face of the dangers presented by the mine, better to be unflinching and even carefree. "When the fuses were lit, we'd hurry away, but not him," recalled don Armando of a fellow miner, "he walked, and slowly, slooowly, sat down beside us. When the blast hit, he had like a little smile."[37] Sticks of dynamite were tossed back and forth. Men sat smoking on the cases of explosives. Carpenters sauntered along high trestles as if they were walking down the sidewalk—the indifference to personal danger was performed in so many day-to-day acts.[38]

And yet, beneath the face of endurance and boldness, there was also recognition and concern. One place that this was enacted was in the many votive shrines that the miners of Cerro de San Pedro carved into the rock. They were built in the immediate vicinity of the mine's mouth, or at the embranchments into secondary tunnels deep within the mines. They varied in size and sophistication from elaborate altars with distinct alcoves for statuary (long since disappeared), candles, and other offerings, to simple, spare, and modest-sized recesses. These were places of day-to-day devotion where miners lit candles, left offerings, and made their prayers. Their emplacement at the mine-mouth, the threshold between the world of the surface and the subterranean world of hard work and constant danger, is suggestive. Were they the occasion to ask for protection on the way in? To give thanks on the way out?

The experience of bodily harm and risk was not, in and of itself, sufficient cause for complaint. It would be an analytical short cut to move simply from the taxing conditions produced by the extractivist complex to the episodes of protests, strikes, or other forms of subaltern revolt. This recalls E. P. Thompson's caution against "spasmodic" interpretations in social history that too-quickly linked privation to protest. Arguably, mine and smelter-workers did not rise up from an incapacity to endure—this would have cut against the prevailing ethos—but out of a refusal of injustice in the field of social and environmental relations. The extractivist complex built by Towne was not only a remarkably exacting socio-ecology but it was also glaringly unequal in its distribution of harms and benefits between the different groups that composed it. Nothing made this more apparent than the environmental segmentation of life at Cerro de San Pedro and Morales.

Remember that the barrios, inhabited by local workers and their families, and the colonias, the retreat of American and other foreigners, were sharply divided. At Morales, the American colonia was stoutly defended by a wall. Two meters thick, built of hewn rock, crenellated, and complete with a set of four round guard towers, it imitated the battlements of an English keep. It was established on a hillock that was both upwind from the chimneys and above the barrios. Inside were two styles of company houses. The first, built under Robert S. Towne's direction, matched the neo-Gothic style of the fortifications: heavy stone masonry walls, arched windows and doorways, high gabled slate roofs. The other, brought in during the ASARCO years, was rather more California ranch: bungalows, creamy stucco siding, low clay tiled roofs, front porches, and porticos. Both were neatly lined up along paved streets and landscaped with small green lawns, flowering plants, and trees. They were wired for electricity, supplied with water lines, indoor plumbing, and telephones. Rent was free. The compound also included tennis courts, an eight-hole golf course, a large recreation hall, and a social club. A hotel catered to visiting staff and their guests. A clinic, staffed by an American doctor and nurses, attended to the colony's health.[39]

Racialized segregation was a hallmark of extractive enclaves and plantations across Mexico and Latin America. It divided people, as is well known, but it also segmented their urban environments. The American colony at Morales may have looked like a keep, but ecologically it was rather closer to a greenhouse. Within its walls an exotic human ecology was produced, one in which water was abundant and available not only for cleaning and drinking but also for watering lawns and plants, where the toxic exhalations of the chimneys breathed away from, rather than toward, its inhabitants, where streets, floors, and the general construction of houses served to insulate human bodies from exposure to ambient parasites and the rigors of the altiplano's climate.

This was at the antipodes of the conditions facing Mexican workers, their families and neighbors. What is striking is the play of distance. The yawning gap between the environmental conditions enveloping American managers and Mexican workers was in fact contiguous in space. From the battlements, the view from the American colonia gave directly down to the 125 workers' cottages—each the standard four meters squared—lined up beneath the lead and arsenic laced slag piles. Immediately beyond them was the large squat that in time became the Fraccionamiento de Morales. This was where a fine rain of lead and arsenic precipitated out of the emissions voided by the smelter, every day, for years; where rates of heavy metal contamination are among the highest in Mexico; where clean water to wash clothing and bodies was scarce; where, in a matter of months, an explosion of hookworms knocked out six hundred workers.

All of which raises the central problems posed by environmental justice: the social production of unequal ecologies, the distribution of environmental health and harms, the slow violence of chronic contamination.[40] The experience of wasting that characterized life in the mining district was quotidian—the struggle to secure clean water, the heavy gasping step of a miner wrecked by silicosis, yet another funerary procession filing behind a child's signal blue coffin. But the immediate, and visible, contrasts with the lives and surroundings experienced by foreigners sharpened experience into a question of justice.[41] The neatly trimmed and deeply watered lawns of the "Yanquis," their spacious bungalows and tennis courts, their emplacement upwind from the mill or plant, their provisioning with plumbing, sewage and clean cool drinking water—all of these distinctions were matters of daily fact, but that did not mean they would be indefinitely endured.

• • • • • •

In 1909, six years after the first stymied protests at Morales and on the eve of the Mexican Revolution, Jesús Silva Herzog updated the situation at the smelter. Silva would go on to become secretary to President Lazaro Cardenas during the nationalization of Mexico's oil industry in 1938, but at the time he was a cub reporter in San Luis Potosí. He had heard that conditions at Robert Towne's Morales plant were terrible. So he began to ride the trolley that people took between the city center and the smelter. Each thirty-minute trip afforded him time to chat with the smelter workers, "and come into contact with their reality. It was true. After some years of labor, the workers would fall ill from inhaling the mephitic airs of the metals. It was silicosis, or something similar. It diminished their energies and their capacity for work. The Company fired them without any indemnification, since ultimately there were always others who could replace them."[42]

The situation that Silva recorded at the Morales plant was but one of a slate of dismal outcomes produced when the industrial mining metabolism interfaced with Mexican bodies. It was the product of an extractive regime accelerating in a context of unequal social relations and racial ideologies. Addressing the problem thus had to come from social—rather than moral—reform. And, contrary to what many foreign observers may have thought (but that Franck, standing beside them, saw more clearly), Mexican workers were hardly the fatalists that they were believed to be. They did not wait for the largesse or moral compunction of company managers. Indeed, it was rather paternalistic when it did arrive, as William French makes clear in his study of the mines of Parral.[43] Building upon their own moral ecology, mine and smelter workers at San Luis Potosí, Cerro de San Pedro, and across Mexico became increasingly organized and politically engaged. Healthy bodies and healthy living spaces were central planks of their political activism. Clean water, sewage, proper housing, adequate medical attention, safe working spaces, corporate acknowledgment, and indemnification; such prosaic and day-to-day things, and yet at the same time, so fundamental to life. And if the harms they suffered were most often chronic in nature it would likewise take great efforts over a long period of time to rectify them. Indeed, while environmental and bodily health animated the core of mine—and smelter-worker politics coming into the Mexican Revolution, the Revolution as such would not give them satisfaction. Their struggle had to push through and past the Revolution until it achieved a basic measure of justice.

12 The Revolution Underground

· ·

When the troubles started in San Luis Potosí on September 19, 1910, they began in the heart of Robert S. Towne's Morales plant. With the night shift over, and the last charges of ore and fuel spent, the operators cut the power to the furnace-blowers. In the early morning quiet that followed, groups of workers began to gather around the machines and on the edges of the sorting yards. Instead of rushing away into their days, the workers from night shift stood around, waiting as their counterparts came filing in. Eventually they numbered hundreds, their banter and talk echoing through the plant now silenced of the normal roar of the blowers and milling machines. The strike was on.

With the furnaces shut down, the flows of ores, coal, and scoria that circulated through the heart of Towne's metallurgical empire were blocked. Shipments of ores from the mines of Cerro de San Pedro and elsewhere could be received but with nowhere to go they would only pile up in the yards. If Towne wanted to get his plant running again, he'd have to talk. And until he did the other main group of plant workers, the *carreteros* (carters), were idled. They were paid piecework, tallied by the tons of ore, coal and scoria that they loaded, unloaded, and otherwise moved in a given week. These movements were now paralyzed. They were on the job but they were not making money. And so they went to have words with the furnace workers.

This was when the commotion began. The carreteros were losing money but the furnace workers were fed up with Towne and his management's refusal to improve working conditions, most particularly their unwillingness to clean up the airs inside the plant. Its "mephitic" atmosphere had been plaguing them for years. They had complained. They had marched and demonstrated. They had walked off the job. In return they were met with silence or the armed intervention of the local constabulary, called in by Towne himself.[1] The meeting between furnace workers and carreteros went badly. Words gave way to catcalls, whistling, and shouting. Then came the shoving, the fisticuffs, and the first volleys of rocks. For the next hours the patios, sorting yards, and buildings of the Morales plant became the grounds for a rolling brawl. Squads of workers gathered behind walls and assailed

one another with hails of ore rocks and leaden scoria. As the battle intensified, the American foremen and managers retreated behind the thick stone
walls of their colonia. It was from there that the call was made into San Luis
Potosí for help. A mounted gendarmerie was quickly gathered and sent trot-
ting toward Morales, followed by small infantry loaded into a set of mule-
drawn trolleys. They all trooped in late in the afternoon. The stones, fists
and tools of the strikers proved no match for horse charges and lined rifle-
fire. Knots of workers were broken apart. The yards and buildings were
gradually brought under control. Ringleaders were shackled, loaded into the
trolleys, and sent off to the San Luis penitentiary. With that the troubles
were snuffed out and the furnaces fired up again.[2] The entire episode lasted
less than twelve hours.

The next day, people outside the plant were still unclear as to what
the commotion was all about. "According to a high-ranking employee of the
Metalúrgica," reported the conservative daily *El Estandarte*, "the malcon-
tents themselves had no idea what they wanted. While some yelled for a
reduced eight-hour shift, others demanded the closure of the works, and still
others asked for an increase in their wages."[3]

The events of September 19 on the outskirts of San Luis Potosí marked
yet another score in a lengthening tally of hundreds of labor conflicts that
erupted in the plantations, mines, smelters, factories, railroads of Porfirian
Mexico.[4] Like the Morales events, they broke out spontaneously, or so it
seemed, and were driven by what, for observers, were obscure motives.
They were punctual episodes, lasting for a matter of hours, at most days,
before they were suppressed.

The strike, the brawl, and the ensuing repression at Morales might be
safely filed away as one more case among so many others if it were not for
where it fell in the calendar of the Mexican Revolution. A week later, September 27, Mexico's Congress ratified the reelection of Porfirio Díaz, now
entering his seventh term as president. For the liberal democratic opposition this was the last straw. Led by Francisco Madero, the *Maderistas* had
galvanized a nation-wide movement against the autocracy and cronyism of
the Díaz regime and had campaigned hard against his reelection. For their
efforts they had suffered imprisonment, repression, and censure during the
months leading to and past the June elections. The last week of September
marked the culmination of the nonviolent phase of this political movement.
Following the elections, the opposition systematically documented electoral
irregularities and submitted their results to the Congress. The Congress,
still loyal to Díaz, found their case without substance and with that the

Maderistas had exhausted their legal options. The only remaining course of action was revolt.

It so happened that Francisco Madero was in the city of San Luis at the time. He had been transferred to its penitentiary in late June, then released on bail after the elections but confined to the city by order of the judge. To plan and lead the insurrection Madero had to be free and so, at dawn on October 6, disguised as a mechanic, he hiked out to an isolated side station north of the city where he boarded the train to Laredo and then exile in San Antonio, Texas. From San Antonio he released the Plan de San Luis Potosí, calling the citizens of Mexico to rise up against the Díaz regime at 6 P.M., November 20, 1910.[5]

In the cities and central states of the republic, the November revolt was generally marked by misfires and the repression or flight of small groups of conspirators.[6] But in the dry pine-covered mountains of northern Mexico the spark of rebellion took and the fires of the Revolution began to burn. In quick succession San Isidro, Santo Tomas, Temosachic, Bachiniva, Matachic, Moris Tomochic took up arms against the regime. These were all small towns and hamlets of highland peasants, timber-workers and, of course, miners. Their isolation, combined with the topographic severity of the Sierra Madre, insulated them from effective military repression from the Díaz regime. Over the next months, from early December 1910 to April 1911, their numbers grew from dozens, to hundreds, and then thousands. A revolutionary conflagration swept across the states of Chihuahua, Sonora, Durango, and Sinaloa.[7] It would not be until March 1911 that the other great hearth of Revolution was lit in the south by Emiliano Zapata and the campesinos of Morelos.

In the north the geography of Revolution and mining coincided. The overlap led a leading Franco-Spanish historian of the Mexican Revolution François-Xavier Guerra, to posit a *Révolution minière*, that is a revolution ignited by a mining proletariat.[8] The suggestion prompted a strong rebuttal from his English peer, Alan Knight, who insisted on the difference between coincidence and causation and who detected nothing particularly revolutionary in the miners' social background. He wrote: "There is nothing to suggest that the pre-Revolutionary experience [of mining] . . . contributed to their politicization." If anything, Knight argued, miners were conservatives, dependent on the mines for their livelihoods and thus structurally inclined to defend them.[9]

This exchange, admittedly a minor thread in the sprawling historiography on the Mexican Revolution, raises a central problem for the history of extraction in Mexico. The debate between Guerra and Knight centers on the question of whether or not mine workers were revolutionary actors. Given the size of this industrial proletariat this is an important question, both for interpretations of the Mexican Revolution and of early twentieth-century revolutions more broadly. But the matter appears differently when viewed from the perspective of the history of the extractive complex. From this angle, the Revolution appears as a key turning point within a much longer story. Revolutionary dynamics coursed into the mining districts where they blew air on to the ember-bed of long-standing causes and political action. But for the longue-durée history of mining in Mexico, the importance of the revolutionary chapter lay in what followed the violent phase of the Revolution. The postrevolutionary order provided mine and smelter workers the openings and political strength to finally advance their platform of occupational and community health, equity in benefits and compensation, and worker control.

As Madero's liberal revolt was transformed into Revolution, mine and smelter worker militancy intensified. At the Cerro de San Pedro, the miners at Towne's mines went on strike early in the new year of 1911. The impasse was broken in March in what reads like an episode out of Zola's *Germinal*. The timber works of the mine were set alight. The fire spread across multiple levels, filling the tunnels with smoke, and triggering numerous cave-ins. Little more is recorded about fire, save that a certain Nicolas Laredo was apprehended for arson. Towne's mines were still smoldering in May when mine workers massed together in the plazas of San Nicolas and the Templo Mayor. Why they gathered is unrecorded and thus open to speculation: to protest the arrest of Laredo? A show of solidarity with the other rebellions now lighting up both north and south of San Luis Potosí? A collective rejection of the new *presidente municipal*, Pedro Estrada, freshly named to the position in what proved to be one of the last acts of the Porfirista governor of San Luis Potosí? To keep the pressure on Robert Towne? All that is known is that the events quickly became violent. The miners overwhelmed the local detachment of *Federales* and then proceeded to level the jail and palacio municipal with dynamite.[10]

In the same month, down in the valley of San Luis, the smelter workers at the Morales plant walked off the job. Unlike the strike of the previous September, they were not met by the army. They marched out to meet it. Leaving the foundry gates they paraded through the city's western barrios until they arrived at the central train station. There they received the train of the Maderista general Cándido Navarro who arrived from the Guanajuato with a battalion of over five hundred armed miners. The strength of this unit had been raised during the uprising at the La Luz and Purísima mines. The Morales workers told Navarro of their long-standing struggle to get the American management to acknowledge and rectify the atrocious working conditions in the smelter. They offered their support to the Maderista revolution in the hope that it would provide remedy where the Porfiriato had given repression. Navarro however, temporized. He promised to take up their complaint directly with Towne and the New York office. In the meantime they should get back to work. Thinking that he had resolved the matter, the General got back aboard his train and went on his way. The foundry workers, however, kept to theirs and maintained the strike. Again this triggered a serious internal conflict with the carreteros and the return of the *federales* to impose peace in the plant.[11]

With the Revolution gathering its energies, the militancy of mine and smelter workers in Cerro de San Pedro, San Luis could only grow. The furnace workers at Morales kept up their pressure. They made formal demands to the new government of Francisco Madero, asking it to force Towne to clean up the airs. They also decried the renewed outbreaks of the hookworm infestations and the increase of "anemia del minero." But on matters of workers' health the Maderistas initially proved little different from their predecessors of the Díaz regime. It was, as Orwell noted, imperative to keep the mines running and not jeopardize such a critical sector of the national economy: "In times of revolution the miner must go on working or the revolution must stop."[12] (Orwell was here writing about the fundamental place of mining in social life. Ideological distinctions were irrelevant. Whether the Bolsheviks, Hitler, or the Papacy, all needed the energy and heat that coal provided.) If anything, Madero's government seemed more willing to apply stiffer forms of state violence to maintain production in a context of spreading social rebellion.[13] It had its hands full. On the heels of the events at Cerro de San Pedro and Morales, four separate miners' rebellions broke out to the north of San Luis Potosí in the district of Real de Catorce. Then another in Villa de la Paz. And then at the ASARCO foundry in Matehuala. Then back again in Morales. And again in Cerro de San Pedro.[14]

Given the historical connections and the mobility of the mine and smelter workers between these districts these detonations were almost certainly connected to one another. There were also other actors in circulation here. In three separate stories between October 1910 and August 1911, the San Luis paper *El Estandarte* noted the presence of "a strange traveller, who goes by the name of Guiseppe Carlo, son of Italians, born in the United States. He says that he has no profession, has traveled the three Americas and is a partisan of Anarchism." He was in the city prior to, and during, the first Morales strike. He then reappeared to the north during the uprisings at the Real de Catorce district where he was seen in the various plazas of the area, distributing socialist propaganda and haranguing workers, until local authorities caught up with him and locked him up.[15]

Whatever his ambitions, it would be a stretch to believe that a single Italian anarchist was responsible for launching this regional revolt of miners. But Carlo's story demonstrates an important connection between anarcho-syndicalism and mine and smelter worker politics. This relationship went back for some years prior to the outbreak of Revolution in 1910. Back in 1895 the miners of Real de Catorce had declared themselves to be the autonomous inhabitants of the *Municipio Libre* of El Refugio. Of anarchist inspiration, the creation of municipios libres was a movement across Mexico to create classless communities with full sovereignty over their lives and collective management of their common property. In El Refugio this meant the mines. Proudhon-inspired mutualist societies were also widespread at the time, in San Luis, and in Parral.

Anarchists and workers were at once strangers and allies to one another. Socially, anarchist organizers were drawn from the middle professions and hailed from the cities or abroad. Mine and smelter workers shifted between industrial wage work and peasant farms and ranches. The former drew from, extensively discussed, and formalized in text, a political ideology largely drawn from continental European anarchism. It provided both a formal critique of capitalist exploitation and state power, and a program of autonomous political life. This was the kernel of the politics that anarchist organizations seeded in the mines and smelters of Mexico. It was received by the more informal moral ecology of mine and smelter workers, with its operational notions of justice, of claims to natural wealth based on the sacrifice of effort, body, and life. Miners' politics followed unwritten ideology, enacted rather than codified. But there existed points of concordance between the two groups. They shared a critique of the liberal mining regime established under Díaz, even if expressed along different registers.

The anarchist critique entered the historical record through their publications, correspondence, and other writings. The Porfiriato had created for Mexico a false progress it argued, one that was highly productive but which only profited the few. Wealth flowed, but not to the people: "Mines there are. Many mines. But who of us has any money in our pockets? Where's the gold that is extracted from the breast of the earth? Where does that river of precious metals flow? The one we see run and run without it staying in the *pueblo*?"[16]

Liberal extraction was recast as dispossession. Supported by Porfirian law and state power, foreign capital had seized possession of Mexico's natural wealth to its profit. If the Mexican people were poor it was because the rich "own the lands, they own the mines, the forests, they own the waters."[17] In August 1911 Ricardo Flores Magón addressed himself to the striking workers of Mexico. He called on them to retake what was rightfully theirs. "Everything the rich own has been made by your hand or is the commonwealth of nature (*un bien natural, común a todos*)."[18] Ultimately, Magón and the anarchists believed in extraction, on condition its benefits returned to the people of Mexico. During the Revolution, they called on miners and smelter workers to continue their work and thereby contribute to the struggle for national liberation.

In general, revolutionary violence spilled into the mines rather than out from them. The mines themselves became strategic targets for various parties. Malacates, power stations and other key infrastructure were destroyed in acts of retribution.[19] The dynamite that miners needed as part of their day-to-day work at the mine-face, but which revolutionary units found just as useful for destroying railroads and bridges, could be found at the mines neatly stockpiled and available for requisitioning. In 1914 one mine in San Luis Potosí has its dynamite magazine seized, restocked, and then seized again three times before the company finally gave up and abandoned operations.[20] Running out of supplies in the mountains of Durango, a Callista unit found their way the ASARCO mines in Tayoltita, Durango. They seized all the provisions they could carry—dynamite, blankets, medication, food—then roused the company doctor to perform field surgery on the wounded in the open-air cinema. The next day they were gone.[21] To keep the mines running meant resupplying stores after a raid but this only refreshed the target. Cleaned out by Pancho Villa's men, the El Tigre mine in Sonora was emptied again only a few months later by his enemies the Callistas.[22] In 1912 rebels raided ASARCO's Velardeña mine and emptied the company safe.[23] Silver and gold bars were seized in transit. Company managers were prevailed upon to extend "loans."[24] American and other foreign employees were occasionally seized and ransomed for much-needed cash.[25] As the Revolution progressed, however, anti-American sentiment among certain parties deepened sufficiently that the complications of successful extortion were simply dropped, and Americans shot on sight.[26]

What really hurt the mining operators was the destruction of railroads, bridges, and other pieces of the infrastructure they needed to keep running. Without fuel, or without means of moving ores and metals, the mines and smelters began to shut down.[27] As they closed, hundreds, and then thousands of workers, were laid off. Some returned home to their farms and ranches but many—the exact numbers are unknown—joined one of the Revolutionary factions. Military service provided arms, food, and even wages. One miner quipped to the American journalist John Reid that at least fighting in the Revolution saved them from having to work in the mines.[28] This was echoed by a columnist for the *Mexican Mining Journal* who wrote that, "some of the miners from the districts have no doubt prolonged their lives by joining one or another of the revolutionary armies."[29]

By late 1916 the worst of the storm was over and the mines and smelters were producing again. As violence of the Revolution tempered, ASARCO seized the opportunity to expand its empire. When Towne died in 1916, the mines of Cerro de San Pedro and the Morales smelter were integrated into ASARCO empire. The period was one of expansion for the Guggenheims, as the company went on to acquire mines in Chihuahua, build up plants and machinery.[30] As the mining sector emerged from the Revolution, the great majority of Mexico's mines and smelters were still owned by foreign companies. Mexican workers continued to labor in conditions that had not changed since the outbreak of the Revolution five year earlier. This to say that the extractive regime built under the Porfiriato had passed through the first phase of the Revolution essentially unaltered. Change would come in time, but when it did, it came from within, from the actions of the mine and smelter workers themselves.

13 Fire in the Mountain

· ·

The Mexican Revolution was first and foremost an agrarian revolution.[1] It reclaimed the lands taken from communities during the Porfiriato. For their part, with few notable exceptions, mine and smelter workers did not claim possession of the mines and smelters of Mexico. What they claimed was their just share of what Magón had called the "commonwealth of nature." It was a share that called into account not only the benefits they should accrue from working the mines but also its costs, most importantly those borne by their bodies and those of their families. Under the Porfirian mining laws of 1884 and 1892, companies were under no obligation to acknowledge, account for, or compensate these damages. Instead, they called in the federales or the local police to discipline workers when they became too insistent in their demands. For the first thirty years of the industrial mining bonanza (1880–1910) its human and ecological costs were never tallied nor were they restituted.

Strike action and labor militancy in general continued through the Revolution. After the strikes against Towne at Cerro de San Pedro and Morales came those of the north of San Luis Potosí state—La Paz, Matehuala, Real de Cartoce—then Cananea, the first big strike since 1906, suppressed by over a thousand soldiers and policemen in July 1911. Within a year there were strikes in most of the mining states of the Republic: Mexico, Aguascalientes, Coahuila, Nuevo Leon, Sonora, Chihuahua, Baja California.[2] This period also saw the foundation of the first national mine and smelter-workers' organization: the Unión Minera Mexicana. It was formed by an assembly of worker delegates from sixteen different districts, gathered at the coal mines of La Rosita, Coahuila. According to Benecio López Padilla, a Coahuila coal miner and later Constitutionalist general, the UMM was the fruit of "a movement that predated the Revolution, one that has walked since the liberal plan of the Flores Magóns and the other revolutionary clubs. The emancipation and progress of the working class has been the sole work of the collective efforts of the sons of labour." When he worked in the mines, he had witnessed the felling of his workmates by the gases; he had "observed how silicosis preyed on those workers with greater seniority." López Padilla felt that only the Unión Minera could defend the miners' "life-rights" (*los derechos de vida*).[3]

There were three main planks to the Unión Minera's 1911 platform. The first was the creation of a national body capable of pushing for mine and smelter workers' rights. The second concerned health. This one had the most specific clauses: the provision of material and moral support to injured or deceased miners and their families; the passing of federal and state laws on accidents and compensation; the creation of a national medical service dedicated to miners' health; cooperatives to provide housing and sustenance at a reasonable price. The third plank was the formation of production and extraction cooperatives and, in a nod to the continued importance of agriculture for many mine and smelter workers, the establishment of agrarian colonies.[4]

Unlike the Díaz government, Revolutionary leaders at national and state levels were more amenable to mine and smelter-worker demands. The need to stitch together some semblance of social peace and to keep the mining sector running made them so. Francisco Madero responded to the Unión Minera in October 1912 by promulgating Mexico's first work safety code for the mines, the *Reglamento de Policia Minera y de Seguridad en los Trabajos de Minas*.[5] Though it would not be seriously applied until the 1920s, the code legislated Mexico's first set of occupational health laws for the mining sector. It included provisions for better ventilation, the monitoring and control of gases, the responsibility of companies to assure medical aid and emergency rescue. In San Luis Potosí the former miner and now Magonista general Eulalio Gutiérrez, issued dispositions aimed at supporting "those unhappy miners whose lives and energy was consumed in the Cerro de San Pedro, in Catorce, in Santa Maria de la Paz and other districts of the state."[6] During his term as governor of Sonora, Plutarco E. Calles supported mine and smelter workers' right to strike for fair wages and legislated companies' obligations to provide them with health care and education.[7] Like those of Madero it is unlikely that these measures were ever truly enforced. In 1915, the Constitutionalist general Venustiano Carranza entered into a pact with the Casa Obrera Mundial (COM) to obtain the support of its members against Pancho Villa. The COM included an important contingent of mine and smelter workers. It would provide "Red battalions" to the Constitutionalist army and would defend the towns against Villa. In exchange it would have entire freedom to organize workers in held territories. Over five thousand workers joined Carranza and were key to Constitutionalist victories in Celaya and Leon against Villa.[8]

For historian Eberhardt Niemeyer this was a key moment because it garnered organized labor an important voice in the 1916–17 Constitutional convention and thereby enshrined labor and social rights in the new constitution. The translation of worker's militancy into the Constitution of 1917 created a fundamental break with the liberal extractive regime dreamed of by Alamán in the 1820s and achieved under Porfirio Díaz in the 1880s on the following points: sovereignty over natural wealth, workers' health, and state arbitration in the relations between labor and capital. The key articles were numbers 27 and 123 of the new Constitution. Both were passed quickly, enthusiastically, and unanimously.

Article 27 reappropriated all lands, waters, carbon fuels, phosphates, salt, precious stones, and minerals "in veins, beds, or fields."[9] This was a return to the patrimonial regime created by the Spanish monarchs. There was one key difference. Instead of forming part of the royal estate (regalia), the resources of nature were now integrated into the patrimony of the Nation.[10] In the language of the Constitution, these different kinds of natural wealth formed a reserve to be exploited "as is necessary for social development."[11] It reserved to the nation the right to impose on private parties for the sake of public interests. "Resources susceptible to appropriation" it read, could be regulated to assure an equitable distribution of public wealth as well as its conservation."[12] Only citizens of Mexico would have the right to ownership in lands or hold exploitation rights over waters and resources.[13] Article 27 represented an important shift in the political economy of extraction. It was now socialized. Some years later, the government of Lazaro Cardenas made this a matter of pedagogy. Schools were to teach that minerals belonged to the citizens of the country and that it "was their duty to bring them to the light of day."[14] The imperative of extraction was maintained even if reconfigured to the new political realities of the post-Revolutionary state.

Article 123 on work and social welfare addressed itself directly to the workers of Mexico, whether they labored in the fields, factories, or mines of the country. It was a ground breaking piece of legislation for the period. Although early twentieth-century Mexico remained a largely agrarian society, the article enshrined as constitutional rights demands that industrial workers had long mobilized for in Europe and the United States but had only been partially satisfied in either administrative regulation or law.[15] Article 123 was restorative in nature and intent. It sought to redress the situation of the majority of Mexicans who were forced to work in conditions that sapped their health. Whether because of extraordinarily long hours or harmful working environments, the existing work regime exceeded people's natural capacity for recovery and health. The result had been the exhaustion and degeneration of the population. According to Cayetano Andrade (delegate for Morelia, Michoacan) it was "withering *la raza* (the people)."[16] The theme of the bodily regeneration ran through the deliberations around Article 123, and its specific provisions—the eight-hour working day and the provisions on health, housing, and hygiene—all aimed at fostering bodily health and vitality.

This was particularly important for mine and smelter workers. Indeed, the subsections on health and sanitation were directly inspired from the platform of the Unión Minera and other militant miners' organizations. Companies were obliged to provide their workers with clean and comfortable accommodations at reasonable rates. They had to establish schools, infirmaries, and other services for the community. Larger operations had to reserve five hectares for public markets, municipal offices, and recreational centers. Employers were made responsible for work-site accidents and occupational illnesses. Consequently, they were obliged to reorganize the worksite in such a way as to "assure the health and lives of workers" and, failing this, to pay out the appropriate indemnities.[17]

Article 123 explicitly framed workers' rights within the relations of labor and capital. Workers gained the right to form unions in defense of their interests and the right to strike. The head of the commission that drafted the article, José Natividad Macías, believed that the state had to provide support to labor and guarantee the orderly resolution of work conflicts. This gave rise to other subsections that protected workers from unfair dismissal, legislated for equitable contracts, and prohibited lockouts.

These articles set the stage for the expansion of unions in the mining and smelting sectors over the next twenty years. In Cerro de San Pedro, the miners established a formal union called the "Liga de Resistencia de los Mineros de Cerro de San Pedro" in 1922. On its banner came, recto, a coat of arms composed of pick, shovel and piston-drill and its motto "Evolución y trabajo," and verso a stark red bend on black background. During strikes the banner was reversed and the assembled miners marched behind the red and the black.[18] They would do so two years later during a major strike action against ASARCO for healthier working conditions and compensation. The strike took Governor Aurelio Manrique's personal intervention for resolution.[19]

The real gains for mine and smelter-worker unions came in the 1930s, after the crisis of the Great Depression had passed. The national mine and smelter-worker union the SITMMBRM (Sindicato Industrial de Trabajadores

Mineros Metalúrgicos y Similares de la Republica Mexicana) was formed in Pachuca in 1934. It integrated over twelve thousand workers from across the country. The smelter workers at the Morales plant came in as Sección 5; the miners of Cerro de San Pedro as Sección 7. The demands echoed those of the Unión Minera: workers' health and safety, decent housing and education, better wages, consumer cooperatives and *granjas agricolas*.[20] The national scope of the SITMMBRM greatly increased its leverage, particularly against large conglomerates like ASARCO. The company had long been refractory to the surge in labor militancy. It had refused the visit of Mexican state health and mining inspectors in its mines and smelters on multiple occasions, and it was not given to negotiating with its workers. The first nation-wide strike for the sector came in 1935. Here the Mexican state, under the leadership of Labor Minister Antonio Villalobos, put its weight behind the workers. Wages went up an average 40 percent in the contracts signed in 1936, and then rose another 25 percent over the next years. By the end of the decade mine and smelter workers were now making more than twice the wages of agricultural workers and 50 percent more than the average wages for industrial workers.[21]

The increase in wages, when placed alongside the gains made in occupational health and safety regulations and state support for health and education in workers' communities, marked the high point for workers in Mexico's industrial mine and smelting industry. It arrived well over twenty years after Francisco Madero fled San Luis Potosí and issued his call to insurgency. Arguably it represented the achievement of the Revolution for mine and smelter workers. It was the culmination of a decades-long struggle against chronic harms and for socio-ecological justice. Such was the cadence of the politics of mining.

But there were other rhythms in motion. The 1930s also marked the beginning of a secular decline in Mexico's mineral production. Back in the 1890s, industrialization and foreign capital had reanimated the abandoned mines of Mexico and brought them to unprecedented rates of extraction. After fifty years, the exhaustion of this cycle was beginning to be felt. Overall output of gold and silver was dropping. So too were the grades, or richness, of the ores. Diminishing ore grades meant rising production costs and this did not bode well for distributing benefits to workers. It might have been easier during the early years of the bonanza but it was too late for that.

In March 1948, the miners of Cerro de San Pedro and the workers at the Morales plant mobilized a new push against ASARCO. This was part of a national strike action, coordinated through the SITMMBRM and linked to national unions in other sectors. It would be active in all of ASARCO's plants and mines, and in most of the other large operations across the country. On March 15, the workers of the Morales plant held their assembly. The complaints were familiar: workplace accidents, dangerous equipment and conditions, inadequate compensation. The atmosphere was described by one journalist as "piping hot." 'When the strike was called today there was much shouting, "Vivas" for the strike and "Down with the capitalists." The first march came the following day. The assembled workers poured out of Morales and paraded for the city center. They followed a great red flag and were led by a vanguard of cyclists. When they arrived in the Plaza de los Fundadores they were met by contingents from the railway workers, the electricians, petroleum workers from the Huasteca and the *CUT en pleno*."[22] It was one of the largest demonstrations in San Luis since the Revolution. Two days later, they all gathered again to commemorate the tenth anniversary of the nationalization of the Mexican oil fields. They were soon joined by the miners of Cerro de San Pedro who took their strike vote on March 21. By then all sections of the SITMMBRM in the country were on strike. "Currently," reported the El Heraldo de San Luis, "two of the largest refining plants in Latin America—Morales and Monterrey—find themselves paralyzed."[23]

The workers at Morales and Cerro de San Pedro had prepared the strike for months. They had formed consumer cooperatives to keep everyone fed and provisioned. In solidarity, a team of local San Luis doctors volunteered their medical assistance free of charge to strikers and their families. Shifts were formed to guard the gates of the operations and keep them shut. They had a large strike fund, capable of disbursing $6,000 a week, and were confident that it would hold for a very long time.[24] Even so, this was not going to be an easy strike to win. The Mexican sub-secretary for labor, Ramírez Vásquez, judged that the situation in the mines was difficult—demand and prices were dropping, and deposits were emptying. He cited the case of the mines at Real del Monte, "whose principal deposits are exhausted and yet must sustain four thousand workers."[25] The Mexican state was no longer the unquestioning ally of the workers. It was first and foremost concerned with maintaining the industry. As for the mine owners' association, they felt that the workers were asking too much. As talks broke down, and the preparations for the strike geared up, the Camara Minera convoked its annual Congreso Minero. After the inaugural speeches by President Miguel Alemán and the secretary of labor, the president of the Camara Gustavo P. Serrano provided a short history of mining in Mexico to put the current conjuncture in perspective. Ever since the times of the colony mining had built cities, roads, artistic monuments, and other good works. But today, "despite the importance that our mining industry occupies in our economy, it is lacking a treatment that would allow its conservation and development as a wealth-spring for the country." Instead, the industry "bears charges far in excess of what is normal in other activities, their magnitude impeding the normal development of mining, causing the disappearance of many mining centers that could have prolonged their life under more favorable circumstances."[26]

Indeed, as all of this was going on, ASARCO was abandoning its plant in Matehuala. It removed all the machinery, leaving only the buildings, now empty shells. On March 19 it dynamited the great central chimney, reducing it to a massive pile of bricks. The company was cutting its losses and consolidating. It chose to keep Morales running, settling with the strikers on the Easter weekend (March 26–28). The new contract improved medical, retirement, and injury and death benefits, as well as a wage hike. The news provoked jubilation among the strikers "who dedicated to celebrating in all conceivable ways."[27]

Easter of 1948 could not have been more different for the miners up at the Cerro de San Pedro. During Holy week people in town had spotted the company carting large flasks filled, it was said, with sulfuric acid.[28] Then, on the Saturday, Easter Saturday, the day of the Harrowing of Hell, while the workers at Morales were jubilating over their new contract, a great fire broke out in the mines of San Pedro. It was enveloped with a great deal of mystery. ASARCO tried to keep it secret at first. It was only a week later that El Heraldo broke the story on its front pages. The conflagration had started in a part of the mine that had not been worked for years, which led to the question of who could have possibly started it. Also: no explosions were noted even though the company announced that the works were mortally polluted with sulfuric gases. Miners interviewed by El Heraldo tried to conduct their own investigation into the causes but ASARCO refused their access to its property. Indeed, the mines were fully paralyzed. The fire had broken out immediately after the night shift had departed which meant that the mines were entirely empty. Since then only engineers and a carefully-selected crew of thirty miners had descended into the tunnels to try to get a handle on the fire. Guards were now posted to control entry into the property.[29] It sounded suspiciously like an anti-Germinal. Rumors began to fly that it was ASARCO itself that had set fire to its own mines as a way of getting out of settling with its workers.[30]

14 The Second Death of Cerro de San Pedro

· ·

ASARCO's shutdown of the mines at Cerro de San Pedro came as a complete surprise to (almost) everyone in town. During what were to be the last weeks and days of the strike against the company, the mood was lively, festive even. The miners relayed themselves at the *plantón* (picket) set up at the company gates, each arriving with his flag, now flipped to show its black-and-red. Their assembled flags snapped proudly in the spring winds funneling down the valley. It was now March, Easter was coming, and the shops were busy. Cash was short on account of the strike, so Mario Martínez's mother sold her dresses and cloths from her shop on credit. The feeling was that, like on all other occasions, the company would come to an agreement, the work would resume, and the bills would get paid. Drinking at the cantinas continued and the Cocinera was as fiery as ever, hardly dampened by either Lent or the continuing strike.[1]

But this was not going to be like all those other times.

Don Mario's memory of what ASARCO did next comes from his father. Blás Martínez was the head carpenter at the mines and a drinking partner and intimate of Martin Tynan, the American mine manager. In don Mario's telling of it, Tynan and ASARCO did the math. The market for silver and gold prices was dropping. So too were the grades of the ore coming out of the mines. On the other side of the ledger, wages and benefits had risen steadily since the 1930s. Strong, well-organized, and with the government behind them, unions had consistently walked away from contract negotiations and strikes with gains for their members. But ASARCO's real worries revolved around the looming costs associated with a workforce whose health was failing. It included miners like don Mario's uncle who, after twenty-five years of work in the mines was, sadly, only forty when he died of silicosis. There were many like him. According to ASARCO, too many. And so, to avoid the forthcoming indemnifications, Tynan gave the orders to blow up the mines. It was now Easter Friday. That night Tynan's crew located an old tunnel that had been sealed to contain a swelling pocket of gases. They set off a charge at the bricked-up wall. It simultaneously released and ignited an amassed cloud of hydrogen sulfide. The resulting explosion set the timberworks aflame and as the fire spread through them, the mines filled with fire and noxious smoke.[2] The *El Heraldo* described the situation: "All its shafts are flooded by gases that are toxic for workers as well as supremely flammable, such as carbon monoxide, carbon dioxide, and methane, all gases that have not been completely evacuated from the mine, and it is these that, at this moment, are burning."[3]

ASARCO declared force majeure and Tynan made good with the saboteurs. They asked for passports, visas, and a ticket to the United States ("If I stay here, they'll kill me"). "Everyone knew everything," Don Mario remembers. Late one night a truck pulled up to his compadre's house. By dawn it was gone, along with the family and everything in the house. "They just took off," he recalls.[4] ASARCO was believed to have paid off the union leadership for their cooperation in getting the rank and file to accept its settlement. It involved mass medical examinations and x-rays for everyone, and then their triage. Young and still-healthy miners were offered jobs in its Chihuahuan operations, mainly in Parral and Santa Barbara.[5] For everyone else it was a lump sum payment of a month's wages and then the company shuttered the operation.

Blás Martínez was kept on for another year to help close the mines. Don Mario, who was ten at the time, remembers his school steadily emptying out of classmates over the span of months: "Well I . . . I didn't have anyone to get together with. There were no more friends because they had all left. And all those empty houses! It was all very sad. It looked ugly. The town emptied out, *¡uuyy, no!* I really felt very sad."[6]

The census figures match his memories. By 1950 over half the town's population was gone. By 1960 only 10 percent remained. People left for other mining towns in Chihuahua, Durango, Coahuila, or they headed back down into the San Luis Potosi's booming industrial quarters.[7]

Here's what others remember:

"The company left overnight. And you might ask: 'So, what happened afterward?' Well, basically everyone was cut out, they were left adrift."[8]

"Others refused the company's offer. They said, 'No way! I'm not going to keep damaging myself for a company that hardly gives you more than to survive.'"[9]

"This community was abandoned, you understand? Abandoned politically. Abandoned in services. There only remained a community deeply wounded by mining . . . Let's say that they were marked by the scars of abandonment and by so many struggles that they had kept up for so, so long."[10]

It had indeed been a long struggle, a fight for a better share of the underground's riches, for a reprieve and compensation from mining's tolls on their bodies. For some, like the Martínez and the Rangels, the struggle reached back through the centuries to their distant ancestors—the Alanises, the Barbosas, the Ramos—people who had stood up against the King of Spain and taken the Cerro de San Pedro as their own commonwealth.[11] But that was all over now. For now, young people like seventeen-year-old Armando Mendoza just wanted out, "to get out of here, out of this mournful place ('*este triste lugar*'), the minute I can."[12]

In 1952 the state congress dissolved the municipality of Cerro de San Pedro (for the second time). The surviving community was reclassified as a "congregation" and incorporated into the city of San Luis Potosí.[13] The villagers lost their resident priest and, when one morning the travelling priest arrived to unlock the doors at the church of San Nicolás, they discovered that thieves had made off with the altar.[14] Robbed of its status, people, and now its soul, there remained only to bury Cerro de San Pedro. "Cerro de San Pedro," wondered one headline, "From Cradle to Grave?"[15] The obituary, written up by the *El Heraldo of San Luis*, was both a lament and a condemnation: "Up there is the ghost town of Cerro de San Pedro, where the crimes for which it is victim remain unpunished, where the winds yell and howl through the empty doors and windows, where wind and memory weep and lament for having been victimized by a foreign company who, enriched by the wealth of Mexico's subsoil, when it was no longer convenient to pay out its benefits, shut off the lights. There, this Mexican land, orphaned of its people, a perpetual skeletal witness, signals a perennial '*J'accuse*.'"[16]

Abandonment and mournfulness ("tristeza") became the theme and feeling of life in the wake of ASARCO's sudden pull-out from Cerro de San Pedro. Abandonment presented a range of aspects: it was an action committed by a company that renounced its obligation and thus suffered by a community that now found itself in need; it was a condition linked to the neglect of others, of the government's or the Church's duty of care, or of San Luis Potosí's duty to remember the cradle of its origins. Abandonment, however, also meant the withdrawal of, and release from, the political ecology of an industrial extractivist regime. It allowed other ecologies, human and otherwise, to remerge.

Notwithstanding the headlines and the occasionally florid recriminations of the newspaper obituaries, a closer look at Cerro de San Pedro, the place and the community, shows that the blow dealt by ASARCO abandonment was ruinous, but not fatal. There was life after the mine. It ran according to slower rhythms now that the intensity of industrial mining had slacked. The local ecology began to shift again. Described in 1923 as completely stripped of trees, the hills and high plateaus around Monte Caldera saw the return of the pines and oaks. By 2006 the area was mosaicked with healthy stands of trees. Elsewhere in the valley, the retraction of grazing (local livestock counts dropped from 22,000 to 7,900 head), allowed the slow return of the native xeric vegetation of cactacea, mesquites, and thorn-brush.[17] Very slow, in fact. The signature species of this ecological community is the *biznaga* or candy barrel cactus (*echinocactus platyacantus*). It needs about a century or so to raise a replacement generation.

The shifts in the human ecology were of a kind. Those who remained, a bit more than a hundred adults at low count, move back to the typical mixed economy of the Mexican peasantry. "I move from one thing to another," said one resident. "I don't rest my hopes on just one thing. I sell things on the markets. I'm a cultivator. I like farming. And then I'm also a rancher. I like ranching. I do a lot of things . . . and so I'm self-sufficient for myself."[18]

They gathered extensively from the growing bounty of the surrounding hills: mesquite seeds, yucca fruit pods, prickly pears, the hearts of the lechuguilla agave, the sweet water of the maguey and a wide variety of roots.[19] Processing this harvest, they made honey, pulque, and charcoal to sell down in the markets of the city. They still raised sheep and goats. Some kept a few horses or some cattle.

That this kind of mixed livelihood would assure the survival of the community in the wake of mining was exactly the argument an earlier generation had made for the collective titling of Cerro de San Pedro as an *ejido*. Ejido tenure would safeguard a local common of gathering, pastoralism and cottage production. The ejido "would shelter them from any sudden paralysis of the arduous work provided by the mines of San Pedro."[20] This was back in 1923. Decades later, after ASARCO's sudden departure, it did exactly what it was meant to do—it kept the community alive during the borrasca.

Nor did ASARCO's abandonment signal the end of mining at Cerro de San Pedro. As in the previous borrascas, the mining bust registered by outsiders mainly concerned capitalist mining operations. The crash, dramatic and thus visible, was essentially a retraction of large-scale and intensive operations.

ASARCO's withdrawal created space for the gambusinos to reoccupy the mines. They had never entirely left, of course. Even at the height of ASARCOs operations, local miners worked stamp-sized claims beside and between the ASARCO concession. They fed the ores into a clandestine refining economy. Pedro Ramos, grand-uncle to Don Mario, bought a defunct mine to cover what amounted to an ore-washing operation. Every morning he purchased sacks of high-grade ore spirited out of ASARCO's mine, claimed that they had come from his mine, trucked them down to Morales and sold the ore back to the company for refining. This was the twentieth-century version of the partido, over 300 years old and still going strong: "The people who worked for the company, many of them would return from nightfall to dawn. They knew. They would enter into an old tunnel over here and knew that they could pop out over there, or there. Whenever they came across a rich vein, they'd say '¡Hijolé!,' and return that night with their sacks, and 'Shhhhhh . . . ,' they'd dig and haul it out. Those were the *buscones*, each with their little lamps. Every night, you'd see those little lights bobbing across the hills from here [points] to there [points]."[21]

When ASARCO left, the work of the gambusinos could continue in peace and keep more regular daytime hours, leaving for the mines at dawn and sleeping with their families during the night. Or they did not—perhaps the day called for a trip to the hills or to the corrals. It depended. ("*I move from one thing to another.*") To legally secure title to the old ASARCO concession, they federated themselves into a cooperative—the Sociedad Cooperativa de Producción Minera (SPCM)—that counted over 130 members at its outset.[22] As one member explained, however, the actual work was not coordinated or administered by the SCPM. It was "every man for himself," each miner working alone or with a few family members and associates. They knew from experience where the rich areas of mineralization were. Working by hand or with minimal machinery, the amounts they brought out were modest—a ton of ore per month per person was a good average—but the grades of this material were high, and they paid well when sold down at the Morales plant.[23] One miner—Aristeo Gutierrez—figured out how to put together a small, closed circuit cyanide leaching operation. It allowed him to refine the ores for himself: "I did this for thirty years. It's very pretty because it's something . . . something like a magic art because you don't see the metals until it's done. You just put in the mineral earths, put in the chemicals, and you don't see anything until it's in treatment, in the smelting. It's very lovely. I liked this very much. And it was even lovelier when you took out the money! [laughs]"[24]

The high grades obtained by the gambusinos eventually enticed mining companies to have another look at the Cerro de San Pedro. Some of these were the large Mexican corporations like Fresnillo, Peñoles, or Minera Mexico, which emerged in the renationalization of the mining sector. Others were American exploration companies such as Geocon or Bear Creek Mining (a subsidiary of the Kennecott mining company).[25] They brought in small crews who tended to one or two rotary drills that slowly bored into the mountain and pulled out three-meter lengths of core at a time. These marbled tubes of rock registered the geochemical composition of the mountain through hundreds of meters, and from many different angles of attack. Such efforts never garnered sufficient prospects to justify more than a few years or months of work, and so they came and went, operating here and there, moving across the flanks of the Cerro de San Pedro and neighboring hills. When they left, they were not heard from again.

The last exploration campaign, that of the Peñoles company, did not so much give up as was swept out. This happened just after midday on June 13, 1990, when a violent rainstorm broke over the valley. The rains only lasted for half an hour, but they were extremely heavy, dumping so much water on the hills that it washed over the bedrock in thick sheets, instantly flooding the arroyos, and rushing down toward town. There was surprise, soon followed by panic as the reservoirs at Los Mendez and beside the church of San Nicolas, filled up, overtopped, and then ceded, sending a great wall of water roaring down the valley. Along the way the flood carried off the entirety of Peñoles's operation: tanker trucks, excavators, pick-ups, compressors, and track drills. They finally came to rest, a great tangle of machinery sealed in the mud, some kilometers down valley.[26]

The following day, Mario Martínez went up to inspect the damages to his family's home and begin the repairs. Having left in 1949 as a youngster, he had returned to the house of his birth in the 1980s and since then divided his time between the Cerro de San Pedro and the city of San Luis. He was not alone. The Rangel brothers—Marcos and Pedro—also returned around this time. "I always aspired to return to the village where I was born," recalled Pedro Rangel.[27] "Bit by bit we went about reconnecting ourselves with our past," said Ana María Alvarado García. Along with her husband, Doña Ana María resettled her grandfather into the family house. He had been among the original petitioners who obtained the ejido title in 1923. He was now the last surviving *fundador* ("founder"), but he was back.[28] So too was Armando Mendoza. So eager to quit the place as a young man, he had since returned with his wife Lola, to herd goats and make honey. A daughter was born whom they named Tonantzín—an old Nahuatl name, "Earth Mother." As Doña Ana María noted in her memoirs, in Cerro de San Pedro the sight of the bougainvilleas spilling over a wall was the signal of habitation.[29] From the 1980s on, the small cascades of pink and yellow flowers were multiplying across the abandoned walls of the village. A couple of entrepreneurs opened up a small restaurant and a ten-bed hotel. The community got the water flowing back into the taps. "It gave us pleasure to see the village returning to life."[30]

Part III **Extractivism, Again**

15 Extractivism, Again (1980s–Present)

This was the portrait of abandonment a generation after ASARCO shut down its operations: Aristeo Guttierez running his grinder, backhoe and pumps around his small processing operation; Mario Martínez fixing his roof; Armando and Lola Mendoza making honey; their neighbors heading into the mines or out to the hills, depending; a few weekenders coming up to escape from the city to soak up the atmosphere of crumbling walls and bougainvilleas; maybe a crew drilling core somewhere up in the hills. The shock of abandonment was long since spent. The bustle and action of Cerro de San Pedro of the ASARCO years was held in memory or put on display in the small museum that don Armando set up for the tourists. It was a small, quiet, rather typical Mexican village, though pretty enough and historically important enough for the state legislature of San Luis Potosí to pass a decree that protected it, and the surrounding valley, as a site of unique ecological and patrimonial value.[1] The local landscape was indeed special. It was full of cacti and mesquites, some like the viznaga exceptionally slow growing and rare. At the same time, it was deeply marked by the past centuries of mining: the lack of soils, the exposed bedrock, the tailings piles blowing their dust across the village. As for patrimony, people remembered, in its abandonment, the fact that Cerro de San Pedro was the birthplace of Potosino society.

What no one saw coming was a new cycle of extractivist mining that would return to Cerro de San Pedro and churn through both its landscape and history one last time. This was *la megamineria*, mega-mining. By the mid-1990s, the eve of high-intensity mining's return, the material organization of mass excavation and processing of very low-grade gold ores had been figured out and deployed across the US West. Cerro de San Pedro was one of the first mines in Mexico to be reorganized in this way. Open-pit gold mining was brought in, not by an American company, but rather a small Canadian "junior" operating out of the Toronto stock exchange. For such companies *la megamineria* opened up a new commodity frontier. Even if it rested on the reworking of old and abandoned deposits, large scale mining was capable of realizing unprecedented rates of extraction. The past fifteen years (2005–20) of Mexican gold output far exceeded the cumulative production of well over four hundred years of colonial and industrial mining (see figure 1).

Open-pit gold and silver mining completely changed the way Mexico's mineral sector was seen. In the seventies, precious metal mining in Mexico was the preserve of national operators—foreign ownership of the country's resources remaining unconstitutional. These included cooperatives and artisanal, gambusino miners, but in production terms the greatest part of the sector was concentrated in the hands of a few Mexican mining and smelting conglomerates like IMMSA or Grupo Mexico. They ran the remaining industrial tunnel-mining operations in places like Charcas, Real del Monte, or Real de los Angeles. Their cumulative production was less than one-third of what it was at the peak of that boom and the overall portrait was of a depressed sector on its last legs. Open-pit mining made the mineral deposits of Mexico look like treasure again. The new generation of gold companies of Vancouver and Toronto could not access these, however, until the Mexican state, under Salinas de Gotari, reformed its mining laws in 1991. Neoliberal reform opened the sector to transnational capital, ushering in a rush of foreign mining companies that snapped up concessions across the nation's mining districts. By 2010 over seven hundred and fifty projects were under way.[2] The concessions under their control covered close to a third of Mexico's surface area. Almost all of these projects were funded by Canadian exchanges that had become, with this neoliberal gold rush, the central hubs for global mining capital. One of these was the Cerro de San Pedro project, run by Toronto-based Metallica Resources.

This last extractivist cycle in the history of Cerro de San Pedro ran from early 1995, the month that Metallica acquired the concession, to 2018 when the company shuttered operations. For a decade of that time, the company could not operate; this was a key part of the story that follows. This overview chapter sets out the context for that story. From a continental perspective, the last thirty years have been defined by the continent's latest great commodity boom. The renovation of the resource sector came to be known as reprimarization and it was felt across all the landscapes of Latin America, in its soya fields, pine plantations, fracking zones, hydroelectric plants and reservoirs, and, of course, its mining districts. The period was also marked by a renewed ideological and political commitment to resource extraction—*neo-extractivismo* or the new extractivism. Both terms, neo-extractivismo and reprimarization, signal the importance of return. Latin American scholars—Svampa, Gudynas, Acosta—all underscore how the contemporary resource boom renews with past scenarios of resource exploitation and dependence that stretch back to the Spanish Conquest. Indeed, it would indeed be impossible to fully understand the last thirty years at Cerro de San Pedro, or in the Mexican precious mining sector more broadly, without recurring to the past. Founded on a colonial bonanza in 1592, Cerro de San Pedro was, four centuries later, about undertake its third mining boom. If the open-pit mine that Metallica built was so huge, it was to overcome the exhaustions left by previous cycles of mining. If it was owned by North American investors it was because, once again, the capital required to fund a project of that scale required access to pools of capital unavailable nationally. That the return of mining Cerro de San Pedro should appear so natural and unquestionable, owed a great deal to the persistent idea that this was, by history and by nature, its destiny.

· · · · · ·

To understand the material origins of contemporary mega-gold mining requires a detour back to Carlin, Nevada, in the mid-1960s. It was there that, in 1965, the Newmont company built the world's first open-pit gold mine: the Carlin Number One mine. This was not the first large-scale open-cast mining operation, as such. This had already been innovated at the turn of the twentieth century to mine massive copper and iron deposits in Bingham Canyon, Utah, and Hibbing Minnesota.[3] Mining gold in this way, however, was new. Gold had already been mined at Carlin in the placers at Maggie Creek back in the 1880s but this work had long since ceased. In the postwar period, it was simply one more of some thirteen hundred abandoned gold mining districts in the United States.[4] Carlin Number One transformed the area into one of the most productive gold districts in the world. It was emerged at the intersection of advances in structural geology, chemical metallurgy, and industrial engineering.

The Carlin ore bodies were first revealed by Nevada state geologists.[5] Their interest was not in gold, but rather in understanding what Bruce Braun called "the inner structure of the landscape."[6] Field observations, sampling, geochemical analysis, and then mapping revealed an interesting pattern at play in the mountains of northeastern Nevada—an alignment of "windows" where the granitic basement layer could be seen peeping through the sedimentary layers of the surface. Seen from above, these granitic extrusions appear as a Northwesterly arc of islands some eighty kilometers long and eight kilometers wide. This was the Carlin Trend or Carlin Unconformity. What was of real interest to the Newmont company's geologists were the large deposits of carbonate rock found here and there on the edges of those islands. They discovered that these bodies of dull gray and finely porous rocks, without shine or crystal, contained an enormous amount of gold. This gold, however, was invisible. It was present in microscopic flecks less than a micron in size evenly disseminated through the carbonate (a micron is one millionth of a meter; most bacteria measure one to ten microns across). One imagines these massive underground bodies of flat gray rock like subterranean cumuli, permeated with the finest mist of gold, a mist so fine that miners had passed it over for generations.[7]

The process of transmuting these low-grade carbonate deposits into a gold mine began in 1961 when Newmont brought in drills to map their structure and composition. The company dispatched the samples to Harry and Clemmie Treweek, a husband-and-wife team considered to be the finest assayers in the world at the time. The managers sensed that they were on to something big and so they insisted the Treweeks work only under the cover of night and in an abandoned assay lab sixty miles from town. There, the Treweeks mixed the sample material with a series of precisely calibrated compounds needed to flux out tiny beads of gold from the carbonate. Fire assay metallurgy showed two important results: that the ore was rich enough to mine and that, given the size of the host deposits, there was a prodigious amount of it.[8]

Newmont rushed to Bechtel Engineering's San Francisco office and put in their order: "We've got to have a mine and we've got to have it built right now!" Three days later, the construction equipment rolled in, and within the year the mine was in operation. "They wanted a gold mine and they figured it would be very profitable," recollected Newmont engineer Robert Shoemaker, "which it was. So they wanted it in a hurry, and they got it in a hurry."[9]

Newmont's application of open-sky mining and processing illustrates how technology and market economics combine to determine the physical dimensions of a gold deposit. This was witnessed in the changes in the estimation of its mineral reserves, as expressed in tons of treatable ore. The difference between what was, and what was not, treatable was more of an economic than technical matter. As open-pit mining dropped the price on per-ton processing costs, and as commodity markets pushed up gold prices, the amount of ore that could be profitably processed increased, and thus the size of the reserves grew. Seemingly immutable and simply "there in the ground," gold deposits are effectively socio-technical, and thus contingent, constructs.[10]

The grades revealed by the Treweeks were too low to be economically mined by existing underground mining operations. "Any time that drilling or exploration encountered deposits of that sort . . . they were passed up as non-commercial."[11] The development open-pit mining techniques allowed Newmont to push the threshold of what constituted an economically viable ore down one order of magnitude, from ores grading between fifty and one hundred grams per ton (of their industrial era predecessors) to ores grading ten grams per ton (in the first days of Carlin's operation) to ores bearing one and a half grams per ton (with the introduction of heap leaching in the 1970s).[12] Coming into the 2010s open-pit gold mines were working deposits with grades of half a gram of metal per ton.

As with the Bourbon revival of the eighteenth century, and then the industrial revival of the nineteenth century, the key to profitably mine lower ore grades was to accelerate the extraction and processing rates. The fundamental sequence and organization of the Carlin Mine differed little from that of colonial or industrial-period mining: ores were broken out of the earth, hauled to the mill and smelter, processed to winnow out metal from ore, and then refined to a state of near-purity. However, to harvest Carlin's "invisible gold," Newmont had to vastly increase the flow of material moving through the system. This was proving to be an ever-more difficult challenge to meet given the operation of Lasky's Law, a rule of thumb in geological economics that holds that the processing rates of ore tonnage increase exponentially as ore grades drop arithmetically.[13] Thus while at the height of industrial-era mining a gold mine had to process ten thousand parts of ore for every part of metal recovered, Newmont's processing ratios were 100,000:1 in the 1960s and then jumped to one million: one in the mid-1970s.[14] Currently there are open-pit mines processing ten million parts dross to one part metal.

In sum, Newmont had to churn through large chunks of the Nevada landscape to achieve profitability. There was no time to plan and drive in elaborate structures of tunnels and shafts. Instead, open-cast mining methods were brought over from the big copper mines to simply rip away the thick overburden of sedimentary layers and disinter the carbonate ore body, en masse. Advanced blasting techniques were used to fracture the rock. New explosive compounds (ANFO, Geldyne, Pentex) were arranged in special arrays and then detonated simultaneously to maximize "brisance," the shattering-capacity of the shock wave that punched into the earth. Ever-larger excavators picked up the fragmented rock and spun around to dump it in the bin of a waiting mining truck. With advances in radial tire design, the size, and thus the payloads, of these trucks were constantly increasing in this period from 10 to 60 tons (1960s) to 210 tons (1980s).[15] They worked in coordinated fleets, assembled head to tail in long lines that snaked their way up and down steep haul roads as fast as their elephantine mass would allow. Blasting, excavating, and hauling like this, Newmont ground its way down into the Nevada desert in a broad circular movement that created the stepped rings that give the open-pit mine its iconic shape.

The scale and rate of open-pit extraction forced corresponding increases in ore-processing. In the first years of operation Newmont used what was a then-standard mill and cyanide plant to reduce ores to fines and then extract the gold in the swirl of the cyanidation vats. It was a big plant but even so its processing rate of twenty-four hundred tons per day was too low. It acted like a bottleneck. The solution was to move ore-processing outside. Instead of being ground inside a mill, ores were simply dumped out of the trucks in large heaps on a leach pad. This last was a large impermeable plane measuring over a kilometer a side, built with a slight incline toward its center. The assembled piles were sprayed with a cyanide sodium solution. As it percolated down through the ore the solution filched out the microscopic flecks of gold as it went. When the enriched solution hit the impermeable floor of the leach pad, it flowed down into a drain where it was pumped to over to a plant that electrochemically precipitated gold from the pregnant solution.[16] Newmont's heap leach pad at the Rain mine (built in the late 1970s) received ninety-seven thousand tons of ore per day.[17] Heap-leaching was less efficient than an industrial mill and plant—only between sixty and 85 percent of the available gold was recovered in this way—but this was easily offset by the significantly greater processing rates.

From the Carlin Number One mine, Newmont expanded, opening up new mines and leach pads along the various carbonate deposits that composed the Carlin trend. In its beginnings this single operation was extracting about fifty-three tons of gold a year. This exceeded the output of the entire US gold industry up to that moment.[18] Newmont's engineers and metallurgists tinkered constantly to increase processing and recovery rates. They tried all manner of additives—lime, Portland cement, even coconut shells—to maximize the recovery rates or to deal with the so-called refractory ores that found themselves in the leach pads. Newmont's efforts led to a quasi-magical paradox in which its reserves continuously expanded even as production levels increased, and ore grades dropped. This inverted the standard pattern in mining in which both ore grades and remaining reserves drop over time as miners work through a deposit. More importantly, however, Newmont's landscape-scale experiment demonstrated that mining gold be very successful at very low concentrations for all manner of material. This suddenly made large and disseminated porphyry deposits economically viable. Even the wastes left over from previous generations of miners could be worked for a profit. Open-pit heap-leach gold mining transformed exhaustion into increase once again.

· · · · · ·

Newmont's experiment in the desert and mountains of Nevada was akin to Sutter's Mill, California in 1848. Open-pit gold mining at Carlin Number One launched the last gold rush in history—a cycle of contemporary extractivism that would, in a matter of decades, draw out more gold than all previous gold rushes—South Africa, the Klondike, Australia, California, Brazil, and Spanish America—combined.[19] The invisible gold rush was not only more intensely productive than its predecessors, its geography was—true to its times—multi-sited and globalized. Open-pit mining spread outward from the Carlin trend to other abandoned gold districts of the US West, and from there to low grade ore deposits around the world wherever they might be found.

Given the rates and scales at which these mines operated, they required large amounts of capital to build and run. If investors chose to place their funds into such projects, it was because the new gold mines were unusually lucrative, with quick turn-around rates, high profit margins, and very large volumes. Large-scale, low-grade gold mines also provided the material anchor for financial speculation around an ever-growing number of small mining companies and for sector-specific derivative trading. An important locus of the invisible gold rush was thus to be found, not at the mine site, but in the joining of mine operators and financiers. In the 1970s and 1980s, the key venue for this kind of congress was the small provincial exchange of Vancouver, British Colombia. It served as home ground for a late twentieth-century commodity frontier driven by the dynamics and imperatives of capital accumulation.

The beginning of Newmont's operations at Carlin attracted a great deal of attention within the mining industry. It was the first major gold mine to be built since before the Great Depression.[20] In 1965 Newmont announced 3.5 million ounces in reserves. Its reserve estimates were regularly revised upwards as the company dropped costs, as gold prices rose, and as greater and greater portions of the deposit became commercially viable. It was a surprising trend given that this was, in principle, a nonrenewable resource but it illustrates how fungible the dimensions of a mineral reserve are. They were just as much defined by markets and technology as by their geology.

Newmont moved quickly to secure land titles to other deposits along the Carlin Trend. It was soon joined by other mining companies looking to repeat the formula. By the end of the 1970s there were fourteen large mining operations strung out along the Carlin Trend, all running open-pit and heap-leach projects.[21] Other open-pit gold mines sprouted up across the US West. As with Carlin, they were developed in historic mining districts first opened during the late nineteenth-century expansion of the frontier, districts whose high-grade stock-work ores had been spent, but whose large low-grade porphyries and other disseminated ore bodies had not been touched. In 1984, the US Bureau of Mines counted eighty new open-pit mines operating in Arizona, California, Idaho, Montana, Colorado, Nevada, New Mexico, and South Dakota.[22]

Low-grade gold mining was a massive undertaking. The largest operations moved up to 190 000 tons of rock a day. The stepwise increase in the physical and energetic scale of the venture, the sharp intensification of its extractive metabolism, required a commensurate rise in costs. Companies compiled these into a cost per ounce produced ratio. In the 1980s, an open-pit gold mine did well if these were kept within a fork of between $120 to $220 per ounce of gold recovered.[23] The gigantic scale of open-pit mining meant that these costs quickly tallied up to hundreds of thousands and even millions of dollars per day. The greatest part of these expenditures—80 percent—was spent in breaking the ore body out of the earth: blasting and hauling.[24] The fuel bills alone were gigantic.

It is worth noting that the mines in the opening decade of this invisible gold rush were not developed by the large mining corporations of the time. Established multinational mining companies such as INCO, Vale, BHP, and ALCAN, and ALCOA focused on the extraction of industrial metals (iron, nickel, copper, bauxite) and coal. The companies behind the new generation of gold mines were small start-up companies. The key exception was Newmont itself, which had been mining gold since the early twentieth century and had the collateral needed to finance the Carlin Number One Mine. For the rest, however, getting an open-pit mining project up and running meant finding financing. In the 1970s and 1980s, this most often meant a trip to Vancouver, Canada. For the mine promoter Robert Friedland, it was "the heap-leach capital of the world."[25]

Vancouver was home to a small provincial exchange that blossomed into the leading source of capital for open-pit mining projects in the US West. In the late 1960s it had received an influx of penny-traders drawn from the world of speculation and wash-trading around the mines of northern Ontario. Since the early twentieth century they had operated unsupervised and unimpeded until a massive trading fraud known as the Windfall affair crashed the market in 1965. This pushed the provincial government to legislate a minimal regulation of the exchange but, for many, this was too much. It pushed "the backbone of the Toronto penny stock market . . . to slide into Vancouver *en masse*," where they promptly resumed their activities.[26] In 1974 an investigation by British Colombia's Attorney General found that 20 to 30 percent of the stocks listed on the Vancouver exchange were under manipulation.[27] Vancouver had become "the new Wild West" for Canadian mining and speculation.[28] Political economists Alain Deneault and William Sacher agree. They document the early history of Toronto and Vancouver, colonial exchanges built around speculation and investments in Canada's natural resources: hydroelectric power generation, timber, and of course, mining. They argue that this changed completely beginning with the open-pit mining boom in the 1980s. Vancouver became the financial base for the global extension of the open-pit gold mining frontier. (It was then replaced in the 1990s by the Toronto stock exchange.)[29]

Permissiveness was a key factor in their rise. The penny traders of Vancouver were willing to invest in the smallest companies, companies with less than a year of experience and less than a million dollars in their portfolio, companies would have been "laughed at it if they approached major exchanges such as New York or London."[30] For promoters such as Robert Friedland, Vancouver was attractive because it was "one the freest and most under-regulated venture capital markets in the world."[31] Friedland was representative of new generation of venture capitalists that rose with the invisible gold rush. They were men who were first and foremost financiers with little to no background in the world of mining.[32] They began their careers as entrepreneurs, accountants, and MBA graduates. (Friedland's experience prior to mining was in trying to launch a commercial apple orchard with Steve Jobs.) They did not know much about mining, but they did know their numbers, and open-pit gold mining's numbers were good.

In 1965, open-pit gold mining was profitable even at a time when gold remained pegged at $35 the ounce. In 1971 the US eliminated the gold standard, allowing the metal to be priced like any other tradeable commodity. In 1980 it reached $850 the ounce before stabilizing in the $300–$400 range for the rest of the decade. This was double or more of the costs of production. Moreover, open-pit operations could be set up quickly. Operators could expect to pour the first bars of gold within less than a year after breaking ground. Compared to large copper mines or underground precious metal mines, this was an extremely fast move to revenue generation.[33] Secondly, the open-pit mines were significantly more productive than their predecessors. This greatly increased yearly revenues. In the early 1980s, Newmont was producing 3.2 million ounces of gold per year, about the same amount as the entire US gold sector on the eve of open-pit gold mining and worth over $2.5 billion dollars. Thirdly, as seen above, open-pit gold mining greatly increased the overall number and size of the reserves that could be mined. Fourthly, the highly mechanized nature of open-pit mining meant significantly reduced labor forces which, when coupled with the growing recurrence to subcontractors, had the effect of dropping the costs that investors associated with unionized labor.[34]

It was a compelling case, and Vancouver's financial community, ever on the look-out for new investment opportunities, was enthusiastic about the new gold mines cropping up in the US West. Friedland arrived on the market in 1983. He needed $26.5 million to get operations rolling at his gold mine at Summitville, Colorado. Within a year he had reaped $300 million in capital.[35] His case was just one of many. From 1978 to 1987, the amount of capital invested in the companies listed on the Vancouver exchange multiplied more than tenfold, from CAD600 million to CAD6.6 billion.[36] The majority of these were mining companies.

It would, nevertheless, be wrong or incomplete to reduce the take-off in open-pit gold mining to the cupidity and risk-taking of Vancouver's traders and financiers. There were structural shifts in the broader political economy to also be considered. The same period saw, with the re-emergence of market liberalism, the move to the deregulation of capital flows.[37] Exchanges around the world saw their capitalization grow rapidly during this period. North American households and employee pension plans began their move toward mutual funds as the privileged means of holding equity. The deregulation of stock markets allowed the institutional investors and pension funds to invest in venture capital markets and start-ups. What was new for Vancouver within this context was the arrival of international capital. It followed the likes of high-profile investors such as Franco-British banker James Goldsmith, and the Saudi-born financier and arms dealer Adnan Khashoggi, people whose arrival and successes on the VSX acted like a calling card for what was a rather parochial exchange in the broader landscape of the world's capital markets. By 1986, well over half of the exchange's capital came from abroad.[38] The VSX was becoming known internationally as the exchange of reference for investing in start-up mining companies.

The elaboration of new investment instruments also helped funnel capital into this renewed gold mining frontier. Gold companies were listed—most notably the Franco Nevada Company directed by Canadian financiers Pierre Lassonde and Seymour Schulich—that held no mines in their assets but were instead based on the royalty agreements they held on producing operators. Barrick Gold, the world's largest gold corporation by the 2000s, built its fortunes on the hedging agreements it offered its investors. It integrated futures trading into the gold market and offered mechanisms that allowed investors to make money even as the price of gold dropped.[39] Finally, the Canadian government itself provided a significant boost to the junior mining market through the fiscal incentive known as flow-through shares. Under this dispensation investors could write off the exploration and development expenses incurred by the mining company in which they had invested.[40] Industry insiders called it the "flow-through boom." Well, they might. It moved billions of dollars into junior mining companies from the late 1970s on.[41] In return, the open-pit gold mines of the US West produced close to thirty billion dollars' worth of gold for the period 1971–90.[42]

Although open-pit gold mining appeared to transform barren grounds into bonanzas, its magic could not escape the problem of exhaustion. If anything, the massive intensification of the extractive process only accelerated the rate at which an open-pit operation ran through a given deposit. The average duration of these projects was now less than a decade in length, a project life span that should be contrasted against the decades and centuries of operation of previous generations of underground mines. And so, by the 1980s, the Newmont Mining Company was already looking beyond Carlin. To maintain profitability as a corporation, it needed to bring new mines into a pipeline of production. Given that options for expansion in the US West were already drawing to a close, Newmont turned its sights abroad to Spain, Indonesia, and Peru.

· · · · · ·

It was in the highlands immediately to the north of the provincial city of Cajamarca, Peru, that one of Newmont's exploration teams surveyed a massive gold sulfide deposit that extended over sixteen kilometers of the Andean cordillera. The deposit was named Yanacocha—"Black Lake"—after the local lake at the heart of a regional complex of marshes, streams, and rivers.[43] Once the wetlands stripped away, the Yanacocha mine would serve as a suitable replacement for Carlin once Carlin began to peter out. In the 1980s, however, Newmont found that it could not move on its prospect. The highlands were still wracked by the conflict between the Peruvian state and rural guerrilla movements. What's more, Peru's subterranean resources, having been nationalized back in the late 1960s under the populist General Juan Velasco Alvarado, remained very much out of bounds to foreign capital. This was true across the continent where the doctrine of national resource sovereignty still held sway.[44]

Newmont's Peruvian prospects only began to improve in 1990, with the election of Alberto Fujimori as president of Peru. To resolve the crisis of the Peruvian economy—laboring under hyperinflation rates of over 7,000 percent and losing GDP at rates of close to 30 percent per year—Fujimori enacted a slate of neoliberal reforms. Within these, the reliberalization of the mining sector occupied a central and strategic position. Newmont's Yanacocha mine became the flagship for Peru's neoliberal extractivist regime. Operations began in 1992, almost immediately after the passage of the mining law 708. It would become the world's largest open-pit gold mine, simultaneously operating six massive open pits and heap-leach pads over a project footprint of over ten thousand hectares.[45] Newmont CEO Tom Conway commented at the time: "We are really blessed with this deposit. We expect gold to just scream out of there."[46] When gold was first poured in 1993, a year after Newmont broke ground, President Fujimori was on site to personally receive the first bar of Yanacocha's gold. It was a national media event, broadcast on television, radio and in print across the country. For Fujimori, Yanacocha was a triumph: "Peru is once again becoming El Dorado."[47]

The opening of the Yanacocha gold mine was a signal event in the broader globalization of the invisible gold rush. In the 1990s, open pit gold mining broke out of North America to tap into disseminated ore deposits across Africa, Asia, Australasia, and Latin America.[48] It was effected by an ever-growing number of mining companies—the majority based out of Vancouver or Toronto—that fanned out across the world to secure extraction rights and to prove up their prospects. The Yanacocha episode illustrates the importance of politics in enabling this new resource frontier. The contemporary boom could not have taken place without the political reordering wrought by post-1989 neoliberalism or without the resource imaginaries that animated these.

El Dorado, again. Fujimori may have been speaking loosely, as a speech maker might, his mind snapping up the first available image for great wealth, but the speaker, the phrase, and the context give pause. What did it mean that one of the leading architects of Latin American neoliberalism should recur to the image of colonial treasure at a time of resurgent extractivism? The question moves us into a field of relations articulated around the commodification of subterranean wealth, one that linked symbolic discourse, territory, and history to political action, Fujimori's reflexive use of the figure of El Dorado linked his neoliberal present to Peru's colonial past. This was hardly a unique or exceptional rapprochement. El Dorado, as well as other symbolic connections to colonial treasure-taking, resurfaced everywhere across the discursive field generated around the invisible gold rush. They were part of corporate heraldics. Aside from the El Dorado Gold Corporation (ELD) the Toronto stock exchange listed an Azteca Gold Corporation (ASG); a Castillian Resources (CT.V); a Conquistador Mines (CMG-V); Maya Gold and Silver (MYA). There was also, for the record, a Golden Predator Corporation (GPY). El Dorado denominated mining projects in the Dominican Republic, Mexico, Peru, and El Salvador. Mining zones in Argentina, Chile, Mexico, Peru, and Venezuela were all described as "new El Dorados."[49] Harkening back to the days of the Conquest became a standard trope in media reporting: "Not since the arrival of the Spanish have outsiders shown so much interest in Andean rocks."[50] "Peru, home of Incan treasures that made Spanish conquerors crave for gold, is seeing a new rush for the precious metal by top mining firms."[51] Returning to the fantastic geographies conjured by Sir Walter Raleigh in the late sixteenth century, a report described the landing strip at the Omai gold mine in Guyana as literally paved with gold: "just one of the massive gold deposits" of the Guyana shield, "an area of 100 000 square miles of largely dense jungle brimming with minerals."[52]

The proliferation of El Dorados and associated forms of colonial imagery cannot be entirely reduced to the hyperbole of promoters and reporters. They accomplished ideological work. They helped build the resource imaginaries that naturalized extractivism in targeted territories.[53] It is a kind of imaginary built through selective foregrounding and silencing. The figure of El Dorado, for instance, signals something more than marvelous and unlimited wealth. It is not cornucopia. It is not mana. It is an emblem of colonial wealth: the story, carried in a symbol, of the encounter with the Other's treasure; an encounter that contains, beneath the immediate sense of excited wonder and astonishment, the barely veiled prospect of acquisition through violence (that is, as Las Casas observed in the sixteenth-century, pillage). Regions targeted by mining companies were stripped of their human and ecological histories and rendered into resource-territories. Yanacocha, for instance, was not represented as Black Lake. No mention is made of millennial history of cohabitation between people, land, water, and other-than-human beings.[54] State and corporate representations of Yanacocha and Cajamarca only moved between two views: the most productive mine in the world or the site of the last Sapa-Inca Atahualpa's ransomed treasure.

In Latin America, the neoliberal resource imaginary deployed a strategic representation of the past that essentialized mining into the history and thus identity of countries and regions. "This is a matter of history," wrote one columnist, "Mexico has been a mining nation since before the Spanish, when exchanges were made between cacao and gold nuggets."[55] "Peru was, is, and will be a mining nation." pronounced Peru's Minister of the Environment, "To say otherwise is an illusion."[56] Throughout Latin America dozens and dozens of small towns and even cities were described as "pueblos mineros"—towns whose vocation and identity were inextricably linked to the activity of mining. Cerro de San Pedro, of course, was one of these.

Argentine sociologist Maristella Svampa argues that these historical discourses encoded an extractivist imperative. She writes, "[they] end up functioning like a historical threshold or horizon, imposing the idea that there exist no other alternatives to extractivist development."[57] They represent a variant of Margaret Thatcher's TINA (There Is No Alternative) deployed in the realm of extraction. Svampa also notes how the imagery of superabundant natural wealth fostered a particular view of the future. The return of El Dorado signaled a return to the idea that resource extraction would enable progress. And it recapitulated the providentialist readings of New World bounty first expressed in the sixteenth century. Consider the following passage of the speech delivered by the CEO of the Barrick Gold corporation before the gathered audience of investors. One hears in it the whispers of the sixteenth-century Jesuit José de Acosta: "The good Lord did not put gold deposits in the middle of Manhattan or Paris. The good Lord picked for some unique and obscure reason to put gold in areas like the middle of the Tanzanian jungle or on top of the Andes mountains in remote communities where the options to escape poverty are nil."[58]

Svampa calls this the "developmental illusion."[59] Mining in the neoliberal period, much like mining in the colonial and national periods, was again positioned as the wellspring of economic and social advancement. The neoliberal resource imaginary animated and legitimated the political and institutional actions of a range of linked actors: corporate managers and employees, industry, and financial promoters, as well as officials within national ministries in Latin America and international organizations such as the World Bank and the United Nations Economic Commission for Latin America. This assemblage produced a "Commodities Consensus," that is the projection of the neoliberal Washington Consensus within the resource sector: the idea that capitalist investment in primary extraction would bring development.

In the early 1990s, the World Bank and the United Nation's CEPAL played a critical role in the creation of the neoliberal mining regime. They guided national policy makers toward a thorough reform of the laws and institutions that governed mining. The experience of the US West in the 1980s, and of Newmont's Yanacocha mine, showed what was possible across the mining districts of the continent: bonanzas of historically unprecedented scales. To realize them, to make El Dorado again, neoliberal policy turned to foreign capital to underwrite the massive costs of large-scale low-grade mining. This entirely shifted Latin America's political economy. Gone was the custodial state of the mid-twentieth century, guarding the nation's natural wealth. In its stead came the transacting state, ready and willing to put that wealth into the circuits of global capital.

The first reforms came in Peru, with Law 708. Mexico followed in 1991 when the mining code was reformed under the government of Carlos Salinas de Gotari as part of the general recasting of Article 27 of Constitution regarding land, labor, and resource rights. Recall that this article represented the key revolutionary achievement of workers and small-hold farmers in the Revolution. The reforms to the article were designed to meet the demands of North America's first free trade agreement. They all adopted the same general formula: the guarantee of foreign property rights and freedom of capital. In the Mexican case, the relevant one for Cerro de San Pedro, mineral deposits could now be held by foreign companies through Mexican subsidiaries. The duration of a mining concession was greatly lengthened—fifty years, with a guaranteed renewal for another fifty. The entirety of Mexico's territory was made available to mining. Concessions would be accorded anywhere without regard for existing claims whether of private property, ejidal lands, indigenous territories, or protected areas. The development of mineral resources was declared to be a public utility, which enabled concession holders to call on the state to evict local landowners. Regulatory oversight of operations was pared down, and the state's principal function was reduced to keeping the country's mining cadaster straight. Royalties on mining operations were dropped to 5 percent of production. With the signing of NAFTA, transnational mining operations further benefited by the lifting of tariffs on imported equipment and supplies. The presence of investor protection clauses guaranteed their profits against any move to stop their operation for environmental or social reasons.[60] The neoliberal mining code updated the basic aims of the late nineteenth-century codes legislated under the government of Porfirio Díaz ("*facility of acquisition, liberty of exploitation, and security of retention*").

By the 2000s, guided by the World Bank and the CEPAL, fourteen Latin American countries had all reformed their mining laws.[61] Latin America's mining reforms contributed strongly to the boom in mining capital. For transnational companies it opened up the mineral deposits of the continent. The moment the laws were passed, foreign companies, the majority of them juniors listed in Canada, rushed to secure concessions across the continent. "This is the closest thing you have to a modern-day gold rush," commented the CEO of the Placer Dome corporation at the time.[62] The number of Canadian owned mining projects shot up. In Mexico they quintupled from 52 to 244 between 1991 and 1995. In the same period for Peru the number went from 3 to 98; in Argentina, 0 to 97.[63] These numbers would only grow. By 2007 the Canadian mine owner's association estimated that Canadian firms were developing over fifteen hundred projects across the continent.[64] Everywhere they counted for the greatest share of all mining ongoing projects. In Mexico, where the greatest number of Canadian projects were developed, 80 percent were held by transnational companies.[65]

A mining project, in the sense adopted by the Mining Association of Canada, covered everything from the possession of a concession to an exploration campaign, to an actual functioning mine. The customary estimate is that only one in ten projects ever becomes an operating mine. But given the greatly expanded reserves that mass open-pit mining made available, these were still excellent odds. Good enough, argue Deneault and Sacher, to incite a massive influx in speculative capital.[66] The amount of global capital flowing into the ever-multiplying number of mining companies greatly increased. In 2001, the Vancouver exchange was bought out by and subsumed into the Toronto Stock Exchange (TSX). It became the TSX-Venture exchange. Together, the TSX and the TSX-V pooled together close to $325 billion in capitalization for mining companies—this was a long way from Vancouver in the 1980s ($6 billion in 1986).[67]

Taken as a whole, the fifteen hundred or more mining projects developed out from Toronto show some clear patterns. The first was a distinct orientation toward gold. Over 80 percent projects had gold as their primary target metal (followed by silver and then copper). The second was the prevalence of mass open-pit techniques launched at large disseminated deposits. This led to the third key characteristic of the neoliberal bonanza in Latin America, the fact that almost all of these projects were developed in territories that had already been mined. This was in many cases the second and even third cycle of brownfield mining. The invisible gold rush was a return to the El Dorados of the past in the hopes of bringing them back for one more round of extraction. "For Canadian companies, Latin America offers the chance of developing elephant-sized deposits in areas where modern mine techniques have not been applied."[68] "Turning old projects opened up by Spanish Conquistadors into new projects is the new style in Mexican mining," reported Toronto's Financial Post in 1997. And, from among this new wave of transnational mining, it singled out the following example:

"Toronto-based Metallica Resources Inc. will revive another gold-silver mine at Cerro de San Pedro, a mine that was worked for 427 years."[69]

16 The Last Mine

∙ ∙

In the broader context of the global invisible gold rush, the Cerro de San Pedro mine was hardly exceptional or unique. It was but one of hundreds of new open-pit gold mining projects being developed across Latin America. Typically for the region, it was not based on a green-field discovery but rather a brown-field return to an ostensibly abandoned and exhausted deposit. Metallica Resources Inc., the junior start-up behind the project, estimated that it would recover ca. one and a half million ounces of gold and fifty million ounces of silver—a level of productivity that was historically unprecedented but one that placed the Cerro de San Pedro mine in the low to middling ranks of the new gold deposits of the late twentieth and early twenty-first century.[1] The company was registered in Toronto, Canada, and capitalized on the Toronto Stock Exchange, though (also typically) its board and management team was based in Denver and other parts of the Americas. In time, it would partner with other Toronto-based companies: Cambior, Glamis Gold, and the mine-royalties company Franco-Nevada—before merging into the New Gold corporation in 2008. The company name resonated with historical symbolism in the mining world, recalling as it did Georgius Agricola's famous sixteenth-century defense and description of mining, the *De Re Metallica*. In Mexico, it operated through its wholly owned subsidiary company Minera San Xavier.[2] The Cerro de San Pedro project was one of the very first transnational open-pit gold mines to be developed in Mexico following the reform of Mexico's mining laws and the opening of the sector to foreign ownership in the early 1990s. It represented, albeit at a reduced scale, a Mexican analogue to Newmont's Yanacocha mine in Cajamarca, Peru.

From a geological perspective, from the chthonic perspective of the mountains and underground bodies of Cerro de San Pedro, Metallica's project represented the last stage in the excavation of the highly mineralized cell that rose up through the earth to create the concentrated ore body that had attracted the attention and efforts of miners since the sixteenth century. It is true that the veins and ores that laced through the upper limestone layers of the local hills were not entirely exhausted. In the wake of ASARCO's abandonment in 1948, small scale miners like Aristeo Gutierrez or the Rangels continued to make their living from these tunnels. They occasionally came across small runs of material of high grades (22 to 28 grams per ton) and, more exceptionally, encountered those really exciting kidney-stones of pure gold. Nevertheless, these leftover ores and micro-bonanzas were hardly enough to attract the attentions of a transnational mining company, even a junior start-up like Metallica Resources.

The company's principal interest was the porphyry, the core and base of the Cerro de San Pedro deposit, that gigantic mass of fine-grained material formed by the slow congealment of the rising cell. Hidden within its bulk were those microscopic particles of gold and silver that were the real target of open-pit miners. Metallica's geologists and engineers estimated that the porphyry held eighty-six million tons of ore, with each ton containing an average of half a gram of gold and twenty-three grams of silver. Before these ores could be removed, however, over a hundred million tons of so-called waste or barren rock had to be stripped off and set aside.[3] The porphyry was not a neatly rounded bulb but rather a polymorphous mass of protrusions that reached up inside the different mountains of the valley. Thus, to excavate the porphyry the company planned to cut out the northeastern slopes of El Barreno and the better part of the Cerro de San Pedro's southern aspect. The town of Cerro de San Pedro would also have to go because it was situated directly atop the largest salient of the porphyry.

The techniques Metallica Resources proposed for the eventration of the Cerro de San Pedro deposit were by then industry-standard. The earthen crust would be broken up by arrays of twelve-meter drill-holes packed with granular ANFO and then detonated by TNT. Three hydraulic shovels would load a small fleet of fourteen mining trucks, four shovelfuls per truck, each bearing about one hundred and thirty tons of material—the equivalent of about a hundred sub-compact cars freighted with every trip.[4] The overburden was to be dumped into one of two arroyos adjacent to the pit while the ores were carted down the valley to the leach pad built on the flats at La Zapatilla and sprayed with a solution of sodium cyanide. The processing plant to treat the resulting solution and smelt the bars of doré would be located immediately beside the leach pad (see figure 5).[5]

The operation was slated to process over thirty thousand tons of ore per day. The entire workforce—including drillers, drivers, mechanics, technicians, and desk workers—numbered about three hundred people.[6] The bulk of the effort required to operate at these material scales was accomplished by machines whose principal cost was energy in the form of diesel—some

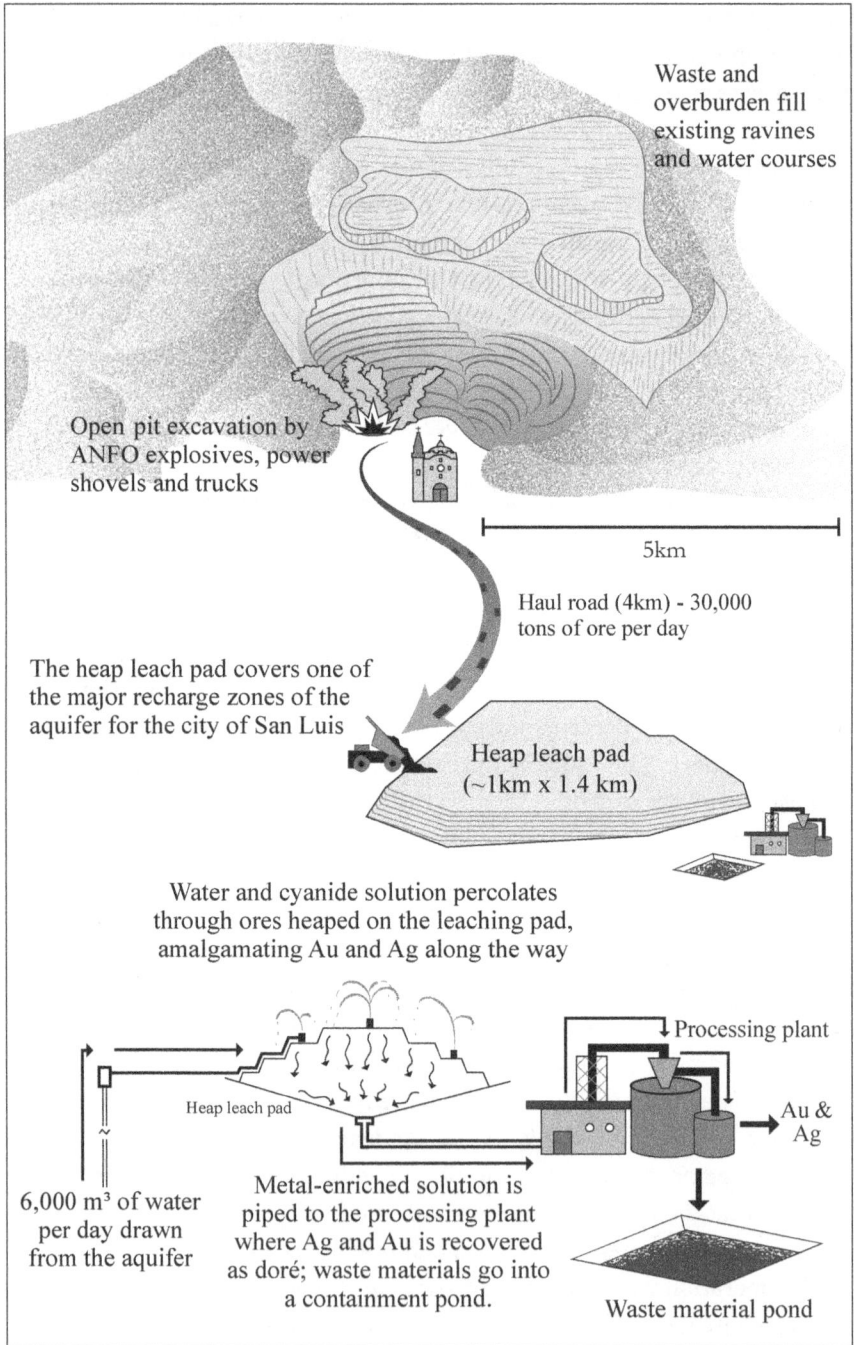

FIGURE 5 The open-pit heap-leach mine at Cerro de San Pedro (illustration by Geoffrey Wallace, 2021, after original by D. Studnicki-Gizbert).

forty-three thousand liters consumed every day.[7] The other critical input for the operations was the water used in the leaching process and to a secondary extent as dust control. Although cyanide leaching is in principle a circuit and thus a portion of the water can be recycled it is not a completely sealed circuit. Leakage and evaporation off the leach pads meant that every day six thousand cubic meters of water had to be replaced (this is the equivalent of three Olympic-sized swimming pools per day).[8] Running two shifts of twelve hours, twenty-four hours a day, six days a week, the company estimated that it would work through the entire Cerro de San Pedro deposit in eight years.[9] Once the operations had ceased the mine would leave behind a crater some three hundred meters deep and about a kilometer in diameter. The missing material would then be found down at the leach pad, piled up in trapezoidal ziggurat of mineralized waste rock a bit more than a kilometer per side.

Metallica Resources' Cerro de San Pedro project represented a stepwise intensification of the extractive process, one necessary to mine the extremely low grades of ore contained in the porphyry. Critics of the mine felt that this was the last mine, the final catastrophic end of over four hundred years of mining at Cerro de San Pedro. The historical pattern of serial reanimation in the district suggests a certain reserve on the question. Still, that same history helps better understand the current conjuncture.

The accompanying chart plots out the colonial, industrial, and open-pit systems that have characterized extractivist cycles of mining at Cerro de San Pedro over the centuries. Each cycle was far more productive than the last. More gold, for instance, would be removed by the open-pit operation than all the mines before it. These gains in output were achieved by intensification, that is, by greatly increasing the flow of the material extracted and processed. This was the working out of Lasky's Law at Cerro de San Pedro as the drop in ore-grades from cycle to cycle entailed exponential increases in the tonnage of material worked in a given year. These shifts were accompanied by corresponding shifts in energy consumption and waste production over time. The other noteworthy characteristic of the longue-durée evolution of mining at Cerro de San Pedro is a kind of temporal compression in which each cycle was of shorter and shorter duration. This was the trade-off of the cyclical intensification of extractivism. It was possible to transcend the limits of physical depletion, but the efforts required could not be sustained. Seen in this light, the last cycle of open-pit mining appears, in metabolic terms, as a paroxysm (see figure 6).

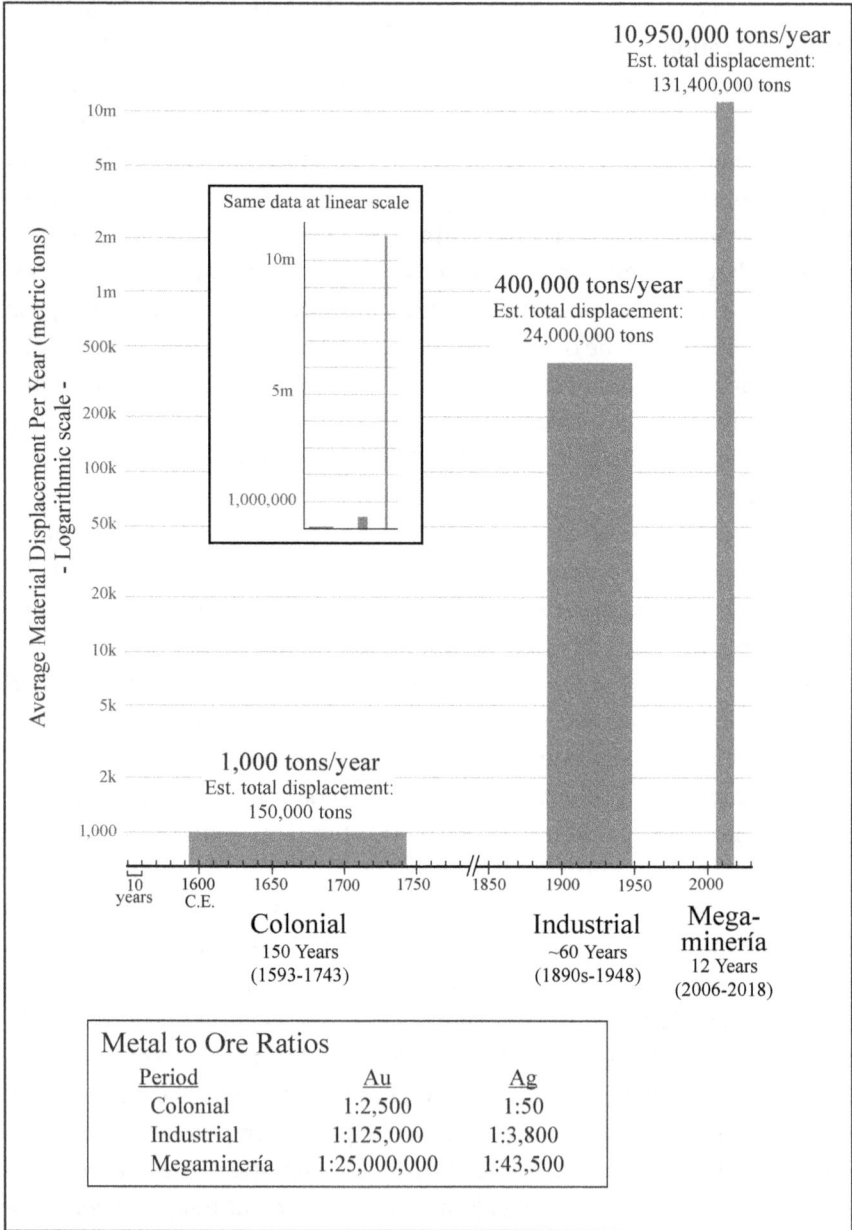

10,950,000 tons/year
Est. total displacement:
131,400,000 tons

Average Material Displacement Per Year (metric tons)
- Logarithmic scale -

10m
5m
2m
1m
500k
200k
100k
50k
20k
10k
5k
2k
1,000

Same data at linear scale

10m

5m

1,000,000

400,000 tons/year
Est. total displacement:
24,000,000 tons

1,000 tons/year
Est. total displacement:
150,000 tons

10 years   1600 C.E.   1650   1700   1750   1850   1900   1950   2000

Colonial
150 Years
(1593-1743)

Industrial
~60 Years
(1890s-1948)

Mega-minería
12 Years
(2006-2018)

Metal to Ore Ratios		
Period	Au	Ag
Colonial	1:2,500	1:50
Industrial	1:125,000	1:3,800
Megaminería	1:25,000,000	1:43,500

FIGURE 6 Cycles of extractivism at Cerro de San Pedro (illustration by
Geoffrey Wallace, 2021, after original by D. Studnicki-Gizbert).

It is also worth drawing attention to the increasing significance of water as an 'object' of impact. By the turn of the twenty-first century, water had arguably become the most ecologically and politically sensitive of life's elements. In the 1990s, when Metallica first promoted its open-pit mine, the politics of water were already moving to the fore in San Luis Potosí. Public water supplies for entire neighborhoods were found to be compromised by fluoride and heavy metal contamination. Much of this was the legacy of past cycles of mining. Rapid urbanization and industrial use were creating scarcity in domestic supply. The overdraft of the aquifer was first measured by university researchers and then became a matter of media coverage and public concern when the dropping water table triggered the collapse of sinkholes.

The scale of the invisible gold rush, the number of companies, the number of opportunities made possible by the new mass-extraction techniques, and the new and swelling flows of capital flowing into the multiplying number of companies that would effect this work, the geographical sweep, the ideological linkages forged anew between mass extraction and the central political projects of modernity and progress and development—all this composed a vast assemblage, a new extractivist regime. Its scale, scope and ideological depth all combined to produce a sense of inevitability—a temporal framing of the future that was itself a key ideological piece of neoliberalism. Having plumbed the size of the gold and silver ores that the Cerro de San Pedro still contained, Metallica Resources' management not only thought that the project's development was certain, but that it would be accomplished quickly. The company's engineers scheduled thirteen months to build the mine infrastructure, break ground, begin leaching, and pour the first ingot of doré (a gold-silver amalgam).[10]

The reality proved to be quite different. Ten years would pass before the company finally began operations. The mine that was built was different from the mine projected in the feasibility report, not for technical reasons but for political ones. Hence the importance of understanding the political dimensions of different extractivist cycles, not only for what they permit, but also for the limits they impose on techno-capitalist projects.

17 ¡Que Viva Cerro de San Pedro!

. . . .

Unlike Newmont's Yanacocha mine, enabled and celebrated by the Peruvian presidency, Metallica Resources' Cerro de San Pedro project was blocked before it even began. Opposition to the project was expressed on the very same day that it was announced to the community, Sunday, September 10, 1995. This was also, coincidently, the feast day of San Nicolás—the patron saint for generations of miners over the centuries. The parish priest had gone door to door convoking the inhabitants of the village to mass and what he described as very important meeting for the future of the community. Everyone showed up—even the atheists and communists. Once mass was concluded, the priest introduced Señor Hugo Gamiño, the director of a mining company called Minera San Xavier. Taking the microphone Gamiño began to explain that the company had decided to build a new gold and silver mine.

What follows is a collation of people's recollection of what happened and what was said at that meeting:

> He said that the community would be reborn like a phoenix . . .
> . . . that there would be hundreds of jobs and that the town would flourish for many, many years.
> . . . that the village would return to being the town that we all knew in our youth, a prosperous town.
> . . . but that there was only one problem: the town had to be evacuated.
> . . . we would be placed on a reservation, far from the danger zone.
> The company had already drawn up the plans. The streets were all laid out and named . . .
> And then El Chencho asked: "Hey, what about the Churches?"
> "They will disappear: everything has to disappear . . ."
> ". . . of course they have to go, because this won't be a traditional mine, you understand?"
> "This is going to be different from the mines that you know, the subterranean mines."
> "What do you mean they'll destroy the Churches? That's impossible!"

That's when people really started getting riled up. They got angry.
"Well then, we don't want this project!"
"But the company is here to help you! We've already given the mayor
a cheque for 20,000 pesos! And another to the priest."
"Well good for them, but what about the people?"
People started yelling; they used words that you don't want to hear in
a church . . .
"Fuck yourself and your project!"
"The project can go to hell!"
"Screw this shit!"
"Now, now, you can't say such things in the house of God."
"Well, you called us in here! Let's take this outside."
And then the priest switched off the microphone. And everyone left.
They were furious.[1]

The anger of the assembled villagers of Cerro de San Pedro surprised. It was something new. No one in Mexico, as far as subsequent observers could tell, had ever rejected a mining project before.[2] This was a community created around the work of mining gold and silver, work that had sustained it through bonanza and borrasca over centuries. It was a community with needs; it had been abandoned. The company promised jobs, hundreds of jobs, for many years. It promised that Cerro de San Pedro would be reborn.

The roots of their rejection were to be found in the kind of mining project this was going to be. As the manager Hugo Gamiño said, the proposed mine would be something entirely different from the sort of mining they had known, the underground operations run by ASARCO and all other miners going back to the sixteenth century. The kind of mine that required the displacement of a village, and its two churches was an open-pit mine. It was a new extractive system that broke up, hauled out, and then processed entire chunks of the landscape.

The furious reaction of the residents at the prospect of the destruction of their churches, their community, and their valley evolved into one the first socio-ecological conflicts that flared up around large-scale transnational mining in Latin America. Others in the mid-1990s included Argentina, Costa Rica, and Panama where hundreds of Ngäbé indigenous protestors blocked and then destroyed the machinery of the Canadian junior mining company Aurum Resources. These were the first conflicts to spark into life along the edge of the renewed commodity frontier engendered by the invisible gold rush. In time, they would multiply, their numbers growing in step with the advance of what came to be known in Latin America as *la megaminería* (mega-mining). They have erupted across the mining belts of the continent and currently number close to two hundred and fifty.[3] The emergence of these conflicts represented a reconfiguration of the politics of mining. The operative moral ecologies of the opposition to mass extraction shifted from an interior struggle over the relations between worker, subterranean matter, and owners. The landscape scale of the operations, coupled with their much-reduced work forces, gave rise to a more exterior struggle over the relations between communities, their territories, and the mining complex as a whole.[4] The new politics of mining addressed the very presence and shape of mining. This emergent politics simultaneously cast backward and forward: backward to the colonial scenarios of extractivist incursion and defense of homeland; forward because it was projected toward the future to the thresholds of what ecologies and communities could bear.

· · · · · ·

In the days and weeks that followed the meeting in the Church of San Nicolas, Cerro de San Pedro was abuzz with talk of the proposed mine. Conjecture mixed with rumor, speculation with gossip. It was a collective conversation probing outwards into an uncertain future. In these exchanges was seeded the conflict that was to cast the tenor of life in CSP for the next decades. This was not exactly a conflict between the community and the company—a rhetorical framing that was much disputed—it was a conflict between opponents and partisans of Minera San Xavier's project or, conversely, a conflict between those who defended life at Cerro de San Pedro and those who participated in its destruction. Some were glad for the return of mining, for the return of the prospects of jobs, for the return of life itself in a community so long abandoned. (*"I'm grateful for the work and I don't reject it; because the mine is a center of life for so many people."*)[5] Some were scandalized that a foreign company should propose to destroy their valley, their community, and their churches and make off with their riches. (*"Minera San Xavier is plundering. It is destroying the mountain that gave us our origins . . . it is a pillage of our homeland."*)[6] Some kept their thoughts to themselves. This included many women in Cerro de San Pedro and the neighboring community of La Zapatilla, also slotted from destruction and displacement to make way for the heap-leach pad and processing plant. They kept their peace. Not all, of course, given that a number of local women became prominent leaders and spokespersons of the opposition.[7]

In its beginnings, the opposition to the open-pit mining of Cerro de San Pedro was all but invisible to outsiders (the significant exception was the company, keenly interested in the question of support and opposition). It probably numbered a few dozen individuals tied to the community. Their prospects were daunting. They faced a corporation with ample funds, political connections, and the endorsement of a state seeking to renew the mining sector in the name of national economic development. They had some experience in collectively organizing to improve the local provisioning of water and electricity, but taking on a transnational mining corporation and its backers was a different matter altogether. The company, at any rate, thought that the local opposition could be safely discounted. A few months after the Church meeting, it announced to its investors that work was advancing at good speed and that operations would begin forthwith with the first pour of gold and silver scheduled within the following year.[8]

When they began, those who fought for Cerro de San Pedro felt like political innocents. *"We did not know what tactics to apply, what strategies to follow. We created and figured these out as we went."*[9] But in creating and figuring things out as they went, the opposition to Minera San Xavier's project achieved something both surprising and possibly unprecedented: they stopped a mine in Mexico for more than a decade. For a junior mining company, trying to grow in the fast-moving waters of a stock market, ten years was an eternity. Over time, however, what happened in Cerro de San Pedro became less and less exceptional. From the mid-1990s on, across Mexico and across Latin America a growing number of community-based organizations emerged to block the expansion of open-pit mining. Their aim was territorial protection through the formal prohibition of large-scale mining extraction. By the 2010s these movements had managed to put an important number of territories out of bounds to open-pit gold mining. They included include entire countries (Costa Rica and El Salvador), sub-national units such as Argentina's chain of Andean provinces, the indigenous Comarca of Panama's Ngäbé and Buglé peoples, and a long list of municipalities that declared themselves "libres de minería" in southern Mexico, Guatemala, Colombia, Ecuador, Peru, and Argentina. Chronicling the evolving opposition to the open-pit mining of Cerro de San Pedro provides an interior view of this larger phenomenon. It shows the life cycle of an environmental justice movement operating within the contemporary extractive zone.

The politics that flared up around Metallica Resources' Cerro de San Pedro mine were not pre-established. They did not lie latent and fully formed, waiting to be activated by the announcement of the project's designs on that

Sunday morning in September of 1995. Both the politics of opposition to the project, and the politics of support for the project, developed over time. Indeed, this is one of the interests in examining the case of Cerro de San Pedro more carefully. As one of a small crop of early socio-ecological conflicts around mining, its story illustrates a process of political emergence. What began as an immediate expression of outrage against the proposed destruction of the community's churches evolved over the course of a struggle that ran for over two decades. Seen from the perspective of the longer political history of mining at Cerro de San Pedro, perhaps it is more accurate to view this as a re-emergence, as long-standing moral ecologies were reconfigured and adapted to the new forms and dimensions of contemporary extractivism.

· · · · · ·

This was, in its beginnings, a movement that declared itself in the positive. It pronounced itself in favor of life in Cerro de San Pedro. It was not, as its members continue to repeat to this day, opposed to mining as such. These were the sons and daughters of miners. Some had themselves worked in mines and smelters. Their activism was directed at the undue exactions that mining imposed on "life," broadly and variously defined. Their first formally constituted organization was named Pro Defensa de Cerro de San Pedro (For the Defense of Cerro de San Pedro). It was made up of an assortment of ejidatarios and their descendants, returnees, and the new arrivals looking to develop some tourism in the town. Hitherto they spent their days tending to their beehives or goats, fixing the roof on the family home, or resetting the stones of their patio. Now, faced with the imminent destruction of the village, they would have to work to save their community.

The founding members of Pro-Defensa de Cerro de San Pedro were under few illusions as to their position. They were isolated, with few political connections to leverage, and even fewer resources. And so, they reached out to Pro San Luis Ecológico, the city's first environmentalist organization. The first meeting between the two groups took place at a public rally in support of the residents of Guadalcazar, an old mining town some eighty kilometers to the north. The people of Guadalcazar were fighting to remove a toxic dump managed by a California-based corporation called Metalclad. Initially described to locals as a forthcoming waterpark, the dump functioned as the Mexican impoundment for the international wastes shipped there by the company. Together, the villagers and Pro San Luis Ecológico convinced the state governor to turn the zone into a protected area and shut the company out. It marked a significant achievement for the incipient environmentalist organization, although one that came at a cost. Metalclad sued the state of San Luis Potosí under NAFTA's investor protection clause—chapter 11—and was eventually awarded over sixteen million dollars in 2001 by a NAFTA tribunal. The decision became *cause notoire* for social movements mobilizing against the Summit of the Americas in Quebec City, April 2001.[10] In Mexico, it cast a cold shadow over subsequent efforts to legislate against transnational forms of environmental injury.

Pro San Luis Ecológico's second major file at the time concerned the chronic heavy metal contamination from the Morales smelter. This was another textbook case of environmental injustice in that the highest concentrations of heavy metals were to be found in the city's working class neighborhoods. The smelter had been processing metallic ores (notably those of Cerro de San Pedro) for close to a century. Under Towne, under ASARCO, now under IMMSA, it had operated without serious emissions control.[11] The soils and airs of its immediate surroundings were saturated with heavy metals and sulfide gases. This was making people of the neighboring colonias chronically sick, but the company denied their harms and the state did not get involved. The impasse motivated a team of chemists, biologists, and medical researchers from the UASLP to examine hundreds of people. Their bodies, especially the childrens', were laced with lead, cadmium, and arsenic at many times the levels set by national and international health organizations.[12] Pro San Luis Ecológico undertook the political work called for by the team's results.[13]

The association with Pro San Luis Ecológico linked the residents of Cerro de San Pedro to a group of elite urban professionals.[14] They included scientists who had participated in the Morales smelter studies, as well as biologists, medical doctors, lawyers, and key individuals within the local pro-democratic Navista movement. With Minera San Xavier rushing to get the mine into operation, the imperative was to slow the project down. It was a contest between the investor-led pressures on the company to move fast, and the strategic need of this emerging movement to secure the time required for a more thorough and inclusive consideration of the project.[15]

In late October 1997 Minera San Xavier published a much condensed and finely printed extract of its environmental impact assessment in the city's papers.[16] Accessing the full document (368 pages plus three volumes of annexes) and thus forcing a more substantive public review of the mining project was the first act of the new alliance. Their petition was initially denied but after a concerted media blitz, a number of public demonstrations ("What are they hiding?"), and the arrival of a new governor, Fernando Silva Nieto, they eventually obtained access to the report. There were conditions, though: only two individuals would be allowed to review it; they could only take handwritten notes as no photocopies would be permitted; they were given eight working days.[17]

What Mario Martínez (son of the Blas Martínez, head carpenter and confidant of ASARCO's managers back in the 1940s) and Dr. Angelica Nuñez (a prominent local biologist) recorded in those harried and hand-cramped days shifted the view of what was at stake in the defense of Cerro de San Pedro. It revealed in detail and measure the scale of the project and its radical departure from previous forms of mining. Specifically, the Environmental Impact Assessment (EIA) signaled the new layers of risk the open-pit mine posed to the area's waters, already an object of political concern given rising rates of industrial contamination and the generalized over-drafting of regional water reserves. Minera San Xavier's mine would draw an additional 2.3 million cubic meters of water per year from an aquifer already unable to replenish itself.[18] This was the equivalent domestic consumption of between 100,000 and 210,000 urban dwellers.[19] The heap leaching pad, with its impermeable liners and dykes, would furthermore block a critical recharge zone for the aquifer.[20] The mass in-filling of the local arroyos with sulfurous waste material would, through the combined action of air, water and a certain class of bacteria, generate acid mine drainage, which in turn would acidify surface water flows and leach out the heavy metals contained in the wastes.[21] To these chronic harms was added the more catastrophic possibility of a breach of the leaching circuit. The scenario of concern was that of a violent rainstorm cutting through the dykes and sending a mucky river of acidic, heavy-metal rich and cyanide-laced material flowing downslope toward the outlying suburbs of the city.

Metallica Resources' coincident announcement in November 1997 that another Canadian mining corporation called Cambior, would enter in as project manager at the Cerro de San Pedro mine set off further alarms.[22] Cambior was the operator of the Omai gold mine in Guyana. Two years earlier its tailings dam had given way, sending over a billion liters of cyanide and heavy metal-laden sludge shooting down the Essequibo River, wiping out all manner of life as it went.[23] Further research revealed that tailing dam failures was an established pattern in open-pit mining, one that significantly began with the highly publicized cyanide spill at Robert Friedland's Galactic Gold Summitville mine in 1989.[24]

For the members of the two organizations, the emerging portrait now surpassed the imminent destruction of Cerro de San Pedro, mountain and town. They saw in the mine a region-wide ecological threat, what they called *un ecocidio*. The project had to be stopped. To do so, they initially followed the clientelist forms and codes of San Luis Potosí's existing political culture. Because a number of the members of Pro San Luis Ecológico were part of the city's professional class, they offered connections that led to a personal meeting with the new Governor Silva Nieto. Better yet, the meeting would take place at Cerro de San Pedro and, better still, the Governor said that he was happy to hike up to the top of the mountain. There, the locals pointed out what they saw as the prospects. An open-pit mine at Cerro de San Pedro would not only destroy the mountain upon which they were standing (the emblem for the state of San Luis Potosí), the village at its feet (a patrimonial site), and the surrounding sierra and valleys (gazetted in 1993 as a Culturally and Ecologically Protected Area), it also put into risk the future water supply of the city of San Luis Potosí sprawling out in the distance below.[25] They asked the governor to conduct an independent and public evaluation of the project. The governor agreed to look into it.[26]

The next person invited to take in this view was the recently elected mayor of Cerro de San Pedro, Baltazar Loredo. Those at that second mountaintop meeting recall working hard to keep up with him as he strode straight up the steep slopes.[27] Tall, thickly mustached, and dressed in the broad brimmed hat and jeans of the ranchero, Loredo was a striking and charismatic local figure known for his straight-talk and independence.[28] After laying out their case, the representatives of Pro Cerro de San Pedro asked him to investigate the land deals put together by the company. The mayor said he would look into it.

Of the two, the mayor was the quicker. He found that his predecessor had overseen a flurry of land deals with the company. And there were questions. The land rental agreement signed between Minera San Xavier and the ejido of Cerro de San Pedro was conducted behind closed doors without the presence of the full quorum of ejidatarios. Those who signed were accused of being outsiders with no ties to Cerro de San Pedro.[29] Then there was a row over the ostensibly abandoned properties. These had been sold to the company as being without owner, but the owners had since been advised and they were furious. Within a month of his hike up the Cerro de San Pedro, Loredo was petitioning the state Congress to conduct a formal judicial review of the land dealings in the municipality.[30] A few months later, at a hotel meeting in the city, he publicly confronted the American manager of the company's operations. Then, the next day, March 21, 1998, Loredo was found dead in his pick-up, a right-handed man shot through the left side of his head. The police declared suicide, but everyone knew better. The responsible party was never identified or apprehended. When later asked about the affair, the CEO of Metallica Resources bristled at the suggestion: "Come on! We're civilized people. We're talking about a mine. You don't kill people over a mine."[31]

The governor reported back more than a year after Loredo's death. In Mario Martínez's account of that meeting, the governor admitted that his hands were tied. He had received phone calls from Los Pinos (Mexico's presidential residence) and from the Canadian embassy. The state university, the Universidad Autonoma de San Luis Potosí, had done a rapid assessment of the company's environmental impact assessment and they cleared it.[32] He could not be expected to go against the experts and would be giving the state's approval for the project.[33]

Clientelist politics having proven a dead-end, and with the shock of Loredo's murder still in the air, the decision was taken to stop the mining project in the courts. The two key issues were the environmental permitting of the project and the land rental agreement signed with the ejido. The first effort was led by a prominent San Luis lawyer working pro bono since the death of his wife, one of the founding members of Pro San Luis Ecológico, now resting in a simple grave on slopes of her "*querido* Cerro de San Pedro." It tested Mexico's young environmental legislation—also passed in the "NAFTA moment" of the early 1990s—specifically its provisions for ecological equilibrium and environmental protection. The federal court's decision annulled the permit given to the company by the Instituto Nacional de Ecologia for not respecting the 1993 designation of the zone as a protected area, as well as for its threats to endemic and endangered species, notably the slow-growing biznaga and snowy cacti.[34] The second legal front was led by a young lawyer active within the local instances of the left-wing Partido Revolucionario Democratico (PRD) party. It brought the matter of the land rental agreement to the Tribunal Agrario (the tribunal, formed with the Revolution, that adjudicated land property). Upon review, the Tribunal disqualified twenty-nine signatories for not being confirmed members of the ejido, thus annulling it.[35]

This is a radically parsed précis of the much bulkier and complex courtroom history of injunctions, appeals and decisions. Throughout, the company and its legal team insisted that they were in the right.[36] But by 2004 the net result was that the company was still not operating. The company's environmental permit was in suspension and the status of the land agreement was unresolved. The Instituto Nacional de Antropología y Historia (INAH), responsible for the area's cultural patrimony, had not cleared the project. The Mexican military withheld the blasting permits. Then, to cap things off, the new mayor of Cerro de San Pedro, Oscar Loredo (the son of the murdered Baltazar) announced that he would not be renewing the company's municipal permits. On June 23, 2004, Metallica publicly announced that it was shuttering the project.[37]

It proved to be only a tactical retreat. After lamenting the fact that a small group of opponents could use Mexico's legal system for political gains, the CEO of Metallica promised that the company would use political and legal channels to lift the blockade.[38] A new director of operations was appointed—a local *Potosino* with close familial and working relations with leading members of the right-wing Partido de Acción Nacional party (then in power, at both the federal and state levels).[39] A bombastic man who publicly modeled himself on Winston Churchill, he proved remarkably adept at assembling the political solution to the company's dilemmas. On his next trip to San Luis, President Vicente Fox insisted that the recalcitrant mayor join him at dinner with the state governor. Oscar Loredo recalls sitting nervously between the two when the President turned to him as said, "Ahhh . . . so you're the youngster whose giving my *compadre* [state Governor Marcelo Santos] such a headache." At first, Laredo recalled:

> He spoke calmly, in an educated way you might say, but then he began
> to raise his voice, loud, in the way that he often does, and said, "Look.
> The fact is you need to approve this project." He wasn't commenting
> anymore, he wasn't asking, he was ordering me: "You need to support
> this project. Because if you don't support this project it's going to
> trigger various clauses of the Free Trade Agreement where Canada
> will submit a consular complaint." Something like that, I don't
> remember the proper term. "A lawsuit because they're investing in a
> legal manner, and if, because of some whim, or whatever you want to
> call it, this investment isn't carried through, then you will be impeding
> a part of the obligations and rights of the Free Trade Agreement."
> Then he slammed his hand on the table, "I need, *need*, you to
> unblock this project. I need this project to be operating, now."[40]

At the following meeting of Cerro de San Pedro's municipal council, Oscar Laredo announced, eyes brimming, an about-face. He would sign off on the permit. Facing the dismay and shock of the assembly he told them that he had no choice. This was bigger than him. He was afraid. Remember what happened to his father.

Resolving things within the ministries was even easier. The SEMARNAT reissued a new environmental permit. A halt was made on the INAH's reclassification of Cerro de San Pedro as a national patrimonial site. The military was enjoined to finally grant the blasting permit. It took some time, but by the spring of 2006 the company was finally in operation and, by April 2007, ten years past schedule, it triumphantly announced the pouring of the first bars of silver-gold doré.[41]

Forms, Tactics, and Emblematics

. . . .

As the company freed itself from its difficulties and began operations, it only gave more energy to those active in the defense of Cerro de San Pedro. By the mid-2000s they already formed a local, if disparate, grouping involving community-members, Pro San Luis Ecológico, former Navistas, student organizations, anarchist collectives, and even the local chapter of the Boy Scouts.[1] Then, as the company began working in earnest, the cause gathered support nationally and internationally. Prominent metropolitan intellectuals such as Carlos Monsiváis, Carlos Montemayor, and the historian Juan Carlos Ruiz Guadalajara, championed it in Mexico City. The case began to get national media coverage. Public forums were organized. Mexico-City based environmental and human rights NGOs lent their support. Delegations traveled across Canada (Kamloops, Ottawa, Montreal, Toronto, Vancouver), where they were received by church groups (KAIROS and the Montreal Social Justice Committee), NGOs (Mining Watch Canada, the Polaris Institute, the Council for Canadians), members of Canadian Parliament (Bloc Québecois, Liberals, NDP), and universities (McGill, UQAM, University of Toronto, York), and solidarity organizations (MISN, CDHAL, QUISETAL). Personally, however, the full measure of how far this opposition movement had traveled came in January 2011, during the jam-packed and heated sessions convened by Panama's National Assembly to debate proposed reforms to the national mining code. Pushing his way through the senators, representatives, professional NGOs, and indigenous Ngäbé, Buglé, and Embera authorities, a *campesino* took the microphone and blazed out from beneath his broad straw hat, punctuating his speech with a sharply pointed finger. He had walked two days through forest and river to get here: "The defense of our lands from the depredations of these [mining] companies must be our maximum objective. We need only consider the Cerro de San Pedro Potosí up there in Mexico, fighting for its life."

As a political creation, the movement for the life of Cerro de San Pedro, escapes easy categorization. It was not a party or a union. It had no formal constitution or memberships. There was, to be sure, an attempt to institutionalize things in 2000, with the foundation of the Broad Opposition Front against Minera San Xavier (FAO-MSX by its Spanish initials for Frente Amplio Opositor a la Minera San Xavier). And it did provide a basic identifier that was useful when it came time to emit a press release or make the call out for allies. But this was a singularly uncoordinated body. Meetings were irregular. Attendance was uncertain. It had no authority or much capacity to order and plan the actions of the multiple groups and individuals trying to stop the company's project. Of all the possible categories applicable to this social and ecological movement, the one that strikes the truest was that proposed by a San Luis-based activist: "The FAO doesn't really exist, there is no organization here. Look at us! The FAO is a bestiary."[2] Indeed. And the fauna of the FAO was of the most diverse. Their colors were on display at the June 2006 forum called to debate what to do about the revival of the mining project: ejidatarios with their broad-brimmed hats and white shirts; young anarchists in their blacks and piercings; professionals and politicians in suits and ties; the pressed jeans and checked shirts of the academics and those working "for an organization"; there was even a contingent dressed in the white spun-cotton blouses and red head bands and beads imagined to be the traditional garb of the indigenous Guachichils.

Within this diversity, however, one can discern a common ideological frame, a moral ecology, now reconfigured to match the scope of low-grade, last-pass mining. Like previous iterations, it was composed in opposition to the negative reciprocities engendered by mining. For the people who grew up and lived in Cerro de San Pedro, these were known as a matter of personal experience and of family history. They were surrounded by the legacies of foreign-owned industrial mining. Better than two on three men in the municipality were still wracked by silicosis and associated pulmonary ailments.[3] Every cough of these cascados, these "husked" workers, was an everyday reminder of the bodily exactions of mining, many years after ASARCO had abandoned the place. The sheer scale of open-pit mining, however, reconfigured the relations between extraction, territory, and people. Whereas in the past mining had consumed human bodies, it was now going to waste the ecological foundations of an entire city.

Judging from the composition of the movement to save Cerro de San Pedro, this framing found a broad resonance. What outraged were the inequities that it foregrounded. On the one hand, a transnational corporation working to enrich its owners, officers, and investors through the extraction of Cerro de San Pedro's last remaining treasure. On the other, the destruction of the mountain and community, as well as the exposure of an entire population to chronic ecological harms and catastrophic risks for generations to come.

People thus came together around the conviction that the project was unacceptable by the most basic terms of social and ecological justice and had to be stopped. They were, with few exceptions, political amateurs. They tried what they could with the time and the resources they could give. One activist characterized it as a kind of "Lilliputian politics" with each groupuscule launching an action like one more cable that might pull down a giant. Surveying their various efforts, one observes the range of political praxis enacted by socio-ecological movements elsewhere in Mexico and Latin America.

Materially speaking, there were attempts to directly paralyze the operations of the mine. These were first deployed in April 2006, when the blasting began, and the trucks began to roll. A small crowd of fifty stopped the company's movement for two days.[4] A week later they were back at it, now with reinforcements from the Zapatista Army of National Liberation.[5] The blockading campaign was kept on and off until the beginning of December when the *plantón* was set upon by a large group of local men armed with rocks and steel bars.[6] It was subsequently abandoned as a tactic. Elsewhere in Latin America, direct action politics were more serious and effective: the Shuar of Ecuador invaded and destroyed the mining camp of Corrientes Resources, the Ngäbe and Buglé blockaded the Inter-American highway, twice, and they too had few compunctions about burning companies out of their territory. In the Panamanian case the government settled the crisis by legislating against mining of any sort in their territories.

Other actions were organized around democratizing Mexico's mining regime. Governance around mining, specifically the decision to allow a mine, was a thoroughly technocratic and hermetic administrative procedure. In only two instances was there a requirement to engage the broader public: by publicizing an extract of the EIA and during a one-day public forum that was held in the same week as Baltazar Loredo was killed.[7] Challenging the lack of democratic input into what they believed to be a momentous decision for Cerro de San Pedro and San Luis Potosí, activists came together to organize a referendum. It was billed as a citizens' consultation. If the organizers are to be believed, they were able to get the ballot out to twenty thousand people in San Luis Potosí and, of those, over 90 percent declared themselves against the project.[8] The company, unsurprisingly, discounted all of this as fabrication.[9] This was an argument heard elsewhere. The first citizens' referenda on large-scale transnational mining were organized in 2007 (Tambo Grande, Peru; Esquel, Argentina; Sipakapa, Guatemala). Dismissed by companies, judged unreceivable by the state, the popular referendum or *consulta*, offered new spaces for community-based debate, decision making, and collective organizing.[10]

Where the movement for Cerro de San Pedro occupied pride of place was in the realm of symbolic politics. They included more established forms such as public marches or *plantones* (encampments) set up in front of the Governor's palace. Project opponents invited themselves into Metallica Resources' annual shareholders' meeting.[11] Graffiti was applied enthusiastically everywhere. Short interventions of protest theatre were staged in front of government ministries, the Canadian embassy, contributing banks, and corporate meetings. The company and colluding politicians were costumed as pigs and rats and held for public trial before the watchful eye of San Luis *justicier*. Pigs and rats made a regular appearance in marches, along with the grim reaper, death, and San Luis himself. For the Canadian Imperial Bank of Commerce, Mexican activists stripped and sprawled naked before the entrance, where they proceeded to be disemboweled (ample pig guts and fake blood) by company miners rooting around for the gold they guarded in their bodies. In Montreal, they stood with representatives from Papua New Guinea, Argentina, Chile, Honduras, Guatemala, and Quebec to stake a concession for an open-pit mine on the city's iconic hill-top park, the Mont-Royal. They went so far as to incorporate a fictious mining company, obtained their prospectors' permit, and formally filed their mining claim with Quebec's ministry of natural resources. Residents of the posh slope-side neighborhood Westmount were advised of their up-coming evictions and invited to a public consultation. It was all show of course. This multi-national coalition only wanted Canadians to understand that what was unimaginable in their city was, elsewhere, a no-less outrageous reality, one orchestrated by their corporations. Quebec's national assembly took no chances and passed a specific motion protecting the Mont-Royal from mining.[12]

There was, in all of this, a great deal of looking back. That is to say that the past itself held an important place within this movement, both as referent and as resource. It had, it's true, its forward-looking politics that identified the risks to future generations of life in the valley and the accretive violence of mining's harms. The mobilization of the past, however, drew upon a centuries-long history of mining as remembered and enacted upon by contemporary activists. The destruction of the Cerro de San Pedro was felt as more than the physical taking apart of a mountain and a village. In the slopes pocked with old mine shafts, in the houses, the caves, in the sparsed detritus was the materialized presence of centuries of mining. The underground space of tunnels, halls, and shafts was for one resident of Cerro de San Pedro, sacralized by past lives and past sacrifices. *"Coming back into these tunnels I'm reminded of all the deaths that occurred in these places, the so many people crushed in the mines. And of how, when they died, one could no longer touch that place, how it became like a place . . . a sacred tomb."*[13] And those ghosts, the people of Cerro de San Pedro, they heard them in their heads: *"What would happen if our ancestors came back and asked, 'So, what did you do? Why did you allow things to come to this?' We wouldn't have an answer for them."*[14]

Faced with the threat of destruction, locals began to reacquaint themselves with the history of Cerro de San Pedro, of its people and of its importance in the broader history of San Luis Potosí. They recovered what they called their ancestral memory, even though, as one recalled, it was "much submerged, which convenienced the government. But it was all here, and now I for one am proud of it."[15] The community-based museum—*El Templete*—displayed murals, objects and photographs assembled to bring the history of the place back into public view and consciousness. They restored the parish archives, "critical for the reconstitution of the history of Cerro de San Pedro and the families that inhabited it over the centuries. . . . They were clear that not only the Cerro de San Pedro [mountain] but also its surroundings had to be preserved."[16] Local memory-making took in the conquest of the Guachichils and the first colonial bonanza. It remembered the role played by José Patricio Alanis and the independent miners of Cerro de San Pedro in the Rebellion of the Barrios in 1767. Indeed, given that the patronym Alanis still survived in the local families, a number of residents claimed a generations-long genealogical tie to the popularly acclaimed King of the Mountain. And, more closely, it took in the days under the Americans and the betrayal of ASARCO in 1948.

Linking these episodes together provided the trajectory and the historical frame into which the contemporary renewal of large-scale mining could be situated. It was a repetition of a cycle: *"Mining has come here in stages. First the Spanish, then the Mexicans, then the Americans, then the Mexican company Peñoles, and now the Canadians."*[17] It represented colonialism, recurrent and renovated: *"Minera San Xavier is plundering. It is destroying the mountain that gave us our origins, our emblem. Well this goes back a long ways. It roots back in the Virreinato, don't you see? And now, now it resurfaces again."*[18] And with this recycling of the colonialist frame came the symbolic rebirth of the Guachichils, key figurants in the colonial antinomy.

Down to its last two survivors in 1674, the return of the Guachichils in the twenty-first century was a re-enactment, a memory-play conducted by a group of urbanites who claimed direct filiation with that Nation and the indigenous resistance they represented. These neo-Guachichils appeared at demonstrations dressed in white slacks and blouses ornamented in beadwork. They blew on conch shells, burnt copal, and performed the circular stamping dance of the *concheros* of Mexico City. When the Sub-Commandante Marcos of the EZLN arrived in Cerro de San Pedro, they took upon themselves to relate the history of their ancestral territory.[19] Activists travelling to Canada and looking for allies, presented themselves as descendants of the Guachichils to the Kanien:keha'ka Mohawks. It was pastiche and performance, clearly, but whatever one's views of this kind of Neo-Indigenismo, it made present the politics of colonialism. Locally, among the villagers of Cerro de San Pedro, the rememorialization of the Guachichils surfaced in less instrumentalized or performative ways: they were the subject of local *cuentos y cantos* (stories and songs) and held as touchstone of history and place: "I love my *pueblo*, I love my roots, I love my Guachichils. It's our identity."[20]

At the same time, the defense of Cerro de San Pedro was framed as the preservation of patrimony. How could the same government seeking UNESCO's endorsement for San Luis Potosí as patrimony for humanity be simultaneously destroying the mountain and community that gave it birth?[21] The figure and emblematics of San Luis himself was dusted off and given new life as a symbol of justice assailed by corporate and state impunity. He was, of course, none other than Saint Louis of France, the patron saint of royal justice and protector of his subjects.[22]

The figure of San Luis perched atop the mountain of Cerro de San Pedro became a standard feature in the movement's emblematics. He was present on banners, as a seal on communiqués, in political cartoons and re-enacted live as a figurant in the FAOs irreverent form of street theatre. As a form of political allegory, the King and the mountain combined to form an iconic visual referent for the assault on justice and identity, on history and territory. It became an emblem for the negative reciprocities churned up by the open-pit project.

19 The Politics of Appeasement

· ·

Of all the tactics and forms deployed by this "Lilliputian" environmental justice movement, it was the work in the Mexican legal system that proved the only means of actually stopping the operations of the mine. In late September 2009, the Federal Tribunal of Administrative Justice ordered the shut-down of the Cerro de San Pedro mine—for the second time. For the activists and supporters, this was justice served, even if they had waited three years for it. The court reaffirmed previous judgements rendered in 2004 and 2005, noting that these were beyond appeal and that "under no condition could an environmental permit be accorded for this project."[1] The judges castigated the SEMARNAT for having illegally reissued a permit to the company in 2006, and demanded that the agency enforce the sentence.[2]

Nothing happened. The drills filled the valley with its percussive drone. The blasting shook the bones of the mountain a bit past three o'clock. The hauling trucks kept to their to and fro between the expanding pit and the leach pads. Back in Canada, the parent company's press releases made no mention of the judgement, and concentrated on sharing the good news of future prospects in Chile. Reality was queerly split between the view of the defenders of Cerro de San Pedro—celebrating the end of impunity and the shut-down of the mine—and that of the company and its supporters— who saw just another legal challenge to be resolved so that business could proceed as usual.

The prospect of the company yet again eluding the conduct of justice galvanized the opposition. They were determined to make it a reality. Strenuous lobbying among Mexican *congresistas* and Canadian parliamentarians produced a four-party bi-national declaration calling on the Canadian company, now New Gold Inc. of Vancouver, to respect the decisions of Mexico's courts.[3] Complaints against the Canadian parent company were formally filed with the British Columbia Securities Commission and the Toronto Stock Exchange for "withholding information of central importance to shareholders."[4] Mexican media reported on the case almost daily, often with front-page coverage. Pressure increased on the incoming Governor of San Luis

Potosí, Fernando Toranzo, and the Mexican undersecretary of the environment, Mauricio Limón, as journalists inquired about their intentions with regards the courts' ruling. Both declared their commitment to upholding the rule of law in the country.[5]

The blasting continued.

In San Luis Potosí, the company insisted that it had all its permits.[6]

On November 19, the head of the Profepa (Procuraduría Federal de Protección al Ambiente), Mexico's environmental enforcement agency, declared to the press gathered in the capital that the project was finished and that the company had been directed to shutter operations and flush the leach pads in view of a final shutdown and remediation. That afternoon, six blasts shook the town of Cerro de San Pedro, exactly on schedule at between three o'clock and three-fifteen, but now captured by the cameras of national media.[7] Meanwhile, down in San Luis Potosí, members of Pro San Luis Ecológico and the media were in the local offices of the Profepa demanding to know why local officials were ignoring their superiors in Mexico City. To calm the row, the personnel eventually walked out to their truck and made the twenty-minute trip up to the Minera San Xavier's main plant and padlocked the gates.[8] After three years of operation, the Cerro de San Pedro mine was shut down.

With its share prices dropping quickly, the Canadian parent company issued its first press release acknowledging the situation but insisting that it was operating in full compliance with the law.[9] Opponents to the mine celebrated and announced further criminal lawsuits against the Mexican general manager of Minera San Xavier for "injuries to the nation and influence peddling allowing the illegal operation of the company in Cerro de San Pedro" as well as the state representative of the Profepa for negligence in the conduct of his public duties.[10]

Up in Cerro de San Pedro, too, the pressure was building. Threats were made against those families who had opposed the mine. Company workers and their families packed the church of San Nicolas for an Advent mass keyed to the message of hope and deliverance and a prayer for "the quick normalization of the mine's operations."[11] As they filed out, however, the mood was less forbearing. "It's really awful that because of those people, of the FAO, people will lose their jobs." "Many people here live from their work; the others, they live off their politics. It's unjust that for the actions of two or three people we lose this project! . . . What will we do with this village [pueblo] when this work disappears? Nothing! NO-THING! What are we supposed to do? Die of hunger?"[12] A week later a large group of people—Amnesty International cited one hundred—set upon known opponents to the mine, showering them with threats and stones as they ran for the cover of their homes. "They wanted to lynch us," one later recounted, "they stoned us really hard, hitting us from behind."[13]

For the company, what moved things forward was not violence against opponents but reaccommodation with authorities. The Toronto office called this the "normalization of the mine's operations," and it was not long in coming. Three days after the attacks on local opponents to the mine, the parent company announced that it had obtained an injunction lifting the court's shut-down order while the merits of the case were discussed. "We are very pleased with the court's decision," announced the CEO, "and view this as an interim step as we continue to work with the government and administrative bodies to find a permanent and mutually-beneficial solution."[14] The solution was for the SEMARNAT to reissue the company a permit for a third time, despite a supposedly final and un-appealable decision to the contrary. The lead lawyer for Pro San Luis Ecológico filed yet another appeal but as he later admitted, his heart was no longer in it. The law was abundantly clear, he said, but the politicians had decided that the mine would go forward and there was not much he could do about that.[15] The company's legal team then coordinated with the state government to pull the problem out by its root. In March 2011, the state of San Luis Potosí brought in a new regional development plan to replace the 1993 plan that had designated Cerro de San Pedro as a zone for cultural and ecological protection. Now its appointed vocation was to be industrial mineral extraction.[16] With the rezoning passed into law, the foundations of the opposition's ten-year legal campaign disappeared. The SEMARNAT and all other permits could build upon new footings. Since the events of late 2009, the company's operations did not cease until 2018 when, after the promised nine-year run of extraction, the company ceased operations and began a two-year flush of the leaching pads.

The company's ability to fend off the 2009 shutdown reveals the local operations of extractivist politics. Like the new moral ecologies enacted by environmental justice movements, this more hegemonic mode followed the movement from the interior contests of work, capital, and the extractives complex, to the exteriorized struggles between open-pit mining projects and the peoples and territories that surrounded them. It was reactive in that it aimed to keep large-scale mining moving forward in the face of critique.[17] Against the activists, that is all those who sought to emphasize the slow violence produced by the mine, its assault on the deeper realms of life like history and the land, the company responded with a praxis and discourse of appeasement. There was no crisis. The mine provided for the community. It had social support. For the mining company, it was the politics of plain vanilla: *"Our interests are to build a simple little mine and employ some people and be a contributing member of that society."*[18] For the French scholar of oil extraction Nicolas Donner, it was a politics of social narcosis, a politics that aimed to appease society's rejection reflexes when faced with the prospects of social and ecological harms, and thereby maintain the support of the ruling institutions of law and government.[19] The new extractivism produced hegemonic relations through anesthesia, normalization, and appeasement.[20]

Minera San Xavier and its different Canadian parent companies engaged with the body politic at different scales: the local world of families and neighbors as well as "higher" worlds of state and national politics. These engagements rejoin the broader shifts in what social scientists call corporate governmentality in extractives. As an object of study by anthropologists, geographers, sociologists and political scientists, its central practices and discourses have now been extensively documented and analyzed with case studies drawn from around the globe. The historical interest here is twofold: to understand its emergence and evolution, and to better see the relations between this renovated form of extractivist politics and those of previous mining cycles. Again, Cerro de San Pedro serves as the local chronicle of this larger "history."

The renovation of extractivist politics can be located in the mid-1990s, in the wake of the large tailings and cyanide spills that the open-pit mining boom provoked around the world. Widely reported and decried by local populations and authorities, these catastrophes revealed that large-scale open-pit mining, with its mass manipulation of toxic materials, was capable of laying waste to entire riverine ecosystems. They showed the invisible gold rush as a runaway experiment in landscape-scale engineering, one that was advancing across the world without much in the way of political control or scientific precaution. And it now involved not only workers but entire populations and territories. Assessing the public's image of the transnational mining industry at the time, a coalition of industry representatives and consultants found that it was "hard to identify any industrial sector (with the possible exception of nuclear power) that features such low levels of trust. . . . Indeed, some polls showed the industry as being held in lower public esteem than the tobacco industry."[21]

It was in the 1990s that mining undertook the considerable task of rebranding itself as sustainable mining—spotted out as a corporate oxymoron by American anthropologist Stuart Kirsch, who shows how it became such a compelling catch phrase in the wake of the Rio Summit of 1992.[22] Among leading Canadian-based corporations, the cause of sustainable mining was championed by corporate officers who saw in it the means of avoiding the imposition of hard-law regulation on their activities.[23] Since its beginnings in the 1990s, this response has greatly developed and institutionalized. Now, the conduct of sustainable mining forms an entire sub-industry within the world of extractives, involving thousands of people, firms, consultants, mercenary academics and government agencies. What sustainable mining or its analogue "Mining CSR" (Corporate Social Responsibility) has consisted of, exactly, depends greatly on the case. But across these some basic ideas emerge: engagement with communities, the management of environmental harms, and the contribution of extraction to capitalist forms of development.[24]

At the beginning of the Cerro de San Pedro project, mining sustainability was not yet fully institutionalized within either Metallica Resources or Minera San Xavier. In the mid-1990s both parent company and subsidiary had just been created and the planning of operations was still at its beginnings. The notion of sustainable mining was still in gestation within the industry. All the same, corporate officers were already clear on the political risks associated with running an open-pit mine. As members of tightly-knit community of corporate miners they would have followed the unfolding of such fiascos as Galactic Gold's Summitville disaster in the late 1980s: the hurried building of the mine, the failure of the retention system to handle a Colorado mountain winter, the killing of aquatic life over twenty-five kilometers of the Alamosa river, the bankruptcy of the company, the legal actions taken against its CEO Robert Friedland.[25] They also knew about others in an even more personal way: Cambior's Omai spill in Guyana, Chemgold's spill at the Picacho mine in California.[26] Recall also that as actors within the neoliberal milieu of the 1990s, these were partisans of deregulation, impatient with the perceived advances marked by environmentalists. The search for unregulated freedom of action informed their choice to develop deposits in Latin America. They were a constituency seeking escape from the "nightmarish bureaucracies in North America, high tax rates, and environmentalists."[27]

Politically speaking, running an open-pit mine in the face of a dedicated opposition was far from plain vanilla. It paid to be attentive to the "social." And so, even as it was surveying the great porphyry beneath the mountain, the company sought to understand the community atop it. A team of contracted anthropologists, sociologists and community consultants was sent to conduct surveys and interviews with the town's inhabitants. They presented themselves as researchers from the state university but the data they obtained was in fact used to produce a social map of Cerro de San Pedro. It made visible to the company management the structure of the communities in the municipality, its factions and fault-lines, its various needs and grievances. Most importantly, it identified who the company could count upon for support, particularly among the families and individuals of influence such as the mayor and the village priest.[28]

It is difficult to clearly establish how many in Cerro de San Pedro actively supported the project at the time. The mining consultancy Behre Dolbear generalized. It reported a community in need and ready to participate in the company's development initiatives (*"they show a great disposition to be incorporated in the projects and programs that mining development can bring"*) but did not release any numbers. Then again, this was not designed as a referendum but as a more technical exercise of identifying and developing strategic partnerships. The surveys showed who might come to an agreement about the sale of outstanding mining concessions and land titles, or the signing of a rental agreement for some three hundred acres of the ejidos' land.

In the event, these arrangements proved the very flash points that ignited the local opposition. With the result that, two years after the announced beginning of operations, the company was being asked to prove itself before the state Governor Torranza's appointed commission of experts. This is when community engagement became political. Or rather, it was instrumentalized to prove the acceptability of the San Pedro mine. As the Canadian activist and researcher Tamara Herman put it, the company began to engineer the production of consent.[29] Initially, the means by which it did so recalled the compensations dispensed by an earlier generation of mine operations. Decent housing, clean water, electricity, roads, education, health care to workers and their families: all of these had been fought for back in the heyday of industrialized mining and institutionalized by the post-Revolutionary state. By the 1990s the companies' responsibility to provide them was a standard convention. For those managers of Minera San Xavier and Metallica Resources who had personally worked in a mining town, these were the measures they reached for to resolve the problem of support. The innovation here was that these compensations were now principally directed at the communities in the project's zone of impact, rather than its workers.

According to the Mexican director of Minera San Xavier, detailing the full list of its compensations would be too long to tell: *"We are supplying the doctor and many of the medicines that they need. We have a program of paving all the roads in the municipality, up in the mountains . . . This will connect all the communities with paved roads. Electricity in each one of them. Sewage and drinking water in each one of them."*[30] It stocked the local schools with desktop computers and internet connections. It paid for the restoration of the villages' two churches. It committed to cleaning up the industrial-era wastes left by ASARCO and the care of endangered cacti. It contracted firms to test local waters and airs for contamination and to monitor the vibrations provoked by the daily blasting. It contributed to employment in the area by providing jobs to local families, over 132 by its own count. It provided a monthly payment to members of the community as compensation for such inconveniences as the mining operation might provoke. And, finally, during Christmas week, it set up a large tent in front of the Church of San Nicolas to hand out everything from children's toys to brand new fridges.[31]

For the company, the costs for all of this were in fact quite modest. At less than half a million dollars, they comprised a fraction of the operation's annual fuel bill.[32] All the same, it proved a most useful solvent against the moral ecology forwarded by the opposition. Investments in the community were offered as proof of a positive reciprocity, as proof of the argument that the project was in fact contributing to the welfare of the community rather than imperiling or harming it.

This argument was unfolded upwards to join larger claims for the new sustainable mining brought by the new transnational mining companies. In Mexico it was perhaps first, and most popularly referred to as "la Mineria Moderna"—a formulation that I find interesting as a historian because of how it rhetorically claims a break with the infelicities associated with previous generations of mining. Large-scale open-pit mining, in this framing, was not the most intensive and potentially damaging form of mining in history, but the one most capable of fulfilling its long-standing promises of development and prosperity. It claimed wide-spread community support and reduced an extensive and broad-based environmental justice movement to "a small opposition. When referring to local opposition, the company stated that this only came from "a handful" or "two or three families."[33] In contrast, the company held up the list of signatories to the much disputed ejido land contract as proof of overwhelming community support.[34] Or it tallied up those who accepted the monthly compensation payments to claim

that "eighty percent back our efforts."[35] Those who organized and participated in the opposition were dismissed as people from outside the community or because they had ulterior motives. *"The people who are against the
mine are politicians, they're people who oppose their own government. Those
are the conflictual ones."*[36] Or they were simply described as congenitally
deranging: they were atavistically opposed to capitalist development and
progress, "a small group of individuals who only raise hell."[37] They were
"minera-fobicos," even "globalofobicos" whose opposition was irrational,
purely reflexive and unconsidered.[38]

Also denied was any responsibility in the various acts of violence suffered by the more visible and local members of the opposition. These included two serious assaults on the lawyer Enrique Rivera Sierra, a machete
attack on Mario Martinez, the drive-by shootings of Armando Mendoza and
his home, and the subsequent rock pitching attacks he endured along with
other local opponents of the mine. The opposition was never able to legally
prove the company's responsibility; the best they could manage was guilt
by association. Likewise, the company disowned any involvement in the
criminalization of its opponents: the incarceration of four student anarchists
for riot during a May Day march against the mine in 2007, or the imprisonment of the community leader Pedro Rebolloso for demonstrating against
the mining project in front of the Gubernatorial palace. Warned of the government's determination to jail him as the intellectual author of these
crimes, and sharply aware of the new decades-long sentences being dealt
to community organizers across Mexico, the lawyer Enrique Rivera Sierra
sought political asylum, in Canada of all places, the same home jurisdiction
as the company at the root of his problems. The irony of the situation was
not lost on the Refugee Board judge who heard his case and ruled to accept
Rivera Sierra as a refugee for political persecution.

As for the environmental harms and risks posed by its San Pedro mine
the company argued that the situation was fully under control. Cyanide—a
substance of mortal toxicity in the iconography and argument of the
opposition—was held to be perfectly safe. "We have put cyanide in its proper
place," lectured the local director of Minera San Xavier, "you can drink [the
leaching solution] right off the pipe and nothing will happen to you. Nothing more than what would happen to you from smoking a cigarette."[39] Nor
did it believe that there was grounds for concern about the more chronic
harms threatened by operations. The mine recycled its waters, and thus did
not draw from the valley's threatened aquifer. The piles of overburden that
now filled the local arroyos to the brim were not sulfidic, nor were they

charged with heavy metals. "The rest that you hear is all lies."[40] In their media interventions, senior management of the company seemed genuinely taken aback that their company should be held as an agent of ecocide.[41] How could it? It had won awards for its environmental management practices. It met its obligations under the industry's International Cyanide Code. It was certified for environmental sustainability by the International Standards Organization under norm ISO-140001.

But perhaps the most remarkable act of denial conducted by the company was the denial that they were physically dismantling the Cerro de San Pedro. With the reduction of the mountain to an open pit some eight hundred meters wide and six hundred meters deep, this took some doing. What was disputed was the very existence of the Cerro de San Pedro itself. The company hired a local historian who researched the issue. He found that there was no connection between the mountain mined by Minera San Savier and the Cerro de San Pedro that featured so emblematically in the history of the region.[42] Close to a decade later, company management was still denying that "the Cerro de San Pedro is the official name for any of the hills in the area."[43] A strange kind of double erasure was at work: an erasure of Cerro de San Pedro from history, an erasure that enabled its physical removal from the landscape.

The last piece to be added in this portrait of the politics of appeasement enacted around the San Pedro mine did not come from a company initiative. It involved, instead, the work of the Canadian state, a state that arguably has contributed more than any other to the representation of transnational mining as sustainable and responsible.[44] On a number of occasions defenders of the Cerro de San Pedro had demonstrated in front of the Canadian embassy, a repeated and favored venue for remonstration and the appeal to the justice of the Canadian state. The response of its consular staff was to try to attempt to reconcile the parties. In the argot of the time it hoped that a win-win solution could be arranged between the stakeholders. In 2013 it was the turn of the newly appointed CSR Counsellor of the government of Canada to see if she could broker an agreement. She met with members of the FAO and explained that she had come because of the notoriety of the case. "Our objective here" she was remembered as saying, "is for you people of the opposition come to an arrangement with the mining company . . . is for the mine to continue working but that it should make commitments, that it should behave better . . ." Her proposal was to hold a number of workshops that would train members of the opposition in how to most productively ask the company for what they were after. ("What you need to say. What you need to ask.") The Counsellor got no further than that modest proposal. People around the table were not interested in this kind of choreographed and paternalistic petitioning to corporate goodwill. They reminded her that their struggle over the previous eighteen years was for the cessation of the project and for the proper conduct of justice.[45] This, the Counsellor admitted, the Canadian state could not deliver. The mediation exercise was brought to an end, and she flew back to Ottawa.[46]

By 2013, the company was almost finished with Cerro de San Pedro. The mine had been in operation—with the exception of those three weeks in the fall of 2009—for seven years. With an announced life of mine of eight to nine years, the end of operations was on the horizon. The majority of the porphyry had been removed to the leaching piles and the greater part of Cerro de San Pedro's remaining stocks of gold and silver had been extracted. As for the opposition, it too was spent. When the company renewed operations after the 2009 shut-down, that is when it achieved "the quick normalization of the mine's operations," many abandoned the struggle as a lost cause. Only a few stalwarts remained in the fight, notably the original core of people from the community of Cerro de San Pedro. And even they fell prey to in-fighting and acrimony, accusing one another of being spies to the company or selling out to cash in on whatever compensatory payments that might still be had.

For Tonantzín Mendoza, the daughter born to Lola Rocha and Armando Mendoza upon their return to Cerro de San Pedro in the nineteen-eighties, it was an abandonment. Her father was now dead, of kidney failure that she suspected was aggravated by the contamination thrown up by the mine. The mountain was destroyed. She did not welcome the continued attention on the case nor did she much care for its status as an emblematic case of environmental injustice. She felt used, made to stand as an exhibit. *"Estamos jodidos, déjanos en paz"* ("We're fucked, leave us in peace").

20 The Third Death of Cerro de San Pedro

When I headed back to Cerro de San Pedro in May 2018, I was admittedly looking for closure. Minera San Xavier—New Gold Inc. was finishing up with the Cerro de San Pedro, the mine, that is. The blasting and the excavations and the constant press of drilling and hauling, all that had ended the year before. There only remained a skeleton crew to tend to the flushing of the heap leach piles, a year-long operation of hosing the now mountainous pile of tailings with clean well-water to wash the cyanide solution out from the rock. In the offices, a small community relations team was working on the implementation of a local tourist development plan. It was part of the company's scenario of a sustainable post-extraction future for the community. My own plan was to spend time looking up and talking with old acquaintances and friends, and even some of our old opponents. The general idea was to absorb the scene at the end of mining's history in Cerro de San Pedro, to reflect on this, to tie up the narrative threads, and generally come to conclusion.

The search for closure, I was quickly reminded, might be a habit of narrative (*how does the story end?*) but not of history. Histories of places and people are not finished so easily, even in those places marked by ruin. Destruction is not the same thing as annihilation. Life—of peoples, of ecologies, and yes, of mountains and the earth—as they say, goes on. I do not mean this in a facile or roseate way. The future dawns assured by the continuation of life at Cerro de San Pedro have hardly been exempted from injustice, sorrow, want, or the invasive touch of contamination. The history of Cerro de San Pedro, that everymine for mining districts across the continent, gives depth and heft to that fact. It shows ruination as a process of the historical longue durée. Begun in the sixteenth century, cyclically renewed, augmented, and reconfigured over four centuries of precious metal mining, ruination also goes on.

The mountain of Cerro de San Pedro is today half consumed. As predicted the impact of mining has been "*adverse, direct, permanent, localized, proximate, irreversible, unrecoverable, without means of mitigation, and with a high probability of presence and critical effect.*" The excavation of the porphyry has cut the mountain in half. The southern flanks of the neighboring peak of Cerro Barreno have been radically excised. As the opposition's iconography had warned, the mount that served as San Luis Rey's pedestal has disappeared. The mines of Cerro de San Pedro no longer serve as his support but instead threaten to pull him into hundreds of meters of free fall down into the abyss of the pit. *"Ya acabarón con el Cerro"* ("They've finished the Cerro off") notes one former activist.[1]

But this is not fully true. Or let's say that the story of Cerro de San Pedro is not fully over. Having been subjected to dismantlement, the history of the mountain continues elsewhere, divided. A minuscule fraction of the mountain's mass—its gold and silver—has left the country. In volumetric terms the 25.5 tons of gold and 637 tons of silver extracted in mining's last run through the mountain would make, assembled, two cubes: the first about the size of a small fridge (1.4 cubic meters), the second the size of two large mining trucks (ca. 60 cubic meters).[2] Where this material is now no one knows for certain. However, if we apply the current proportions for precious metal consumption we can get a rough idea: 19 percent into jewelry, eight into manufacturing applications, and a full 73 percent into bullion. Having been extricated from earth's matrix, concentrated, refined, and fully commoditized, this last slice of the mountain's treasure finds itself back underground in guarded vaults. There immobilized, it serves to support speculation and trade in leveraged investment instruments.

A much greater mass of the mountain, its top third of limestone capping, now lies rubble as waste material heaped up in great slanted piles, flat-topped and filling the arroyos to the brim. The remainder, the largest part, the material of interest treated on the leach pads, now constitutes a new mountain a few kilometers down valley. The company has even Christened the new formation with a new name, "El Cerro del Porvenir" (literally "Prospect Mountain" or "The Mountain of the Future"), though it is unclear whether it has yet been gazetted into Mexico's official toponymy. The material prospects of this mountain-sized pile of waste, was of course, a central issue in the polemics of the last decades. This is not inert material. Packed with sulfides and heavy metals, exposed to air and precipitation, it is geochemically alive and going somewhere. It will metabolize and leach and circulate, albeit extremely slowly. The biochemical and geochemical processes of acid-generation and heavy metal leaching have tremendous temporal inertia. It is a slow catastrophe that will be an ongoing source of sterilization and contamination for millennia. It is true that there currently exist filtering and treatment systems that can assure that this massive and slow release of contaminants does not enter into local waters. The real design problem, however, is time. As Houston Kempton and his colleagues. write, what needs to be crafted is a perpetual water management system, something that can last as long as the ten thousand-year Clock of the Long Now but do more than just keep time.[3]

The company always put a lot of store in the capacity of its system of dykes and membranes to keep this material out of the city's aquifer. The temporal specifications of their design are unclear: years? decades? Perhaps. But no further than that. Already local waters are beginning to gnaw away at these enclosures. The rains, when they come, strike with particular intensity in this part of Mexico, gouging out and churning through the gravelly material that makes up the retention walls and dykes. In 2014 a particularly strong rainstorm dumped enough water as to over-top the dykes and send a first flush of escaping material out of its impoundment.[4] At the time Minera San Xavier had a full complement of workers on hand to tidy up and shore the dykes back up. In a year or so, the last worker will leave and the waters will have free rein.

The overall point is that this story, this story of matter, chemistry, and the elements, in still in motion. It projects a storyline of contamination's slow violence: at first undetectable, then endurable, and eventually unbearable. The sterilization of the land will last for a long time to come. The miners' mark on the land is temporally speaking a deep one. The play of time and water will determine how far this sterilization process will extend over local area. But it is in motion; it is in play.

Minera San Xavier fully shuttered its operations in 2019 and with that the last chapter of over four hundred years of mining in the Cerro de San Pedro was closed. Whether this proves to be the final or simply the latest chapter remains an open question. The struggle to preserve the churches and community of Cerro de San Pedro forced the company to redesign its contours to maintain a small peninsula of land upon which the village is currently perched. This has also preserved a good part of the porphyry. This can be seen quite clearly now that the mountain has been cut in half: the bulk of the porphyry extends diagonally down beneath the two churches and the community. There remains a treasure to be disinterred. The concession for the deposit will remain in force for another thirty years after which time it can be easily renewed for another fifty. That buys the mining industry ample time to find a company willing to take on the task, for new techniques to develop to make it feasible, or for scarcity to deepen sufficiently to make it worth it. "We all know that mining is cyclical," commented a functionary working for the SEDECO (Mexico's ministry of economic development), "as an engineer, I can tell you that there remains a great deal to exploit. Other interested parties will come. Cerro de San Pedro was born a mining town and will die a mining town."[5] Considering these possible storylines, a San Luis-based environmentalist thought that the industry simply needed the time for the opposition to die out. "The people have most strongly defended the village are the oldest people and they're all dying now."[6]

In that week in May 2018, it looked like the fight was over. Whenever I raised the subject of Cerro de San Pedro among former activists it felt like putting a match to wet wood. Years earlier, when the opposition was most mobilized and active, the smallest episode or the most incidental question would have sparked a fiery round of talk, speculation and proposals. The core work of the movement had always been to activate concern about the invisible harms of chronic environmental risk and damage. But while my friends and former compañer@s acknowledged that these were still very much in play, they themselves had trouble mustering themselves to the struggle against this slow violence. They were swamped in the urgencies of the no-holds-barred narco-war. Not the soft politics of appeasement, but the extreme and performative violence that shocks people into terror and submission. Like too many other places in Mexico, the last years have transformed San Luis Potosí into a theatre of violence of the most gruesome kind. Dismembered bodies have been hung from highway overpasses in full view of the morning traffic. Corpses have been dumped in the plazas and public halls. The dusty country road out to Cerro de San Pedro was now a dumping ground for victims, thrown into quickly excavated pits and not always buried. "If the mining company had arrived now, we wouldn't even be talking about it," commented Enrique Rivera Sierra, "I hate to say it but I feel as if this is the end of the story of all of this."[7] He may be right. Given the overlapping geographies of mineralization and drug cultivation, dozens of transnational mining companies currently operate in the midst of the narco-zones and strongholds of the Sierra Madre Occidental. It is far from clear how they manage their relations. The only thing that is clear is the almost perfect silence concerning their operations.[8]

Up in Cerro de San Pedro a less disquieting tranquility had returned to the community now that operations were over. The ceaseless drilling and roar of the haul trucks' diesel electric engines was gone. There would be no roar, concussion-blast, and billowing dust cloud at three o'clock in the afternoon. Those most active in the fight against the mine found themselves working alongside former supporters, joining in the same meetings of the *ejido*. It was cohabitation. Relations are cordial, though the lingering memories of assault by rock, machete, and gunfire mean that true reconciliation are still unattained.

One of the founders of the movement for the life of Cerro de San Pedro felt that that struggle for justice in extractive politics had flown elsewhere. He himself was involved in the formation of a national coalition of communities in resistance to mining (REMA—Red Mexicana de Afectados por la Minería). He toured the country from Baja California to Oaxaca relaying what the FAO had learned over the years—the strategies, the pitfalls, and so on. Members of the REMA were working hard on the root of the issue. They sought a serious reform of Mexico's mining code, the legal and institutional regime erected in the 1990s to give life and support to large-scale transnational mining. With members of the Mexican congress and senate, a coalition of civil society organizations, including a vestigial representation of the FAO, was drafting various versions of a citizen's mining code. "The problem is fundamentally one of sovereignty," he noted. "We've given up close to half of the nation's territory to the mining companies. They can do what they want, wherever they want, so long as they get the metals out."[9] The proposed mining code would challenge the position of mining as the ruling imperative of Mexico's political economy. Mineral extraction would be embedded within a frame of human rights, environmental protection, and democratic governance. Without claiming responsibility or authorship, he placed the socio-ecological conflict around the Cerro de San Pedro at the beginning of a historical and nation-wide trajectory.

Seen in these broader terms, it is clear that the history of struggle around contemporary extractivism in Mexico, and across Latin America is far from finished. The continent, indeed the world, remains in the thick of the resource boom that began in the 1990s. The movements for territorial protection, for local control or sovereignty over "our natural commons" have lost none of their dynamism or vitality. The referendum movement, so powerful in the indigenous heartlands of Mexican and Guatemalan Mesoamerica, is currently sweeping across Colombia, with municipality, corregimiento, and cabildo organizing plebiscites on whether or not to accept large scale mining in their territories. On the other hand, the industry's efforts to expand the frontiers of extraction are likewise unabated. Efforts to push past depletion through new configurations of capital, technology and energetics continue. The fully automated mine, with the massive equipment run remotely from centralized control stations in Toronto or Sydney or Santiago is currently being assembled.[10] The exigencies of scaling up and acceleration have made the problem of energy supplies as keenly felt as ever. In response, the mining industry has developed a serious interest in boosting its traditional carbon-based fuel package with massive solar arrays and hydro-electric development. Finally, after decades of pushing the frontiers of intensification, the industry has returned to an older mode of pushing mining's frontiers outward into the new spaces of deep-sea trenches and even nonterrestial spaces of asteroids and the moon.

· · · · · ·

So the multiple threads of the story refused to be tied off. They extended and ramified into the future and Cerro de San Pedro, whether mountain, mine, or place, was not ready for a eulogy. Its history, on the other hand, was evident everywhere one cared to look. Against the brooding backdrop of the Cerro de San Pedro were the signs of past cycles of mining, along with an almost constant evocation of past scenes and people: the Guachichils and Miguel Caldera, the King under the Mountain, the Revolutionary dynamiting of the municipality, ASARCO's fire, the corporate insignia of New Gold. Jumbled and layered atop one another they composed what Mumford had called mining's mark on the land.[11] Here are, to close, four views of that scene as it appeared in May 2018.

The first comes from the track leading to the neighboring village of Monte Caldera, looking north. The scene is crowded by rubble, mainly the remnants of some abandoned houses. One arched doorway still survives but the walls around it are half gone, broken up by piles of quarried stone. The last time these walls would have been raised would have been about a hundred years ago. When the stones were first dressed I do not know, but it's not unimaginable that some might have been cut in the first bonanza of the early seventeenth century. They've fallen down and been picked up a few times since then. Today, people are using concrete to hold things together. The place has been emptied of people for long enough that it is now overgrown by mesquites, nopales, and fluted cacti. It's a resurgent ecology but of the kind able to root itself in a land whose soils were washed out by centuries of erosion that have followed the removal of the trees from the area. The bared sheets of blue limestone peep out here and there. Behind all of this the pit face and the massive piles of waste rock fill the horizon.

The word that kept coming to mind was ruination. Ruination is the term adopted by the likes of Shannon Dawdy, Laura Ann Stoler, and Gastón Gordillo (following Walter Benjamin) to describe the process by which capitalism and colonialism create ruin.[12] The scene shows the ruinations of mining as a recurrent process, a sedimentation as newer formations layer themselves upon, draw from, and reuse, the remains of older formations.

The second view looks northeast across the plaza of San Nicolas. On the right is the facade and atrium of church of San Nicolas. To the left is what used to be Perico Rangel's house. Since his death a couple of years ago it was acquired by an entrepreneur from San Luis who has remade it into a restaurant—"La Ultima Mina" ("The Last Mine"). Between the two, peeking out above a ruined wall are the waste piles again, and to the side a small edge of the pit face.

"The Last Mine" is not the only public reference to the long history of mining in the village. There is "La Victoria," named after the adit begun after the Independence war and finished by Towne mines. There is La Conquista and in front of it, a full muralized treatment of Guachichil leader Gualinamé's surrender to Miguel Caldera. For the "insumisos" there is "Pueblo Bravo" and "El Charapé," the last the name of the fortified pulque drunk by Tarasco miners in Spanish times.

As for the plaza in front of the church of San Nicolas, it is quiet now except for the birds. But throughout its past it was one of the central spaces of the political history of Cerro de San Pedro. It has served as a space of propitiation and succor in the moral ecology of miners. It was here that they assembled before descending into the valley to take the city of San Luis in 1767. It was here that in May 1911 the miners crowded together to face down the detachment of federales sent to bring them to order and then proceeded dynamite the palacio municipal. It was here that in September 1995 the company and the priest convoked the community assembly that touched off over twenty years of clash and polemic: the visit of the EZLN, Carlos Monsiváis' public address, the assaults and attacks, the mass of salvation organized by the company to pray for the reopening of the mine. For all its silence, the plaza still holds all those acts and passions.

The third is a panorama, from one of the few viewpoints in the valley where one can take in the full extent of mining at Cerro de San Pedro. It looks west from the eastern mojonera or boundary stone of the municipality. It is dominated by the pit face. You can't see the bottom of the pit but one does see the larger sweep of the mine face cutting a jagged circle across both El Barreno and San Pedro. Again, the waste piles and, in front of them, the surviving peninsula of land upon which perches the village and its two churches.

The giant dissection effected by the pit reveals in its strata, colors, and texture the geological drama that took place in this not particularly particular Mexican mountain all those millions of years ago. One clearly sees the edge of the explosive contact between the cell of hot mineralized and acidic material and the cool basal bed of limestone. It is marked by an angry red band tracing a long arc across the pit face. Above it the stratified synclines and anticlines created by the Laramide orogeny. Below it, the reddish material of the porphyry, rough-grained, arenaceous, and constantly sloughing off. Here too is revealed in all its phases, the long history of extraction. At different points on the pit-face one can see small inky-black holes. These are the points where the pit face has sliced across the tunnels excavated by miners since the seventeenth century. They include the rather larger and well-cut rectangle of the Socavón de la Victoria, first designed to access the deepest veins of the mine but now emerging into the light of day more than a hundred meters above the pit-bottom. When compared to the sweep of the pit face, it's all one needs to see to understand the material differences between hundreds of years of tunnel mining and a bit more than ten years of mass-extraction.

The final view looks westwards out from the steps of the church of San Nicolas. This was pretty well exactly where I sat with Juan Carlos, back in December 2001, chatting about the early history of Cerro de San Pedro and its new prospects under Metallica Resources. The view now is a portrait of absence. Behind the spire and dome of the parish church but dominating the scene are the mottled dun—green slopes of the Cerro Barreno, the smoothly rising curve of its skyline cut suddenly by the eastern edge of the pit. One does not see the pit. Nor can one any longer see the full rounded arc of the Cerro de San Pedro. But what the extending skyline suggests, memory fills. The loom and shade of the mountain is gone, replaced by the empty blue desert sky, but I can see it even as I draw. It provokes a strange oscillation, a flitting between what was and what is no longer. It is the view produced by the meeting of history and extraction.

Acknowledgments

This book could not have been written without the contributions of all kinds from many people. It gives me great pleasure to thank them here, at last.

The book roots back to the first years of my deep and long-lasting friendship with Juan Carlos Ruiz Guadalajara. It was while we were both doctoral students doing research at the Archivo General de Indias in Seville that we decided to work together. From that decision came the move to San Luis Potosí, the encounter with the full-scale of contemporary extractivism, and the motivating concerns of the present book. Juan Carlos presents an unusually precious alloy of political engagement and scholarly erudition. He can be fierce, but he is also one of the most loyal and great-hearted people I know. For all these qualities, for all the conversations we have had, for all his help in finding and sharing documents (including many stints in his exceptional library), and for his unreserved friendship, I am deeply grateful. All of these have animated and sustained the project over the twenty years of its development.

In the fall of 2006 I met David Schecter, another key figure in the origins and early development of the book. At the time Dave was in his first year as an undergraduate at McGill. In a lecture, I had relayed the observation made by Spanish officials in Zacatecas that work in the district might cease, not for the depletion of metals but for lack of fuelwood for the smelters. That same afternoon he was in my office asking for more details, which I did not then have. From that point, and for the next years, we worked together on many of the key foundations of this book. We coresearched an article on the historical links between colonial-era mining and deforestation. Dave also played a critical role in pulling together the student-based research collective (MICLA) that documented the activities of Canadian mining corporations in Latin America and fed its results into a broad range of venues: Canadian parliament, public fora, and the media. Throughout, he was a key organizer in different solidarity efforts, for the specific case of Cerro de San Pedro, of course, but for many others besides. It was with Dave that we began to discuss and sketch out a history that would give sense to what was unfolding before us. It was in the course of these exchanges that the basic outline of the present book was set. We never realized our project for coauthoring the book that we had originally imagined. In the time that it took to research and write for what is printed here, Dave went on to work on Parliament Hill, finish a law degree, and embark on a career as a Crown prosecutor. Even from a distance, however, he continued to keep tabs on the project and was a vital interlocutor as it progressed. Dave may not have written this book, but his spirit and intellect are here throughout.

Others, in their moment and in their way, have also been important contributors. Over the ten years of its work (2006–16) MICLA ran on the energy and brainpower of numerous undergraduate students at McGill University. They include Claire Lyke, Alix Stoicheff, Arthur Phillips, Cleve Higgins, Mary Roberts, Bronwyn Lira Dyson, Sean Phipps, Ella Myette, Aidan Gilchrist-Blackwood, Kathleen Whysner, and Jason Hirsch. While their collective efforts largely surpassed this particular project, their findings helped sound and frame the contemporary precious metal boom that is the subject of the third part of this book. Alix Stoicheff took the photograph that appears on the book cover. Two of my doctoral students produced high-quality historical research on subjects linked to the early history of mining at Cerro de San Pedro: Laurent Corbeil, on the indigenous migration to the region during the first mining boom of the 1590s–1650s, and Saul Guerrero, on the chemical and material history of precious metal refining in the pre-industrial period. Their work and findings stand alone. I was very glad to be able to draw upon some of their results here.

As a history born at the conjuncture of research and activism, one of the most fertile environments for the development of this book were the numerous meetings and exchanges organized to understand, and push back against, the current extractivist boom. These spaces brought together people who experienced the incursions of extractivism directly, organizers from civil society, and publicly minded researchers. I would like to thank here, for their insights and our conversations, the following colleagues and compañer@s: Gerardo Aiquel, Pierre Beaucage, Bonnie Campbell, Alain Deneault, Lorena Gil Barba, Tamara Herman, Veronica Islas, Jamie Kneen, Marie-Dominick Langlois, Ugo Lapointe, Lazar Komforti, Marie-Eve Marleau, Jen Moore, Isabel Orellana, Dawn Paley, Pedro Reygadas, Enrique Rivera Sierra, Marta Rivera Sierra, Etienne Roy-Grégoire, William Sacher, Stephen Schnoor, Vivianne Weitzner, and Anna Zalik. Wide-ranging conversations with those who fought for the Cerro de San Pedro drew me into the deeper dimensions of what was at stake in this last chapter of its history. Ana Maria Alvarado García, Hector Barri, Mario Martínez, Armando Mendoza (q.e.p.d.), Tonantzín Mendoza, Gabriel Muñoz, Marcos Rangel, Sergio Serrano: gracias por su confianza.

I am also, of course, tremendously grateful to my university colleagues. I begin by thanking my colleagues at McGill. For a good number of years, chairs and colleagues at the Department of History and Classical Studies not only tolerated my excursion into the work of public interest research and activism, but in fact supported it. Catherine Desbarats, Elizabeth Elbourne, Catherine LeGrand, Nicolas Kosoy, Jason Opal, Leah Temper, and Gavin Walker all read early sketches and drafts and encouraged me to continue. Leah Temper and Ismael Vaccarro provided personal introductions to the indiscipline of political ecology. Over the years, different pieces of this book, some very preliminary and provisional indeed, have been presented at a number of academic meetings. I would like to thank the organizers and participants at the following venues: the American Society for Environmental History (Tallahassee meeting, 2009 and Madison meeting, 2012), the Colegio de San Luis Potosí, the Johns Hopkins University, the Post/Extractivism Workshop, the Rachel Carson Center (Munich), Saint Mary's University (Halifax),

Simon Fraser University, the University of British Colombia, the Universidad de Panamá, York University (Toronto), and Yale University.

Once the book was assembled as a complete manuscript, it passed through a number of iterations until it reached its final shape. The first draft was submitted, prematurely it turned out, to an academic press and firmly rejected. All the same, I did want to thank the two anonymous readers for their work in reviewing the manuscript and giving their frank response. It helped me see more clearly what was missing. Then followed readings by Alan Greer, François Furstenberg, Ray Craib, and Bob Whitney (twice!). With each reading, more reconsideration, more rewriting, progress. Thanks to you all. But the best, perhaps, was for the last. On Bob Whitney's recommendation (thanks again, Bob), I submitted the manuscript to the inestimable Elaine Maisner at the University of North Carolina Press. In her friendly and no-nonsense way, she pushed me to figure out the main line through the material. The manuscript was then sent to the external readers. Both delivered constructive reports. I must, however, give particular thanks to one reader who provided an extraordinarily fulsome and sharp-eyed review. Reading their comments, and going through their remarkably thorough mark-up of the text, was an epiphany. The book had to be entirely revised one last time, but now I knew what needed to be done. I'm pleased with the results, and while, of course, they have no responsibility for whatever shortcomings that may persist here, their input has been invaluable. Anonymous reader one, you have all my gratitude. At the UNCP I would also like to thank Andreína Fernández for fielding my queries and assuring a smooth movement toward production. Geoffrey Wallace, a prize-winning environmental historian in his own right, produced the maps and line drawings that accompany the text.

Finally, my deepest thanks to Mylène, Emile, Marike, and Sacha who found themselves sharing their lives with a very long-term project that was not their own. They have done so, for many, many, years now without ever begrudging it. Now it is done. Now we can have a summer free of *le livre*!

Notes

Introduction

1. Marcos Rangel in Valladolid, "Muerte y vida de Cerro de San Pedro."

2. Interview with Aristeo Gutierrez, March 2008.

3. Aschmann, "The Natural History of a Mine," 175–77.

4. Langue and Salazar-Soler, "Bibliografia minera"; Langue, "Bibliografía minera colonial."

5. Brown, *History of Mining in Latin America*; Dore, "Long-Term Trends in Latin American Mining."

6. Ontiveros, *Transformaciones del paisaje*; see also Ontiveros and Merodio, "Neocolonialismo y minería."

7. Keeling and Sandlos, "Ghost Towns and Zombie Mines."

8. Goody, *Metals, Culture and Capitalism*; Salazar Soler, *Anthopologie des mineurs*.

9. Tepaske and Brown, *New World of Gold and Silver*, 26.

10. Gunder Frank, *ReOrient*; Wallerstein, *The Modern World-System*; Tutino, *Making a New World*; Moore, "Silver, Ecology, and the Origins of the Modern World."

11. Lasky, "How Tonnage and Grade Relations Helps Predict Ore Reserves."

12. TePaske and Brown, *New World of Gold and Silver*, 140, 135; INEGI, *Estadísticas Históricas*, 470; USGS, *Mineral Commodity Survey*, 2015, 2018.

13. Jonsson, *Enlightenment's Frontier*, 167–87 and "Origins of Cornucopianism"; Mukerji, "The Great Forest Survey of 1669–1671"; Warde, "Fear of Wood Shortage"; Wing, "Keeping Spain Afloat."

14. Bellamy Foster, "Marx's Theory of Metabolic Rift."

15. Martínez Alier, "Ecology and the Poor"; Muradian and Martínez Alier, "Global Shifts in Social Metabolism"; Stoll, "A Metabolism of Society"; Clark and Bellamy Foster, "Global Metabolic Rift"; Schott, "Uban Environmental History"; Walter, *Toxic Archipelago*.

16. Ye, van der Ploeg, Schneider, and Shanin, "Incursions of Extractivism"; Gudynas, *Extractivisms*.

17. Braudel, *Civilization and Capitalism*, vol. 1, 337–340; Smil, *Energy in Nature and Society*.

18. Parsons, "Gold Mining in the Nicaragua Rain Forest" and "The Settlement of the Sinu Valley of Colombia"; West, *The Parral Mining District*; *Colonial Placer Mining in Colombia*; and *Sonora: Its Geographical Personality*; on emissions and effluvia from silver mining, Robins, *Mercury, Mining, and Empire*; Guerrero, *A Chemical History of Silver Refining*; on working environments, Brown, "Workers' Health and Colonial Mercury Mining"; on energetics Chantal Cramaussel, "Sociedad colonial y

depredacion ecologica"; Craig, "The Ingenious Ingenios"; Dore, "Long-Term Trends in Latin American Mining"; on mining-driven agrarian change, Scott, "The Contested Spaces of the Subterranean"; Moore, "Silver, Ecology, and the Origins of the Modern World."

19. French, *A Peaceful and Working People*; Wyman, "Industrial Revolution in the West"; Azucena Rodríguez López, "Andamos en las entrañas de la tierra"; Jock McCulloch, "The Politics of Silicosis."

20. Garibay et al., "Reciprocidad negativa en el paisaje minero"; earlier, Steven Bunker, *Underdeveloping the Amazon* proposed the concept of "unequal ecological exchange" to describe the ecological dimensions of colonial and neocolonial forms of resource exploitation.

21. Klubock, *La Frontera*, 5. See also Varese, "Ethnopolitics of Indigenous Resistance," for an earlier use of the concept.

22. Thompson, "Moral Economy of English Crowd."

23. Wakild, "Environmental Justice and Environmental History"; Martinez Alier, "Environmentalism of the Poor"; Carruthers, *Environmental Justice in Latin America*.

24. Nash, *We Eat the Mines*; Taussig, *The Devil and Commodity Fetishism*.

25. On the joining of the social and ecological histories of labor, see Soluri, "Labor, Rematerialized"; Peck, "The Nature of Labor"; McEvoy, "Working Environments"; Barca, "Bread and Poison"; Rogers's excellent monograph study of the sugar fields of Brazil, *The Deepest Wounds*; and Santiago, *The Ecology of Oil*.

26. Observatorio de Conflictos Mineros de América Latina, https://www.ocmal .org/; EJ Atlas—Global Atlas of Environmental Justice, https://ejatlas.org.

27. Urkidi and Walter, "Dimensions of Anti-Gold Mining Movements in Latin America," 684–85.

28. Building on older struggles over the land, these constitute what Enrique Leff calls the new politics of territory, *Saber Ambiental*.

29. Bebbington and Bury, *Subterranean Struggles*.

30. "Declaración de Huancayo," Huancayo, Peru, July 23, 2011.

31. Guha and Martínez-Alier, *Varieties of Environmentalism*; earlier formulation in Martínez-Alier, "Ecology and the Poor"; overview and intellectual history of the concept in Martínez Alier, "The Environmentalism of the Poor: Its Origins and Spread"; as related to mining conflicts, in particular, Martínez Alier, "Mining Conflicts, Environmental Justice, and Valuation."

32. Li, "Relating Divergent Worlds," 411.

33. Asamblea El Algarrobo, *Por vida de Andalgalá*.

34. Riofrancos, *Resource Radicals*, 55.

35. Gudynas "Ten Urgent Theses about Extractivism."

36. Acosta, "Extractivism and Neo-Extractivism," 62; Bebbington and Bury, *Subterranean Struggles*, chap. 1; Killoran-McKibbon and Zalik, "Re-Thinking the Extractive/Productive Binary," 538; Sawyer, *Crude Chronicles*, 39; McKay, "Agrarian Extractivism," 200; North and Grinspun, "Neo-Extractivism and the New Latin American Developmentalism," 1484; Gudynas, *Extractivisms*, chap. 1.

37. Coronil, "Speculations on Capitalism's Nature," 75.

38. C. H. Haring, *The Spanish Empire in America*, 258–59.

39. Betancor, *Matter of Empire*; Bigelow, *Mining Language.*

40. Carlos Montemayor, "Amenaza ambiental con Minera San Xavier," *La Jornada*, October 18, 2004.

Chapter 1

1. Aiton, *Antonio de Mendoza*, 182, 183–85, 190.

2. "Relación de Antonio de Mendoza a Luis de Velasco," 40.

3. Sempat Assaoudrian, "La mita'a minera del Virrey Toledo," 39–43.

4. *Recopilacion de Leyes*, Tomo III, Titulo 8, Ley 1.

5. Wey Gómez, *Tropics of Empire*, 40.

6. Cited in Pastor Bodmer, *The Armature of Conquest*, 154.

7. Von Humboldt, *L'histoire de la géographie du nouveau continent*; Sauer, *The Early Spanish Main*; Bauer, *Alchemy of Conquest*, 179–81; Gerbi, *Nature in the New World*, 13, 16, 291. Wey Gómez, *Tropics of Empire*, 411ff.

8. Columbus, *Select Letters*, 4.

9. Cited in Gerbi, Nature in the New World, 16.

10. Cited in Gerbi, Nature in the New World, 16.

11. Columbus, *Diario.*

12. Tepaske and Brown, *New World of Gold and Silver*, 29.

13. Sauer, Early Spanish Main, 28.

14. Columbus, *Diario*, 71.

15. Columbus, *Select letters*, 69.

16. Columbus, *Select letters*, 69.

17. Columbus, *Select letters*, 72.

18. Bridge, "Resource Triumphalism," 2149, 2154.

19. Moffitt Watts, "Prophesy and Discovery."

20. Bauer, *Alchemy of Conquest*, 137–38, 179, 183.

21. Sauer, *Early Spanish Main*, 77–83.

22. Varela and Aguirre, *El juicio de Bobadilla*, 153–54, 158.

23. "Carta patente de los Reyes de Castilla al Almirante Cristóbal Colón," in *Cedulario de tierras*, 105–6.

24. CODOIN, 1st series, vol. 31, 41. 108–9, 176–77, 217.

25. Wing, "Keeping Spain Afloat"; in England Linebaugh, "Enclosures from the Bottom Up," 19; on beaches and rivers in Spain, *Las Siete Partidas*. Part III, Tit. XXVIII. Law V and Benton, *A Search for Sovereignty*, 57–59.

26. Siete Partidas, Part. III, Tit. XXVIII, Law XLV.

27. MacLachlan, *Spain's Empire in the New World*, 15–18; Warsh, "Political Ecology in the Early Spanish Caribbean," 518; Wing, *Roots of Empire*, 97; Wing, "Keeping Spain Afloat," 122.

28. Ramos Perez, "Ordenación de la minería en hispanoamerica."

29. "Instrucciones a Ovando," CODOIN, 1st series, vol. 31, 20; TePaske and Brown, *New World of Gold and Silver*, 30.

30. TePaske and Brown, *New World of Gold and Silver*, chap. 1.

31. TePaske and Brown, *New World of Gold and Silver,* 32, 33.

32. Though in Colombian and Andean regions pre-Columbian gold mining operations were more intensive and required coordination of labor West, *Colonial Placer Mining in Colombia,* 54; Berthelot, "L'exploitation des metaux précieux au temps des Incas," 952.

33. Fernández de Oviedo, *Historia general y natural,* vol. 1, 185; Lane, *Quito, 1599* for coastal Ecuador; and West, *Colonial Placer Mining* for Chocó—Pacific Colombian lowlands.

34. Las Casas, *Historia de las Indias,* vol. 2, 336–37.

35. "Leyes de Burgos," 38.

36. Sauer, *The Early Spanish Main,* 155–56.

37. Scott, "Contested Spaces of the Subterranean," 11; Málaga Medina, "Las reducciones en el Virreinato de Perú"; Jeremy Mumford, *Vertical Empire.*

38. "Leyes de Burgos," 38–39.

39. "Leyes de Burgos," 35, 38–39.

40. Ramos Pérez, "Ordenación de minería en hispanoamerica," 375–77.

41. Whitehead, "The Crises and Transformations of Invaded Societies," 868.

42. Las Casas, *The Devastation of the Indies,* 31.

43. Whitehead, "The Crises and Transformations of Invaded Societies," 70ff.

44. Recopilación De Las Leyes De Los Reynos De Las Indias (1681), Libro Lib IV, tit 22, ley 1 pg 75.

45. West, *Colonial Placer Mining in Colombia,* 66–67; Espinosa Baquero, "Metales preciosos en Nueva Granada," 489; Parsons, "The Settlement of the Sinu Valley," 71–72.

46. Ramírez, *The World Upside Down,* 121–52.

47. Personal communication Nicole Couture, Dept. Anthropology, McGill University, February 2009.

48. Kris Lane, *Quito 1599,* 120–22.

49. Brunhes, *Human Geography,* chap. 5.

50. Sánchez Crispín, counts 143 districts (each containing many mines) for the late sixteenth-century New Spain in his "Territorial Organization of Mining in New Spain," 163. Add to these 91 mines identified in TePaske and Brown, for Española, Puerto Rico, Cuba, Panama, Nueve Reyno de Granada, Ecuador, Peru, and Chile in their *New World of Gold and Silver,* 31, 37–38, 40, 146, 147, 157.

51. Van Buren and Presta, "Inka Silver Production in Porco."

52. Abbott and Wolfe, "Pre-Incan Metallurgy"; Cruz and Absi, "Potosí antes y durante el contacto," 95; Lane, *Potosí,* 10–29.

53. Hernán Cortés, "De los descubrimientos de minas," 35–37.

54. Cited in Lane, *Quito 1599,* 116.

55. Jara, *Guerra y sociedad,* 30.

56. Altman, *The War for Mexico's West.*

57. TePaske and Brown, *New World of Gold and Silver,* 20.

58. Álvarez Nogal and Chamley, "Debt Policy under Constraints," 198; Lynch, *Spain 1516–1598,* 79–82, 201.

59. *Recopilación de Leyes de Indias,* Lib III, Tit. 8, ley 1.

60. *Informe de Antonio de Mendoza a Luis de Velasco,* in Colección de documentos ineditos para la historia de España, 288.

61. Craig, "Silver Beneficiation at Potosí," 272, 282n2; Bakewell, *Miners of the Red Mountain,* 24.

62. Aiton, "Antonio de Mendoza," 181.

63. Moore, "Silver, Ecology."

64. Cole, *Potosí Mita,* 4; Brown, *Mining in Latin America,* chap. 2.

65. Lynch, *Spain, 1516–1598,* 174ff.

66. Cited in Molina Martínez, "Legislación minera colonial en tiempos de Felipe II," 1026, footnote 18.

67. Molina Martínez, "Legislación minera colonial en tiempos de Felipe II," 1019.

68. Lib. IV, Tit. XIX, Ley i, "Que permite descubrir y beneficiar las minas a todos los Espanoles, e Indios, vassallos del Rey," and Ley xiiii "Que los Indios puedan tener y labrar Minas de oro y plata como los Españoles," *Recopilación De Las Leyes De Los Reynos De Las Indias,* 68, 71.

69. Molina Martínez, "Legislación minera colonial en tiempos de Felipe II," 1019; Barrera-Osorio, *Experiencing Nature,* 65–68.

70. Lib. IV, Tit. XIX, Ley xv, *Recopilación De Las Leyes De Los Reynos De Las Indias,* 71.

71. Lib. IV, Tit. XX, Ley i, *Recopilación De Las Leyes De Los Reynos De Las Indias,* 72.

72. Scott, "The Contested Spaces of the Subterranean."

73. Francisco de Toledo, *Memorial y Ordenanzas,* 34.

74. Cole, *Potosí Mita;* Mumford, *Vertical Empire;* Wernke, "A Reduced Landscape," 52.

75. Gade, "Landscape in the Post-Conquest Andes."

76. *Recopilación De Las Leyes,* 308, 311.

77. Cole, *Potosí Mita,* 23–25; Robins, *Mercury, Mining, and Empire;* Brown, "Workers' Health at Huancavelica."

78. Informe de Antonio de Mendoza a Luis de Velasco, n.d. in CODOIN, vol. 26, 288.

79. Scott, "Contested Spaces of the Subterranean," 15.

80. Randolph, "The Lakes of Potosí," 529–36.

81. Craig, "Colonial Water Mills at Potosí."

82. Acosta, *Natural and Morall historie,* 171.

83. Guerrero, *Silver by Fire,* 135.

84. Full treatment of Huancavelica in Robins, *Mercury, Mining and Empire;* Pearce, "Huancavelica: History and Historiography."

85. Robins, *Mercury, Mining, and Empire;* Brown "Workers' Health at Huancavelica"; Gade, "Landscape, System and Identity"; Moore, "Silver, Ecology and the Making of the Modern World"; Scott, "Contested Spaces of Subterranean"; West, *Parral;* Cramaussel, "Sociedad colonial y depredación ecologica"; Folchi, "Los hornos y los bosques."

86. Castañedo Delgado, "Los trabajos forzados en las mines," 837. Both Galeano and Robins draw on Santo Tomas's master metaophor: Galeano, *Open Veins of Latin America;* Robins, *Mercury, Mining and Empire,* 74–75.

87. Capoche, *Relación General de Potosí.*

88. Cited in Robins, *Mercury, Mining and Empire*, 179.

89. Castañedo Delgado, "Los trabajos forzados," 882.

90. Castañedo Delgado, "Los trabajos forzados," 881.

91. Cruzat, *Dialogo sobre el comercio*, 155v, 163r.

92. Castañedo Delgado, "Trabajos forzosos," 886–87.

93. Castañedo Delgado, 819.

94. Betancor, *Matter of Empire.*

95. Hanke, "The Just Titles of Spain," 6; CODOIN, vol. 13: 425–69.

96. Pellicer y Osau, *El Comercio Impedido*, fol. 2r.

97. Acosta, *The naturall and morall historie of the Indies,* 204, 207–8. I have modernized the seventeenth-century translation.

98. Pedro Camargo to Philip II, 1596, cited in Heidi Scott, "Contested Territories," 179; Garcia de Toledo *Dominio de los yngas en el Peru.*

99. Bernabé Cobo, *Historia del Nuevo Mundo*, vol. 1, 292; *Relacion de Las Minas de Çimapan*, 102; Langue, "Formas de trabajo en las minas zacatecanas," 466.

100. Barrett, "World Bullion Flows," 225.

Chapter 2

1. Cox, *The Elements*, 72.

2. Cox, *The Elements*, 109.

3. Cox, *The Elements*, 9, 12.

4. Tardy et al. "La estructura de la Sierra Madre Oriental."

5. Snoke, "North America—Southern Cordillera."

6. Winterburne, *Alteration and Mineralization*; Petersen et al., "Geology of the Cerro San Pedro Gold-Silver Deposit."

7. M. Tardy et al. "La estructura de la Sierra Madre Oriental."

8. Petersen et al., "Geology of the Cerro San Pedro Gold-Silver Deposit."

9. del Campo, *Reseña del Mineral del Cerro de San Pedro,* 4, 6.

10. Winterburne, *Alteration and Mineralization*, 43.

11. Winterburne, *Alteration and Mineralization*, 65.

12. Del Campo, *Reseña del Mineral del Cerro de San Pedro,* 4.

13. Personal communication, Peter Tarasoff, curator of mineralogy, Redpath Museum, Montreal, October 2014.

Chapter 3

1. Altman, *War for Mexico's West.*

2. Bakewell, *Silver Mining and Society*, 6–8.

3. Bakewell, *Silver Mining and Society*, 26–27.

4. On the geography of frontier settlement in Mexico see Bakewell, *Silver Mining and Society*, 20–24; Swann, *Tierra Adentro*; Powell, "Genesis of the Frontier Presidio"; Sánchez Crispín,"Metallic Mining in New Spain," 160.

5. Fernán González de Eslava, *Coloquios espirituales y sacramentales y canciones divinas* (Mexico, 1610), cited in Powell, "Scourge of the Silver Frontier," 315.

6. Santa Maria, *Guerra de los Chichimecas*, 193.

7. Powell, "Chichimecas," 331.

8. Santa Maria, *Guerra de los Chichimecas*, 195.

9. Santa Maria, *Guerra de los Chichimecas*, 191, 193, 195; Powell, "Peacemaking on North America's First Frontier," 221.

10. Velásquez, *Documentos para la historia de San Luis Potosí*, vol.1, 452.

11. Powell, "Scourge," 316.

12. de la Mota y Escobar, *Descripción geografica*.

13. Bakewell, *Silver Mining and Society*, 27.

14. West, *The Parral Mining District*, 86, 87.

15. Ruiz Guadalajara, "Miguel Caldera y la frontera Chichimeca," 52.

16. Velasco Murillo, *Urban Indians in a Silver City*, 5, 64.

17. Powell, "Caldera of New Spain," 325, 331.

18. Ruiz Guadalajara, "Miguel Caldera y la Frontera Chichimeca," 42.

19. Ruiz Guadalajara, "Miguel Caldera y la Frontera Chichimeca," 49.

20. From Relación de Servicios rendered to the King Philip III, cited in Ruiz Guadalajara, "Miguel Caldera y la Frontera Chichimeca," 23.

21. Sego, *Aliados y adversarios*.

22. Corbeil, *Indigenous Migrants on the Urban Frontier*; Sego, *Aliados y adversarios*.

23. Behar, "Visions of a Guachichil witch," 122; De La Torre Villar, *Instrucciones y memorias de los virreyes novohispanos*, vol. 1, 234; Sego, *Aliados y adversarios*.

24. Feliciano Velazquez, *Historia de San Luis Potosí*, vol. 1, 514–15, 517–18.

25. Sego, *Aliados y adversarios*, 16--66.

26. *Relación de Servicios de Pedro Arizmendi de Gongorrón, 1619*; AGI, Patronato 87, no. 3, r .1.

27. Cited in Velázquez, *Historia de San Luis Potosí*, vol. 1, 498–99.

28. Powell, "peacemaking" provides the fullest English-language treatment to date of the *paz por compra* strategy.

29. The full accounts are contained in AGI, Contaduria, 851. The legajo and its organization is discussed in Powell, "Peace-making," 249–50. Total tallies are given in Powell, *Mexico's Miguel Caldera*, 84, 216–19.

30. Powell, "Peace-making," 231.

31. Powell, *Soldiers, Indians, and Silver*, 18–19.

32. Powell, *Mexico's Captain Caldera*, 217.

33. Rivera Villanueva, *Los Tlaxcaltecas: Pobladores de San Luis Potosí*.

34. Powell, *The Taming of America's First Frontier*, 181.

35. Behar,"Visions of a Guachichil Witch," 124.

36. *Indios caciques contra indios Chichimecas salteadores que mataron a dos indios hijos de uno de estos caciques*, 19.02.1594—AMESLP, AM A-44 1594; *Pleito en contra de Cristobal de Solana*. AMESLP, AM 1623 (2), exp. 40; *Informes del caudillo Bernardo Garai sobre los castigos a indios chichimecas sublevados en el Nuevo Reino de*

León. 24.12.1633, AMESLP, AM 1633 (7), exp. 13; *Carta de fray Juan Bautista de Mollinedo,* 1623, AGI, Mexico, 301; see also Sego, *Aliados y adversarios,* 178.

37. *Sublevación de indios Chichimecas,* 01.09.1645. AMESLP, AM, 1645.3, exp. 8.

38. Behar, "Visions," 123.

Chapter 4

1. *Informaciones de oficio y parte Cristóbal Gómez de Rojas,* 1600. AGI, Audiencia de México, leg. 223, n. 13, f. 3r.; Montoya, *Población y sociedad en un real de minas de la frontera,* 107.

2. Corbeil, *The Motions Beneath.*

3. Montoya, *Población y sociedad en un Real Minas,* 107.

4. *Memorial de Lucas Fernández Manjón,* 1v.

5. Vázquez de Espinosa, *Description of the West Indies,* 174.

6. Monroy, *Documentos para la historia de San Luis Potosí,* 122.

7. *Memorial de Lucas Fernández Manjón,* 3v.

8. Garcia de Llanos, *Diccionario de las minas,* 115–16.

9. AHESLP Alcaldia Mayor 1608.1, exp. 11.

10. Langue, "Formas de trabajo en las minas zacatecanas," 478–79; Ladd, *Making of a Strike.*

11. Anon., *Informe sobre el mineral de San Pedro,* 2.

12. Serrano Hernández, "Contrabando y evasión fiscal," 39; AMESLP, AM 1630(3), exps. 1, 2; AMESLP, AM 1629(7), exp. 24.

13. Bros, "Mineral de Cerro de San Pedro."

14. Barba, *El Arte de los Metales,* 3–4, 7; Bros, "El mineral de Cerro de San Pedro."

15. Escamilla González, "Un metalurgista germano en Guanajuato," 111.

16. Bros, "El mineral de Cerro de San Pedro."

17. Salazar, *Las Haciendas,* 86–88.

18. Guerrero, *Silver by Fire,* 55ff.

19. *Venta de una hacienda de sacar oro para azogue,* AMESLP, AM, 1673(3), exp. 2; *Poder a Fernancisco Bernal Lobo para venta de azogue,* AMESLP, AM, 1681 (1), exp. 14,

20. Gilliam, *Travels 1843,* 170; Bakewell, *Silver Mining and Society,* 150ff.

21. AMESLP, AM 1607, exp. 2, 17.01.1607; Salazar, *Las Haciendas,* 274; Urquiola Permisan, *Agua para los ingenios.*

22. Gamboa, *Mining Ordinances of New Spain,* vol. 2, 198.

23. Guerrero, *Silver by Fire,* 55–59.

24. Serrano, *Articulación económica de San Luis Potosí,* Appendix VII.f, 429–34.

25. Serrano, *Articulación económica de San Luis Potosí,* 379, 383.

26. Salazar, *Las Haciendas,* Appendix 2, 485.

27. Salazar, *Las Haciendas,* 282–83.

28. Salazar, *Las Haciendas,* 290.

29. AMESLP, AM 1609.1 exp. 6, 07.01.1609; Feliciano Velázquez, *Historia de San Luis Potosí,* vol. 2, 108; Salazar, *Las Haciendas,* 274.

30. Salazar, *Las Haciendas*, 274; Serrano, *Articulación económica de San Luis Potosí*, 434.

31. Salazar, *Las Haciendas*, 79.

32. Guerrero, *Silver by Fire*, 63.

33. Guerrero, *Silver by Fire*, 63; smelting of ores at Antonio Arizmendi Gogorron's hacienda, 1610 AMESLP, AM C.934, exp. 29.

34. Peak silver production was achieved in 1599; peak gold production in 1618, Serrano, *Articulación económica de San Luis Potosí*, 310, 318. Averages drawn from production figures for 1618–23 in appendix VII.f, 429–34.

35. Serrano calculates that San Luis Potosí registered just shy of 87 percent of New Spain's gold production for the seventeenth century. Serrano, *Articulación económica de San Luis Potosí*, 341, 348; see also TePaske and Brown, *New World of Gold and Silver*, 35–36.

36. Tutino, *Making a New World*.

37. Sempat Assadourian, "Producción de la mercancia dinero"; Semo, *History of Capitalism in Mexico*; Colmenares, *Cali: Terratenientes, mineros y comerciantes*.

38. Le Goff, *Saint Louis*, 644–57.

39. Ruiz Guadalajara, "Vestigios de un prodigio."

Chapter 5

1. Marx, *Capital*, chap. 1, section 4.

2. Velásquez, *Historia*, vol. 2, 135.

3. AGI, Audiencia de México, leg. 223, n. 13, f. 3r.

4. Smil, *Energy in Nature and Society*, 119–37, 174–79.

5. Smil, *Energy in Nature and Society*, 119–38, 155–60, 180–87.

6. The estimate is based on Serrano's average annual production figures, multiplied by the average 2 percent silver to ore ratio given by Guerrero.

7. Salazar, *Haciendas*, 325.

8. Guerrero, *Silver by Fire*, 59, 65; Salazar estimates 17,250 kg/furnace/yr, *Las Haciendas*, 80.

9. *Informaciones de oficio y parte Cristóbal Gómez de Rojas*, 1600. AGI, Audiencia de México, leg. 223, n. 13, f. 3r. Montoya, *Población y sociedad*, 107.

10. Salazar, *Las haciendas*, 112.

11. AMESLP, AM 1607, exp. 2, 17.01.1607; Salazar, *Las haciendas*, 274.

12. West, *Parral*, 43; Bahre, *Legacy of Change*, 145, 147.

13. Studnicki-Gizbert and Schecter, "A Colonial Fuel Rush," 112–13; see Guerrero, *Silver by Fire*, for revision of these estimated ratios.

14. Guerrero, *Silver by Fire*, 92.

15. 100:1 lead to silver ratios at an estimated average ore grade of 2 percent. Guerrero, *Silver by Fire*, 84, 85.

16. Corbeil, *The Motions Beneath*, details the provenance and patterns of settlement and work for these migrants.

17. Salazar again provides the overview, *Las haciendas*, 213–16, 225–32, 329–30.

18. Sempat Assadourian, *El sistema de la economía colonial*; Brading, *Merchants and Miners*; Tutino, *Making of New World*.

19. Birrichaga Gardida, "Hidraulica colonial."

20. Most of the material that follows is drawn from a fuller discussion in Studnicki-Gizbert and David Schecter, "Environmental Dynamics of a Colonial Fuel Rush."

21. Cook, *Demography and Ecology of Teotlalpán*, 30, 32.

22. Studnicki-Gizbert and Schecter, "Environmental Dynamics of a Fuel Rush," 95.

23. Studnicki-Gizbert and Schecter, "Environmental Dynamics of a Fuel Rush."

24. *Ciudades, Villas y Lugares de Mechoacan*, MS 1106-C3, f. 133v and MS 1106-A, 47r.

25. *Informes y licencia de los pobladores de Tlaxcalilla sobre las lluvias* AMESLP, AM 1626 (3), exp. 21, 21.08.1626; *Petición para reparar gran parte de la iglesia que se cayó a raíz de las lluvias* AMESLP, AM 1626 (3), exp. 23, 21.08.1626.

26. *Información sumaria sobre la abundancia o escasez que en esta siudad hay* 20 Oct. 1739, AMESLP, AM 1739(1), exp. 23.

27. Capitán Manuel Pascual de Burgoa, *Plano de la ciudad de San Luis Potosí*, 1794. Original in AGI, Mapas y Planos; Juan Mairano de Vildosola, *Plan horisontal de la ciudad de San Luis Potosí con todos sus pueblos y barrios*. 1799. Reproduced and analyzed in Salazar, *Las haciendas*; see also Guerrero, *Silver by Fire*, 62.

28. Salazar, *Las haciendas*, 473.

29. Velázquez, *Historia de San Luis Potosí*, vol. 2, 144–45.

30. Borah, "Un gobierno provincial de frontera," 537–38.

31. Martínez, "Minas de la Unidad San Pedro," 375.

32. C. R. Boxer, *Brazil's Golden age*, 184.

33. Martínez de Viedsma, "Informe sobre las minas de Taxco," 36.

34. Ulloa, *Noticias Americanas*, 219–20; Gamboa, *Mining Ordinances of New Spain*, vol. 2, 321.

35. Gamboa, *Mining Ordinances of New Spain*, vol. 2, 194.

36. Guerrero, *Silver by Fire*, 69.

37. AHESLP, AM 1635.1, exp. 6, 16.01.1635.

Chapter 6

1. Fabry, *La Rebaja del Azogue*, 187, 188–89.

2. Domínguez de la Fuente, *Leal informe político legal*. Cited in Bustamante, "Trabajo en las minas de Guanajuato," 71.

3. Mourelle, "Minas de Guanajuato," 33, 38, 43–44.

4. Brown, *History of Mining*, chap. 4; Cornbilt, *Oruro*; Castro Gutierrez, *Nueva Ley y Nuevo Rey*; Noblet Barry Danks, "Labour Revolt of 1766"; Tutino, *Making of New World*, 235–36, 490; Brading, *Miners and Merchants*, 276; Higgins, *Licentious Liberty*.

5. Toledo, *Ordenanzas*, Ord. VII: 277; Gamboa, *Mining Ordinances of New Spain*, vol. 1, chaps. 5, 9–12.

6. Reygadas, *Vetas del Lenguaje Minero*, 24.

7. Peter Bakewell, *Silver Mining and Society*.

8. Goodrich, *The Miner's Freedom*, 16, 19.

9. Orozco, *Documentos sobre el conflicto de trabajo*, 27; Danks, "The Labor Revolt of 1766 in Real Del Monte," 153–55.

10. Serrano Hernández, "Contrabando y evasión fiscal"; AGI-Escribania, 868A, exp. 3.

11. Oral history interviews with Marcos Rangel, February 2008, Mario Martínez, December 2014; French, *Peaceful and Working People*, 127–28, 130; Smale, *Labor Activism in Bolivia*, 16–17, 34.

12. Barragán, "Mining Labor and Popular Economy."

13. *Cedula Real 20.11.1599* AMESLP, AM, C-2, exp. 1.

14. *Interrogatorio de Diego de Zamora, ca. 1600*, AMESLP, AM, C-934.6, exp. 6.

15. *Declaración de Diego Muñoz, 06.06.1601*, AMESLP, AM, leg. 1601, exp. 8.

16. *Denuncia de Juan Bautista, 06.11.1621*, AMESLP, AM, 1621.5, exp. 32.

17. *Orders of Pedro de Salazar, Alcalde Mayor, 10.11.1613*, AMESLP, AM, C.925, exp. 23.

18. *Orders of Juan Camacho Jayna, Alcalde Mayor, 30.05.1683*, AMESLP, AM, C.928, exp. 13.

19. *Manifestación de fuelles, 16.04.1651* AMESLP, AM, leg. 1651.2, exp. 3; *Manuel Ortiz de Santa María and Martín José de Iraizós, condiciones para los trabajadores de las minas*, 08.1767, AMESLP, AM, leg. 1767.2, exp.12; *Testimonio de Ignacio de Orgovosa*, 14.06.1791, AMESLP, AM, leg. 1791.2, exp. 3.

20. West, *Placer Mining in Colombia*, 88; Minaudier, "Barbacoas, 1750–1830"; Restrepo, *Minas de Oro y Plata de Colombia*.

21. Higgins, *A Brazilian Gold-Mining Region*.

22. Barragan, "Silver for the World"; Stern, "Feudalism, Capitalism and the World-System in the Perspective of Latin America," 41; Zulawski, *Work and Social Change in Colonial Bolivia*, 86.

23. Tandeter, "Forced and Free Labour in Late Colonial Potosi," 134; Margaret Rankine states that half of the ore extracted Guanajuato's Valenciana mine in the early nineteenth century was "lost" to the miners, "The Mexican Mining Industry in Guanajuato," 35.

24. French, *Peaceful and Working People*, 128ff; Navarro, *Cultura politica de Cerro de San Pedro*; Thompson, "The Moral Economy of the English Crown," 188ff.

25. Ramírez Reynosos, "El conflicto de Pachuca y Real del Monte."

26. Ramírez Reynosos, 'El conflicto de Pachuca y Real del Monte," 557, 566; Danks, "The Labor Revolt of 1766," 143–65; Boorstein Couturier, *The Silver King*.

27. Chavez Orozco, *Documentos sobre los mineros de Real del Monte*, 27, 28.

28. See also Taussig, *The Devil and Commodity Fetishism*; and Nash, *We Eat the Mines*.

29. Ramírez Reynoso, "Conflicto de Pachuca y Real del Monte"; Ladd, *Making of a Strike*; Galvez, "Instrucción particular."

30. Gálvez, *Informe sobre el estado de Mexico, California, Sonora*, 419.

31. Gálvez, *Informe sobre las rebeliones populares de 1767*, 32.

32. Granados y Gálvez, *Tardes americanas*, 446.

33. Galvez, *Informe sobre las rebeliones*, 37, 39; Pérez Navarro, *Cultura politica de Cerro de San Pedro*, 5,6, 84.

34. Pérez Navarro, *Cultura politica de Cerro de San Pedro*, 133.

35. Castro Gutiérrez, *Nueva Ley y Nuevo Rey*, 125–26.

36. Castro Gutiérrez, *Nueva Ley y Nuevo Rey*, 134, 142.

37. Granados y Gálvez, *Tardes americanas*, 447–48.

38. Galvez, *Informe sobre las rebeliones*, 32, 35.

39. Pérez Navarro, *Cultura Politica en Cerro de San Pedro*, 95.

40. Granados, *Tardes Americanas*, 449.

41. AHESLP, 1767.2, exp. 12.

42. Carbajal López, *La minería en Bolaños*, 233–35, 277.

43. Castro Gutiérrez, *Nueva Ley y Nuevo Rey*, 141.

44. Brading, *Miners and Merchants*, 277.

45. Cornblit, *Oruro*; Hernández, "Disciplinar la frontera," Minaudier, "Barbacoas 1750–1830," 95; Junho Anastasia, *Vassalos Rebeldes*; Brown, *History of Mining in Latin America*, chap. 4.

Chapter 7

1. Fay, *Glossary of the Mining Industry*, 94, 97; *Diccionario de Autoridades*, vol. 1; Gamboa, *Mining Ordinances of New Spain*, 41.

2. TePaske and Brown, *New World of Gold and Silver*, 54, 59, 114, 127. These figures do not record the unknown but significant amount of metals appropriated by the miners' share.

3. Garner, *Accounts of the Cajas Reales*; TePaske and Brown, *New World of Gold and Silver*, 128.

4. Salazar González, *Haciendas de San Luis Potosí*, 327.

5. Del Campo, "El Mineral de San Pedro."

6. Muster of indigenous population of SLP *Reunión de los Alcaldes de los barrios*, January 24, 1695, AMESLP, AM, 1695(1) exp 8; *Auto sobre la obra del socavón en el Cerro de San Pedro*, n.d. 1695, AMESLP, AM 1695(3) exp 23; loan extended by Crown Carta del virrey conde de Galve al Rey, May 25, 1694, AGI México, 61, r.1, n.11; AGI - México, 63, r.1, n.1; description of remnant pit in *Demanda de Maria Theresa Dominguez*, November 2, 1727, AMESLP, AM 1727, exp. 30.

7. Basalenque, *Historia de la Provincia de San Nicolas*; Villaseñor y Sánchez, *Theatro Americano*, vol. 2, 48.

8. Guerrero, *Chemical History of Silver Refining*, 54.

9. Gamboa, *Mining Ordinances of New Spain*, vol. 2, 198.

10. Guerrero, *Chemical History of Silver Refining*, 27–30; Dahlgren, *Historic Mines of Mexico*, 15.

11. Guerrero, *Chemical History of Silver Refining*, 51.

12. Humboldt, *Kingdom of New Spain*, vol. 3, 278.

13. *Petición de Jhoan de Paz*, March 1614, AMESLP, AM, leg. 1614.2, exp. 2, 2r.

14. *Gastos de la Hacienda de Juan de Valle*, March 1602, AMESLP, AM C.114, exp. 1; *Cuentas de la hacienda del Cp. Miguel Maldonado*, June 1612 to January 1614, AMESLP, AM C.146, Exp. 13; *Memorias semanales de la Hacienda de Francisco de Mora*, February 1767 to October 1768, AMESLP, AM C.929, exp. 7.

15. "Informe del Real de Cerro de San Pedro (1774)," in López Miramontes and Urrutia, *Minas de Nueva España*, 139.

16. "Informe de Antonio de Mendoza a Luis de Velasco," n.d., in CODOIN, vol. 26: 288; *Ordenanzas de Minería*, Titulo XIII, art. 13 and 14.

17. Wing, *Roots of Empire* and "Keeping Spain Afloat."

18. Hammersley, "Charcoal Industry and its Fuel," 613.

19. Tutino, *Capitalism in the Bajío*, 311–12; Barrera de la Torre, "Paisaje de Real de Catorce."

20. Salazar González, *Las Haciendas*, 81.

21. *Denuncia criminal contra Andres Monzón y otros*, February 25, 1651, AMESLP, AM 1651 (1), exp. 4.

22. *Denuncia de Pedro Sánchez Matias contra ladrones de minas*, November 22, 1689, AMESLP, AM 1689 (2) exp. 24; *Domingo y Miguel Martín por el despojo de una mina*, December 22, 1708, AMESLP, AM 1711 (1), exp. 3; *Despojo de una mina*, August 8, 1759, AMESLP, AM 1759 (2), exp. 17.

23. Villaseñor y Sánchez, *Teatro Americano*, vol. 2, 48.

24. Pérez Navarro, *Cultura politica en Cerro de San Pedro*, 27, 30.

25. Avalos Lozano, "Paisajes Mineros en el Altiplano"; Montejano y Aguiñaga "Real de Minas de Catorce"; Barrera de la Torre, "Paisaje de Catorce."

Chapter 8

1. *Parecer of Benito Novoa Salgado*, 1691 reported in Gómez del Campo, *Mineral del Cerro de San Pedro*.

2. Villaseñor y Sánchez, *Theatro Americano*, 48.

3. Fabry, *La Rebaja del Azogue*, 82, 128, 164–65.

4. Galaor, *Minas hispanoamericans del siglo XVIII*, 14 ff.

5. Dobado *Minas de Almadén*.

6. Dobado and Marrero, "Mining-Led Growth of Bourbon Mexico," 867, 868–69.

7. Bakewell, "Periodización de la producción minera," 42–43.

8. Humboldt, *Kingdom of New Spain*, vol. 3, 203.

9. Brading, "Revival of Zacatecas," 666.

10. *Reales Ordenanzas*, Tit. XII, Arts. 3, 4, 8, 13.

11. Charles I, *Pregmatica sobre Vagamundos*; Baca, "Mining Legislation in Mexico," 527.

12. Mourelle, "Minas de Guanajuato," 43.

13. Brading, "Revival of Zacatecas," 671.

14. Gálvez, *Informe sobre el estado de Mexico, California, Sonora*, 155–56.

15. Output rose from 2 million to 4.7 million kilograms of silver produced for decades 1730 and 1810. TePaske and Brown, *New World of Gold and Silver*, 115–16.

16. Ca. 55 percent of world total between 1790 and 1810. TePaske and Brown, *New World of Gold and Silver*, 110, 140.

17. Humboldt, *Kingdom of New Spain*, vol 3, 138–39.

18. Carbajal López, *Minería en Bolaños*.

19. TePaske and Brown, *New World of Gold and Silver*.

20. Humboldt, *Kingdom of New Spain*, vol. 3, 118, 119–37, 333, 334–35.

21. Humboldt, *Kingdom of New Spain*, vol. 3, 334–35.

22. Tutino provides the most recent synthesis in, *The Mexican Heartland*, 150ff.

23. Ward, *Mexico in 1827*, vol. 1, 399–400.

24. Elhuyar, *Influjo de la Minería*, 57.

25. Marucci, "American Smelting and Refining Company," 29.

26. Ward, *Mexico in 1827*, vol. 1, 399; St Clair Dupont discusses capital flight from Mexico in 1820s and 1830s in *Métaux Précieux au Méxique*, 391–93.

27. Haber, Maurer, and Razo, *Politics of Property Rights*, 36ff.; Marucci, "American Smelting and Refining Company," 41.

28. Elhuyar, *Influjo de la Mineria*, 60.

29. Elhuyar, *Influjo de la Mineria*, 57.

30. A "capitalist vanguard" according to Mary Louise Pratt, it included writers such Francis Bond Head (Argentina and Chile), Charles Cochrane (Colombia), Joseph Andrews (Bolivia and Peru), Richard Burton (Brazil). See her *Imperial Eyes*, 146–55.

31. Rawson, *Future prospects of the Mexican Mine Associations*, 14.

32. Roberts Poinsett, *Notes on Mexico*, 203, 222.

33. Dalhgren, *Mines of Mexico*, 190.

34. Alamán, *Autobiografia*, 12, 14, 18.

35. Charles Hale, *Mexican Liberalism in the Age of Mora*.

36. Arnaiz y Freg, "Don Fausto Elhuyar," 80, 84.

37. Elhuyar, *Influjo de la minería*, 72.

38. Rawson, *Mexican Mining Associations*, 25.

39. Eguía, *Memoria sobre la utilidad de la minería*, 19.

40. Elhuyar, *El Influjo de la Mineria*, 72.

41. Rawson, *Mexican Mining Associations*, 41–43.

42. Eguía, *Memoria sobre la utilidad de la minería*, 11.

43. Alamán, *Memoria presentada al Congreso General*, 149.

44. Dalhgren, *Historic Mines of Mexico*, 190.

45. Dawson, *First Latin American Debt Crisis*.

46. Alamán, *Autobiografia*, 13, 21, 24; Gilmore, "Henry George Ward," 38–39.

47. Abdulrahman Nzibo, *Relations between Great Britain and Mexico*, 167.

48. Rippy, "Latin America and British Investment," 128.

49. Humboldt, *Kingdom of New Spain*, vol. 3, 279–80.

50. Nzibo, *Relations between Great Britain and Mexico*, 183.

51. Rankine, "Mining in Nineteenth Century Guanajuato."

52. Dalhgren, *Historic mines of Mexico*, 191–92.

53. INEGI, *Estadísticas Históricas*, 469.

54. Serrano records ca. 700 kg/yr, *Articulación económica de San Luis Potosí*, 318.

55. TePaske and Brown record an average annual silver output of 488,000 kilograms, *New World of Gold and Silver*, 113; INEGI records an average of ca. 244,000 kilograms for 1873–74, *Estadísticas Históricas*, 469.

56. Marucci, "American Smelting and Refining Company in Mexico," 35.

57. Lurtz, "Developing the Mexican Countryside," 433. For parallel developments in Peru and Chile see, Himley, "Extractivist Geographies"; Vicuña Mackenna, *Edad de Oro en Chile*, chaps. 13 and 14.

58. Ramírez, *Riqueza Minera de Mexico*, 59.

59. Ramírez, *Riqueza Minera de Mexico*, 711.

60. Ramírez, *Riqueza Minera de Mexico*, 709.

61. Ramírez, *Riqueza Minera de Mexico*, 710.

62. Ramírez, *Riqueza Minera de Mexico*, 715.

63. Ramírez, *Riqueza Minera de Mexico*, 698.

64. Ramírez, *Riqueza Minera de Mexico*, 709.

65. Ramírez, *Riqueza Minera de Mexico*, 707.

66. Haber, Maurer and Razo, *"Politics of Property Rights*, 240–41; Kortheur, *Compagnie du Boleo*, 81, 82–83; Ramírez, *Rigueza minera de Mexico*, 744–46.

67. Ramírez, *Apuntes para un proyecto de Código de Minería*, Titulo I, art. 1, 3, 15–18, 35–36; Titulo VI; Titulo X, art. 2, 9, 14–15, 17; and Martínez Baca, "Historical Sketch of Mining Legislation in Mexico," 551, 553, 559.

68. Chism, "Mining Laws of Mexico," 5.

69. Bernstein, *Mexican Mining Industry*, 27–28.

70. Chism, " A Synopsis of the Mining Laws of Mexico," 5, 8, 40; Martínez Baca, "Historical Sketch of Mining Legislation," 559.

71. Cited in in Martínez Baca, "Historical Sketch of Mining Legislation," 559.

72. Dumett, *Mining Tycoons in the Age of Empire*, 10–11, 16, 17–18.

73. Cosío Villegas *Historia Moderna de Mexico*, 1091, 1103.

74. Bernstein, *The Mexican Mining Industry*, 75.

75. Rickard, *History of American Mining*, 132; Emmons, *Geology and Mining Industry of Leadville, CO.*

76. Rickard, *Interviews with Mining Engineers*, 139.

77. Wyman, *Western Miners and Industrial Revolution*, 3.

78. Conversion based on Smil's estimations in *Energy in Nature and Society*, 158.

79. Rickard, *History of American Mining*, 133, 437; Wyman, *Western Miners and Industrial Revolution*, 102–3.

80. Wyman, *Western Miners and Industrial Revolution*, 104–5.

81. Wyman, *Western Miners and Industrial Revolution*; Francaviglia, *Hard Places*; Hardesty, *Mining Archaeology in the American West*.

82. Rickard, *History of American Mining*, 101.

83. Rickard, *Recent Cyanide Practice*, 256ff.

84. Manahan, *Mining Operations of ASARCO*, 9, 14.

85. Manahan, *Mining Operations of ASARCO*, 7.

86. Dahlgren, *Historic Mines of Mexico*, 199.

87. Dalhgren, *Historic Mines of Mexico*, 199,

88. Manahan, *Mining Operations of ASARCO*, 7.

89. Southworth and Holmes, *Mining Directory of Mexico*, 8.

90. Southworth and Holmes, *Mining Directory of Mexico*, 14, 15.

91. Harvey, *The New Imperialism*.

92. West, *Sonora Its Geographic Personality*; Langue, "La minería en Sonora"; del Rio, "Real de la Cieneguilla, Sonora."

93. "Map Supplement: The Mines and Railways of Mexico."

Chapter 9

1. Grossman provides the full treatment of the professionalization of US mining engineers. See her *Mining the Borderlands*, 44–66.

2. Bernstein, *Mexican Mining Industry*, 22.

3. J. W. Malcomson, "Sierra Mojada"; Van Law, "Aerial Tramway of the Real del Monte Company," 214; Dominguez, "District of Parral," 473; Managahn, *Mining Operations of ASARCO*, 3.

4. Romero, *Geographical and Statistical Notes on Mexico*, 28; Hahn, "Silver Smelting in Mexico," 233.

5. Gamez Rodriguez, *Movilización de los mineros*, 51.

6. *Contrato celebrado con el C. General Carlos Pacheco para exploración y explotación de minas de toda especia y contruccion de cinco haciendas metalúrgicas*, 20 de marzo, 1890, AHESLP.

7. Del Campo, *Mineral de Cerro de San Pedro*.

8. Bros, "El Mineral de Cerro de San Pedro."

9. Martínez, *Minas de la Unidad San Pedro*, 374–76.

10. Hahn, "Silver Smelting in Mexico."

11. Matías Romero, *Geographical and Statistical Notes on Mexico*, 28.

12. Gómez Serrano, *Minería y metalurgia en Aguascalientes*.

13. Bernstein, *Mexican Mining Industry*, 145.

14. Guerreo, *Chemical History of Silver Refining*, 228, 288.

15. Romero, *Geographical and Statistical Notes on Mexico*, 28.

16. *International Mining Manual*, 263.

17. *International Mining Manual*, 252–86.

18. Malm, "Water to Steam in the British Cotton Industry."

19. Hart, *Silver of the Sierra Madre*, 160–61; Marucci, *The American Smelting and Refining Company*, 42–43.

20. Bernstein, *Mexican Mining Industry*, 43.

21. Bernstein, *Mexican Mining Industry*, 44.

22. *Decreto sobre la Hacienda Metalurgica 1891*—AHESLP—SGG manuscritos, leg. 1890.

23. *Alvarez Land and Timber Co. General Ledger, 1898–1953*.

24. Poinsett, *Notes on Mexico*, 222.

25. *Alvarez Land and Timber Co. General Ledger, 1898–1953*.

26. Montes de Ocá, "La fundición de San Luis Potosí."

27. Humboldt, *Kingdom of New Spain*, vol. 3, 243; Poinsett, *Notes on Mexico*, 225.

28. Ramirez, *Riqueza Minera de Mexico*, xiv–xv, 148.

29. Bernstein, *Mexican Mining Industry*, 35–36; *International Mining Manual*, 235–36, 267, 273, 274; see also, Vergara, *Fueling Mexico*.

30. INEGI, *Estadísticas Históricas*, vol. 1, 472.

31. Bernstein, *Mexican Mining Industry*, 35.

32. Hahn, "Silver Smelting in Mexico," 289.

33. Hahn, "Silver Smelting in Mexico," 290.

34. Minero Mexicano notes.

35. Chism, "Sierra Mojada," 569.

36. Juan Martínez, "Minas de la Unidad San Pedro, de la Compañía Metalurgica Nacional, en San Luis Potosí," Boletin Minero (Marzo, 1927), 376.

37. Gamez Rodriguez, *Movilización de los mineros*, 139.

38. At the height of the Bourbon boom the total workforce of Mexican miners and workers was thirty-five thousand. Weyl, "Labor Conditions in Mexico," 53.

39. Martínez Baca, "Mining Legislation in Mexico," 522.

40. Torres Pares, *La Revolución Sin Frontera*, 39.

41. Martínez Chavez et al., "Procesos historicos y ambientales en Cerro de San Pedro," 213.

42. Interview Marcos Rangel, Cerro de San Pedro, February 2006.

43. Medina Esquivel, "Vida cotidiana en Cerro de San Pedro," 41, 102; Martínez Chaves et al. "Procesos historicas y ambientales en Cerro de San Pedro," 217.

44. Interview Marcos Rangel, March 2006.

45. Interview Mario Martinez, July 2008.

46. Medina Esquivel, "Vida cotidiana en Cerro de San Pedro."

47. Palacios Garcia, *La Fundición de Morales*, 68–70; Calvillo Unna, *La fundición de Morales*, 72–74; Hahn, "Silver Smelting in Mexico," 292.

48. Walker, *Toxic Archipelago*, 71ff.

Chapter 10

1. Myrna Santiago, *The Ecology of Oil*; Vergara, *Fueling Mexico*, and "How Coal Kept my Valley Green."

2. This estimate based on the transfer of per ton emissions rates from a comparable ASARCO smelter at El Paso to the average production rates at Morales. Daily emissions for the ASARCO plant were 1386 kgs Pb, 680 kgs Zn; 1.3 kgs As. Shapleigh, *Towards a Brighter Future*, 8, 10.

3. Aragón-Piña et al., "Emisiones Industriales en San Luis Potosí," 9, 12.

4. "El MINEM tiene registrados 6,847 pasivos ambientales mineros."

5. Change and Tjalkens, "Neurotoxicology of Metals," 491.

6. Fowler et al., "Arsenic, Antimony and Bismuth"; Jacobs, "Lead"; Grandjean and Yorifuji, "Mercury"; Chang and Tjalkens, "Neurotoxicology of Metals"; Davidson et al., " Molecular Mechanisms of Metal Toxicity."

7. Medina Esquivel, "Fiestas y diversiones en Cerro de San Pedro," 7–8.

8. Gamez Rodriguez, *Movilización de los mineros,* 139; Cárdenas Garcia, *Empresas y trabajadores en la gran minería mexicana,* 160; Luna Morales, *La fuerza de trabajo en la mineria de Chihuahua,* 119–22.

9. McNeill, *Mosquito Empires;* Sutter, "Nature's Agents or Agents of Empire?"

10. Bogitsh, Oeltmann, and Carter, *Human Parasitology,* 340–43; Ettling, "Hookworm Infection."

11. de la Garza Brito, "La anquilostomasia en el estado de Hidalgo," 141.

12. "El microbio de la holganzaneria," *El Pais,* October 1, 1905.

13. "Uncinaria o gusano de la pereza," *El Informador,* May 1,1931.

14. Gamez Rodriguez, *Movilización de los mineros,* 139. Such rates of infection were significantly higher than those obtained in mining districts in northern Europe—between 13 and 25 percent—and comparable to those of Italy and Spain. See International Health Board, *Bibliography of Hookworm Disease,* xv; Martínez Ortiz and Tarifa Fernández, *Medicina social, demografía y enfermedad,* 245 and appendix 25.

15. Bernstein, *Mexican Mining Industry,* 86; Rebollar "Mal que aflige a los mineros," 49–50; Manuell, "Anchylostomiasis in Mexico," 202; Garza Brito believed that almost all the miners of Pachuca were infected though only a portion were symptomatic. Garza Brito, "La anquilostomasia en el estado de Hidalgo," 141.

16. Álvarez, "Ankylostomiasis in Mexico," 1388.

17. Bogitsh et al., *Human Parasitology,* 344.

18. Álvarez, "Ankylostomiasis in Mexico," 1388.

19. Manuell, "Anchylostomiasis in Mexico," 202.

20. Instituto Medico Nacional, *Informe sobre la distribucion de la uncinariasis.*

21. Instituto Medico Nacional, *Informe sobre la distribucion de la uncinariasis.*

22. Garza Brito, "Anquilostomasia en Hidalgo," 140.

23. Martinez, "Minas de la Unidad San Pedro."

24. Interview Armando Mendoza, July 2008; Interview Gabriel Muñoz, February 2006.

25. Balliet, "Wanted—Light and Air," 8.

26. Hughes, "Accidents from Falls of Rock or Ore," 263.

27. Weyl, "Labor Conditions in Mexico," 12–17.

28. Colonel Tepetate, "Safety First," 314.

29. Menéndez Taboada, *Minería y enfermedad en Zimapán,* 75.

30. S. C. Hotchkiss, "Occupational Diseases in the Mining Industry," 134–35.

31. Rodríguez López, *"Andamos en las entrañas de la tierra,"* 217, 226.

32. Menéndez Taboada, *Minería y enfermedad en Zimapán,* 79.

33. Gillespie, "Accounting for Lead Poisoning," 303–5.

34. Gamez Rodriguez, *Movilización de los mineros,* 73.

35. Interview Marcos Rangel, February 2008.

36. Hotchkiss, "Occupational Diseases in Mining," 138.

37. Camacho Bueno, "El trabajo mata," 166.

38. Leung and Tak Sun Yu, "Silicosis."

39. Testimony transcribed in Menéndez Taboada, *Minería y enfermedad en Zimapán,* 151.

Chapter 11

1. Anderson, *Outcasts in Their Own Land*, 331–33.
2. *El Estandarte*, August 2, 1903; transcribed in Appendix 2 of Gamez Rodriguez, *Movilización de los mineros*, 172.
3. *El Estandarte*, June 2, 1892; October 29, 1892; May 20, 1893. *Periodico Oficial del Estado*, January 27, 1903; Gamez Rodriguez, *Movilización de los mineros*, 74–75.
4. Gamez Rodriguez, *Movilización de los mineros*, 87, 88–89, 90.
5. Gonzalez, "Copper Companies and Labour Conflict in Mexico"; Truett, *Fugitive Landscapes*; Sariego, *Historia social de los mineros de Cananea*; Romero Gil, "La Revolución en el socavón."
6. For US comparison, see Peck, "Nature of Labor," 216.
7. Soluri, "Labor, Rematerialized."
8. Gamez Rodriguez, *Movilización de los mineros*, 72.
9. Nixon, *Slow Violence*.
10. Medina Esquivel, *Sobrevivir en un pueblo minero*, 116.
11. Ralph Ingersoll, *In and Under Mexico*, 119.
12. Martínez, *Minas de la Unidad San Pedro*, 376.
13. Calvillo Unna, *La fundición de Morales*, 81.
14. Bureau of Labor Statistics, "Industrial Accidents [notes,]" July 1927, 1259.
15. Bureau of Labor Statistics, "Industrial Accidents and Safety [notes]," May 1933, 1070.
16. Camacho Bueno, "El trabajo mata," 163–65.
17. Franck, *Tramping Through Mexico*, 65–66, 69; Ingersoll, *In and Under Mexico*, 119.
18. Ingersoll, *In and Under Mexico*, 119.
19. Rogers, "Character and Habits of the Mexican Miner," 701; French, *A Peaceful and Working People*, 23.
20. Rogers, "Character and Habits of the Mexican Miner," 701.
21. Ingersoll, *In and Under Mexico*, 117, 120; more along these lines for Batopilas, Cananea, and Parral in French, *A Peaceful and Working People*, 51–55.
22. Tays, "Present Labor Conditions in Mexico," 624; Rogers, "The Mexican Miner," 700.
23. Weyl, "Labor Conditions in Mexico," 12–13.
24. French, *A Peaceful and Working People*, chap. 3; Kortheuer, *Santa Rosalia and Compagnie du Boleo*, 226–27.
25. Ingersoll, *In and Under Mexico*, 121.
26. Franck, *Tramping*, 70, 100.
27. Franck, *Tramping*, 69, 70, 71, 110.
28. Calvillo Unna, *La fundición de Morales*, 71.
29. Calvillo Unna, *La fundición de Morales*, 58.
30. Calvillo Unna, *La fundición de Morales*, 81.
31. Medina Esquivel *Vida cotidiana en Cerro de San Pedro*, 301.
32. Cited in Menéndez Taboada, *Minería y enfermedad en Zimapán*, 141.

33. Alvarado García, *Mí pueblo viejo*; see also Flores Clair "Trabajo, salud y muerte," 13.

34. Calvillo Unna, *La fundición de Morales*, 71.

35. Interview don Gabriel Muñoz, February 2006.

36. Medina Esquivel, *Sobrevivir a un pueblo minero*, 266–69.

37. Interview don Armando Mendoza, March 2008.

38. Franck, *Tramping*, 66; Colonel Tepetate, "Safety First," 314.

39. Palacios Gomez, *Fundición de Morales*, 68–70.

40. Wakild, "Environmental Justice and Environmental History" and David Carruthers for definition in early twenty-first century Latin American context, *Environmental Justice in Latin America*.

41. On the visibility of injustice in the industrial mining districts: Cárdenas Garcia, *Empresas y trabajadores en la gran minería*, 165–66.

42. *Obras* cited in Gamez Rodriguez, *Movilización de los mineros*, 75.

43. French, *A Peaceful and Working People*, chap. 3.

Chapter 12

1. Gamez Rodriguez, *Movilización de los mineros*, 73, 75, 76, 88–89.

2. Gamez Rodriguez, *Movilización de los mineros*, 108.

3. *El Estandarte*, September 20, 1910, transcribed Appendix 3 in Gamez Rodriguez, *Movilización de los mineros*, 174.

4. Anderson, *Outcasts in Their Own Land*, 302–3, 331–38.

5. For overview and analysis of liberal democratic Maderista opposition, see Knight, *The Mexican Revolution*, vol. 1, 55–77; narrative account and details on Madero in Cumberland, *Genesis Under Madero*, 112, 115–18; Estrada, *La revolución y Francisco I. Madero*, 289–90.

6. Knight, *Mexican Revolution*, 172–74.

7. What Friedrich Katz called the "Chihuahuan Revolution," see his *The Life and Times of Pancho Villa*; Knight, *The Mexican Revolution*.

8. Guerra, "Une révolution minière?"

9. Knight, "Révolution minière ou révolution serrano?"

10. Medina Esquivel, *Vida cotidiana en Cerro de San Pedro*, 295.

11. Gamez Rodriguez, *Movilización de los mineros*, 109.

12. Orwell, *The Road to Wigan Pier*, 29.

13. Gamez Rodriguez, *Movilización de los mineros*, 113–14.

14. Gamez Rodriguez, *Movilización de los mineros*, 105–6, 113–14, Appendix 1.7.

15. Gamez Rodriguez, *Movilización de los mineros*.

16. Anakreón, "El progreso material," *El Colmillo Público*, no. 127, February 11, 1906.

17. Junta Organizadora del Partido Liberal Mexicano, *La Bandera Roja en Tamaulipas* (5 de octubre de 1911).

18. Junta Organizadora del Partido Liberal Mexicano, *A los huelguistas y a los trabajadores en general* (15 de agosto de 1911).

19. Hall and Coerver, "La frontera y las minas en la revolución mexicana," 394–95.

20. Bonney, *Supplement to Commerce Reports*, 5.

21. Lane, *Mine Doctor's Wife in Mexico*, 29.

22. Romero Gil, "La Revolución en el socavón," 31.

23. French, *A Peaceful and Working People*, 161.

24. Meyer," Pancho Villa and the Multinationals."

25. Hall and Coerver, "La frontera y las minas," 396, 397–98.

26. Meyer, "Pancho Villa and the Multinationals," 358–59; Hall and Coerver, "La frontera y las minas," 399.

27. Romero Gil, "Sonora: La Revolución en el Socavón, 1910–1918," 28.

28. Reed, *Insurgent Mexico*, 64.

29. Tepetate, "Safety First," 313–14.

30. Manahan, *Mining Operations of ASARCO*.

Chapter 13

1. Knight, *Mexican Revolution*.

2. On Cananea Cárdenas García, *Empresas y trabajadores en la gran minería mexicana*, 219, 222–23; see also Leal, *En la Revolución*, 155, 161.

3. Cited in Villarello Vélez, *Revolución mexicana en Coahuila*, 221.

4. Leal, *En la Revolución*, 156–57.

5. Leal, *En la Revolución*, 166.

6. Gamez Rodriguez, *Movilización de los mineros*, 144; citation on 146.

7. Cárdenas García, *Empresas y trabajadores en la gran minería mexicana*, 249.

8. Niemeyer, *Revolution at Querétaro*, 25.

9. Mexico, *Constitucion de 1917*, Art. 27.

10. Ferry, *Not Ours Alone*, 200, 201.

11. Niemeyer, *Revolution at Querétaro*, 139.

12. Mexico, *Constitucion de 1917*, Art. 27: "La Nación tendrá en todo tiempo el derecho de imponer a la propiedad privada las modalidades que dicte el interés público, así como el de regular el aprovechamiento de los elementos naturales suceptibles (sic) de apropiación, para hacer una distribución equitativa de la riqueza pública y para cuidar de su conservación."

13. Niemeyer, *Revolution at Querétaro*, 147.

14. Cited in Bernstein, *Mexican Mining Industry*, 182.

15. Niemeyer, *Revolution at Querétaro*, 101.

16. Cited in Niemeyer, *Revolution at Querétaro*, 106.

17. Mexico, *Constitucion de 1917*, Subss. xii, xiii, xiv.

18. Medina Esquivel, "Sobrevivir en un pueblo minero," 296.

19. Medina Esquivel, "Sobrevivir en un pueblo minero," 295.

20. Camacho Bueno, "El trabajo mata," 153–54.

21. Bernstein, *The Mexican Mining Industry*, 198.

22. "Hoy a las diez horas puede empezar la huelga de los metalúrgicos en Morales," *El Heraldo de San Luis*, March 15, 1948.

23. "Todo indicaba anoche que la huelga minera se prolongará más tiempo del que se creía," *El Heraldo de San Luis*, March 16, 1948; "Habrá un mítin obrero para

conmemorar la expropiación," *El Heraldo de San Luis*, March 18, 1948; "Formal ofrecimiento de los mineros al presidente," *El Heraldo de San Luis*, March 21, 1948.

24. "Preparativos de huelga" *El Heraldo de San Luis*, February 26, 1948; "La secretaría del Trabajo trata de evitar la huelga minera," *El Heraldo de San Luis*, March 12, 1948; "La huelga minera se prolongará más tiempo del que se creía," *El Heraldo de San Luis*, March 15, 1948.

25. "Magnos esfuerzos para resolver el Amago de la Huelga Minera," *El Heraldo de San Luis*, March 8, 1948.

26. "La Industria Minera es uno de los más fuertes sostenes de nuestra economía," *El Heraldo de San Luis*, March 9, 1948.

27. "A los de Morales les hacen más concesiones que a los de Avalos," *El Heraldo de San Luis*, March 28, 1948.

28. As recounted in René Medina Esquivel's, *Sobrevivir en un pueblo minero*, 311.

29. "Estalló un incendio en el mineral de San Pedro que tiene en peligro y paralizado a este fundo"; "Inexplicablemente mantuvo eso en secreto la empresa minera," *El Heraldo de San Luis*, April 4, 1948.

30. "Se teme que el mineral de San Pedro, importante centro de trabajo, pare," *El Heraldo de San Luis*, April 7, 1948.

Chapter 14

1. Interview with Mario Martínez, December 2014.

2. Interview with Mario Martínez, December 2014.

3. "Se acordó prorrogar el plazo para que estalle la huelga en San Pedro," *El Heraldo de San Luis*, April 7, 1948.

4. Interview with Mario Martínez, December 2014.

5. Medina Esquivel, "Vida cotidiana en Cerro de San Pedro," 312, 321.

6. Interview with Mario Martínez, December 2014.

7. Martínez Chavez et al., "Procesos historicos y ambientales en Cerro de San Pedro," 218.

8. Interview with Pedro Rangel, March 2008.

9. Interview with Pedro Rangel, March 2008.

10. Interview with Padre Marguerito Sánchez Grimaldo, May 2008.

11. Interview with Mario Martínez, December 2014. The genealogical ties between local families to the tumultosos of the eighteenth century was corroborated by research undertaken in the parish archives in the late 1990s and 2000s. Email correspondence with Margarita Villalba, February 20, 2018.

12. "Cerro de San Pedro: El pueblo fantasma," *El Heraldo de San Luis*, June 11, 1961.

13. Paula Andrea Chavez et al. "Processos historicos y ambientales en Cerro de San Pedro," 220.

14. "Cerro de San Pedro: el Pueblo que se niega a morir," *El Heraldo de San Luis*, March 6, 1978.

15. "Cerro de San Pedro: ¿De cuna a sepultura?" *El Heraldo de San Luis*, October 24, 1971.

16. "Cerro de San Pedro: El pueblo fantasma."

17. Chavez et al., "Processos historicos y ambientales en Cerro de San Pedro," 211, 225.

18. Interview with Aristeo Gutierrez Chavéz, March 2008.

19. "Cerro de San Pedro El pueblo fantasma."

20. Chavez et al., "Processos historicos y ambientales en Cerro de San Pedro," 211, 225.

21. Interview with Mario Martínez, December 2014.

22. Chavez et al., "Processos historicos y ambientales en Cerro de San Pedro," 218, 219.

23. Interview with Aristeo Gutierrez Chavéz, March 2008.

24. Interview with Aristeo Gutierrez Chavéz, March 2008.

25. Chavez et al., "Processos historicos y ambientales en Cerro de San Pedro," 219; Metallica Resources, *Technical report*; "Cerro de San Pedro: ¿De cuna a sepultura?"; interview with Tonantzín Mendoza, March 2008.

26. "Violenta Tromba en Cerro de S. Pedro," *Pulso*, June 14, 1990.

27. Interview with Pedro Rangel, March 2008.

28. Cited in Espinosa, "Victoria revertida," *Proceso*, January 7, 2010.

29. Alvarado García, *Mi Pueblo Viejo—Cerro de San Pedro*.

30. "Victoria revertida."

Chapter 15

1. Estado de San Luis Potosí, *Decreto del 24 septiembre de 1993*; and Estado de San Luis Potosí, *Plan de ordenación*.

2. Direccion General de Minería, *Proyectos mineros de capital extranjero*.

3. LeCain, *Mass Destruction*; Manuel, *Taconite Dreams*.

4. The US Bureau of Mines estimated that only 2 percent of these were of sufficient grade to be mined at a profit. Bureau of Mines, *Production Potential of Known Gold Deposits*.

5. Roberts, *Autobiography*, 55, 83.

6. Braun, "Producing vertical territory," 16.

7. Heitt, "Newmont's Reserve History on the Carlin Trend," 35; Coope, "Carlin Trend Exploration History."

8. Shoemaker, *Recollections of Frank Woods McQuiston Jr.*, 164.

9. Shoemaker, *Recollections of Frank Woods McQuiston Jr.*, 165, 166, 168.

10. Bridge, "Mapping the Bonanza," 416; Moore, *Capitalism and the Web of Life*, 44.

11. Shoemaker, *Recollections of Frank Woods McQuiston Jr.*, 174.

12. Heitt, "Newmont's Reserve History at Carlin," 35.

13. Lasky, "How Tonnage and Grade Relations Helps Predict Ore Reserves," 81–85.

14. Heitt, "Newmont's Reserve History at Carlin."

15. Today the average payload is a bit more than three hundred tons, up to four hundred tons in the case of the colossal Terex trucks.

16. Chamberlain and Pojar, "Gold and Silver Leaching Practices in the United States."

17. Shoemaker, *Recollections of Frank Woods McQuiston Jr.*, 176.

18. Seventy million ounces of gold were recovered over 37 years of operation at Carlin, D. Heitt, "Newmont's Reserve History"; annual US gold production from Kelly and Matos, *Historical Statistics.*

19. Müller and Frimmel, "Historic Gold Production Cycles."

20. Shoemaker, *Recollections of Frank Woods McQuiston Jr.*, 162.

21. Eggert, *Metallic Mineral Exploration*, 56.

22. Chamberlain and Pojar, "Gold and Silver Leaching Practices in the United States," 17–18.

23. Worstell, "Precious Metal Heap Leaching in North America."

24. In 1992 these costs were estimated at 64 percent for loading and hauling; 15 percent for drilling and blasting. See O'Hara and Suboleski, "Costs and Cost Estimation," 422.

25. Harvey Enchin, "Ex-Hippie's Gold Bug Still Glowing," *Globe and Mail*, November 9, 1984.

26. Cruise and Griffiths, *Fleecing the Lamb*; Deneault and Sacher, *Imperial Canada Inc.*

27. Kryzanowski, "Misinformation and Security Markets," 123.

28. Heffernan, "Viola Heffernan: From the Ground Up."

29. Studnicki-Gizbert, "Canadian Mining in Latin America."

30. Michael Bernard, "Vancouver Trade Volume Swells. Record Potential for Global Stocks Seen Changing Face of Exchange," *The Toronto Star*, December 16, 1986.

31. Roger Moody, "The Ugly Canadian: Robert Friedland and the Poisoning of the Americas," *Multinational Monitor*, 15 (1994): 10.

32. Others include Pierre Lassonde, Seymour Shulich, Peter Munk, Ian Telfer, and Frank Giustra.

33. On the longer time cycles of large-scale copper mining see, Schmitz, "The World Copper Industry, 1870–1930."

34. Dan Worstell, "Precious Metal Heap Leaching in North America"; "Summitville: Leaching at Its Peak"; Wallace Immen, "Heap Leaching Is Used to Extract Gold," *Globe and Mail*, September 2 8, 1984.

35. Jennifer Hunt, "But What Would His Guru Say?" *Globe and Mail*, March 18, 1988; Enchin, "Ex-Hippie's Gold Bug still Glowing."

36. Albert Sigurdson, "Continuing Record Growth in 1980 Is Expected at Vancouver Exchange," *Globe and Mail*, January 5, 1980; Michael Bernard, "Vancouver Trade Volume Swells," *The Toronto Star*, December 16, 1986; Staff, "Vancouver Exchange Set Records in 1987 Despite October Crash," *Toronto Star*, January 4, 1988.

37. Harvey, "History of Neoliberalism."

38. Bernard, "Vancouver Trade Volume Swells."

39. Rhodes, "Barrick Gold Corporation," 63.

40. Denault and Sacher, *Imperial Canada Inc.*, 35–36, 138–39, 176.

41. Thompson, "A Brief History of the PDAC," 10, 14–15.

42. US gold production for 1971 to 1990 was 1,676 metric tons or 59,119,160 ounces. The estimated market value for the metal based on the average, in deflated $2000 of $502 per ounce. Kelly and Matos, *Historical Statistics*.

43. For description of local ecology prior to operations at Newmont's mine see, Li, *Unearthing Conflict*, 80–83.

44. Riofrancos, *Resource Radicals*, chap 1.

45. For the physical and material dimensions of Yanacocha mine, see Bury, "Livelihoods in Transition," 78.

46. Cited in Pratap Chatterjee, "Conquering Peru: Newmont's Yanacocha Mine Recalls the Days of Pizarro," *Multinational Monitor* 18, no. 4 (1997): 21.

47. Cited in AP Staff, "Peru Gold Rush Turns Huge Profits," *Hamilton Spectator*, August 16, 1996; on context of Fujimori's neoliberal reforms in and around the mining sector, see Li, *Unearthing Conflict*; Bury, "Mining Mountains," 222–23.

48. For Australasia and Oceania see Stuart Kirsch, *Mining Capitalism*.

49. For Argentina: Staff, "Canadian Companies Discover Gold in Argentina," *Gazeta Mercantil*, April 30, 1997; for Chile: Nancy Yañez and Sarah Rea, "The Valley of Gold," *Cultural Survival Quarterly* 30, no. 4 (Winter 2006); for Mexico: Chris Aspin, "Looking for Leftovers in Mexico's Old Gold and Silver Mines," *The Financial Post*, January 31, 1997; for Peru: Sally Bowen, "Conquistadores Fight over Peru Gold," *The Australian*, March 18, 1993; for Venezuela: Rod Nutt, "Venezuela Turning to Gold for Placer," *The Vancouver Sun*, March 9, 1994.

50. "South American Mining: The New El Dorado," *The Economist*, September 2, 1995.

51. Mary Powers, "Top Mining Firms Rush to Peru for Gold," *Journal of Commerce*, May 27, 1994.

52. Jason Nisse, "The New Klondike," *The Daily Mail—The Mail on Sunday*, April 17, 1994.

53. Bridge, "Resource Triumphalism," 2154, 2162; see also Ferry and Limbert, *Politics of Resources and Their Temporalities*.

54. Li, *Unearthing Conflict*.

55. Ignacio Ramírez, "Riqueza Minera en Controversia," *El Universal* (Mexico), July 7, 2001.

56. Cited in Li, *Unearthing Conflict*, 7.

57. Svampa and Viale, *La Argentina del extractvismo*, 17.

58. Frigon, *Mirages d'un Eldorado*.

59. Svampa, "Resource Extractivism," 125; on the quasi-magical agency of oil extraction in twentieth-century Venezuela, see also Coronil, *The Magical State*.

60. Estrada and Hofbauer. *Impactos De La Inversion Minera Canadiense En Mexico*; Francisco Cavioto, "La legislación minera vigente en México." An overview of mining reform in other Latin American countries is given in Studnicki-Gizbert, "Canadian Mining in Latin America." Gavin Bridge shows that Latin America was part of a global reform of some ninety mining codes across the world, "Mapping the Bonanza."

61. Sánchez Albavera, Ortiz, and Moussa, "Mining in Latin America in the Late 1990s," 14, 17–18.

62. "Canarc raises $3.25 billion for Venezuela gold projects," *Business Wire*, January 23, 1993.

63. Lemieux, "Canada's Global Mining Presence."

64. Mining Association of Canada, *Facts and Figures 2007*.

65. Dirección General de Minas "Proyectos mineros operados por compañías de capital extranjero."

66. Deneault and Sacher, *Imperial Canada Inc.*

67. Mining Association of Canada, *Facts and Figures 2007*.

68. Allan Robinson, "Canadian Miners Strike It Rich: Hunt for Gold in Latin America Grabs Attention, Cash of Speculators Worldwide," *The Globe and Mail*, February 14, 1994.

69. Staff, "Looking for Leftovers in Mexico's Old Gold and Silver Mines," *The Financial Post*, January 31, 1997.

Chapter 16

1. By 2013 the estimated reserves had grown to 2.5 million ounces of gold. This put it in position 324 on 580 gold deposits inventoried in Natural Resource Holdings, "Global 2013 Gold Mine and Deposit Rankings," http://www.visualcapitalist.com/global-gold-mine-and-deposit-rankings-2013/. It is worth noting that deposits of between one and five million ounces represent close to 70 percent of the overall number of deposits, another measure of Cerro de San Pedro's typicality.

2. Minera San Xavier S.A. de C.V. (Sociedad Anomima de Capital Variable).

3. Winterburne, *Alternation and Mineralization*; Petersen et al., " Geology of the Cerro San Pedro Porphyry"; Metallica Resources, *Technical Report*.

4. Behre Dolbear, *Manifestación de Impacto Ambiental*, 38–39.

5. Behre Dolbear, *Manifestación de Impacto Ambiental*, 30.

6. Behre Dolbear, *Manifestación de Impacto Ambiental*, 50.

7. Behre Dolbear, *Manifestación de Impacto Ambiental*, 54.

8. Behre Dolbear, *Manifestación de Impacto Ambiental*, 55.

9. Behre Dolbear, *Manifestación de Impacto Ambiental*, 64.

10. Behre Dolbear, *Manifestación de Impacto Ambiental*, 16.

Chapter 17

1. Collated from: Interview with Marcos Rangel, February 2008; Interview with Padre Marguerito Sánchez Grimaldo, May 2008; Interview with Mario Martínez Ramos, December 2014; testimonies given in "Muerte y vida de Cerro de San Pedro" Dir. Juan Carlos Valladolid.

2. Interview with Sergio Serrano Soria, December 2014.

3. In March 2018, the Santiago-based Observatorio de Conflictos Mineros documented two hundred forty-five conflicts across the continent. https://mapa.conflictosmineros.net/ocmal_db-v2/.

4. On this shift from the politics of labor to the politics of territory associated with the renewal and transformation of mining in the late twentieth century, see

Li, *Unearthing Conflict*; Kirsch, *Mining Capitalism*; Sawyer and Gómez, *The Politics of Resource Extraction*.

5. Interview with Aristeo Gutierrez, March 2008.

6. Marco Rangel cited in *Muerte y vida de Cerro de San Pedro*, dir. Juan Carlos Valladolid.

7. Annand, *How One Mexican Town Is Navigating a Canadian Mine*.

8. Business Wire, "Metallica Resources Inc.—Third Quarter Report—November 1995."

9. Interview with Mario Martínez, December 2014.

10. Steve Mertl, "NAFTA Ruling under Attack: Protestors Join Federal Lawyers in Outrage over Case of Mexican Toxic Dump," *Montreal Gazette*, February 20, 2001.

11. IMMSA or Industrial Minera Mexico S.A. is part of the Mexican mining conglomerate Grupo Mexico that also took control over the problematic ASARCO smelter in El Paso, Texas. See Perales, *Smeltertown*; Hampton and Ontiveros, *Copper Stain*.

12. Calderón et al., "Exposure to Arsenic and Lead in Mexican Children"; Carrizales et al., "Exposure to Arsenic and Lead of Children."

13. Pedro Medellín Milán, Fernando Díaz Barriga, and Luz María Nieto Caravego, "IMMSA debe presentar su plan de cierre y las autoridades deben exigírselo," *Pulso—Diario de San Luis*, October 19, 2000.

14. On the "scaling up" of anti-extractivist movements, see Uriki and Walter, "Dimensions of Environmental Justice."

15. It exemplified what anthropologist Stuart Kirsh called the politics of time in the conflicts between mining corporations and their critics. Stuart Kirsh, *Mining Capitalism*, chap. 6.

16. "Manifesto de Impacto Ambiental—Proyecto Minero-Metalurgico Cerro de San Pedro," *Pulso*, October 22, 1997.

17. Interview with Mario Martínez, December 2014; interview with Sergio Serrano Soria, December 2014.

18. Behre Dolbear, *Manifestación de Impacto Ambiental*, 55.

19. Drawing from Daniel Jacobo-Marín's estimated range of per capita water consumption of between thirty and fifty liters/person/day. See his *Agua para San Luis Potosí*, 61.

20. Behre Dolbear, *Manifestación de Impacto Ambiental*, 150.

21. Behre Dolbear, *Manifestación de Impacto Ambiental*, 164.

22. Metallica Resources, Inc. "Press Release: Cambior Inc. and Metallica Resources Inc. Announce a Joint Venture for the Cerro de San Pedro, Mexico," November 21, 1997.

23. Staff, "Cyanide from Gold Mine Spills into Major River in Guyana," *Montreal Gazette*, August 22, 1995. Cambior would fend off a precedent-making extra-territorial lawsuit laid in Montreal, Quebec by Guyanese organizations, "Guyana Said to be Place for Suit vs. Cambior," *Montreal Gazette*, August 18, 1998.

24. Dan Noyes and Rick Young, "The Road to Summitville, a Gold Mining Debacle," *New York Times*, August 14, 1994.

25. Estado de San Luis Potosí, *Plan de Ordenación de San Luis Potosí.*

26. Interview with Mario Martínez, December 2014.

27. Interview with Mario Martínez, December 2014. Also recounted in Ortiz's *Lucha para Cerro de San Pedro.*

28. Interview with Padre Marguerito Sánchez Grimaldo, May 2008.

29. Herman, *Extracting Consent*, 58.

30. Ortiz, *Lucha para Cerro de San Pedro*, 72.

31. Cited in Oakland Ross, "Tarnish on a Miniature El Dorado," *Toronto Star*, April 16, 2005

32. Comisión de Medio Ambiente de la USALP, *Opinión técnico-científica sobre proyecto Cerro de San Pedro.*

33. Interview with Mario Martínez, December 2014.

34. TFJFA Sentencia D.A.24/2005-3011, October 5, 2005.

35. Registro Agrario Nacional, March 15, 2000. Herman, *Extracting Consent*, 8.

36. González Márquez and Montelongo Buenavista, *Minera San Xavier. Grado de cumplimiento.*

37. Staff, "Metallica Suspends Mexican Mine," *Toronto Star*, June 24, 2004.

38. Metallica Resources Inc. Press Release, *Metallica Resources Suspends Construction Activities at Cerro San Pedro Project, Mexico*, June 23, 2004.

39. Some years later the documentary evidence was published that laid out these lines of relation between Jorge A. Mendizabal, the local director of the Minera San Xavier, Angel Candia, the company's legal counsel, and Marcelo de los Santos, the Governor of San Luis Potosí, see Patricia Rodríguez Calva, "Destapan compadrazgos de la Minera San Xavier," *La Jornada,* March 2, 2009.

40. Annand, *How One Mexican Town Is Navigating a Canadian Mine.*

41. Herman, *Extracting Consent*, 134.

Chapter 18

1. Interview with Mario Martínez, December 2014.

2. Personal communication Martha Rivera Sierra, July 2008.

3. Behre Dolbear, *Manifestación de Impacto Ambiental*, 280.

4. H. Becerra, "Amenazan opositores a MSX con tomar dependencias del gobierno," *Jornada de San Luis*, February 21, 2006

5. H. Becerra, "Con barricadas y persecucioines, activistas obligan a la MSX a suspender labores," *La Jornada de San Luis*, April 12, 2006.

6. E. G. Ruíz, "Vocera de MSX deslindo a la empresa de hechos occuridos la fin de semana," *La Jornada de San Luis*, December 12, 2006.

7. Herman, *Extracting Consent*, 14, 127.

8. Frente Amplio Opositor, *La consulta ciudadana Cerro de San Pedro—MSX.*

9. R.E. Pedraza, "Minera San Xavier descalifica consulta pública organizada por miembros del FAO," *La Jornada De San Luis*, October 25, 2006.

10. Nolin and LaPlante, "Consultas and Socially Responsible Investing in Guatemala"; Schnoor, *Governmentality and the Spirit of Extraction.*

11. Veronica Islas, "Meeting Crashers. Anti-Mining Activists Confront Shareholders at AGM," *The Dominion*, July 14, 2008.

12. Tobi Cohen, "Activists Protest Open-Pit Mines by Stalking Claim to Mount Royal," *Globe and Mail*, May 12, 2009.

13. Armando Mendoza, cited in *La mutilación de San Pedro*, dir. Oliva Portillo.

14. Interview with Pedro Rangel, March 2008.

15. Interview with Marcos Rangel, February 2006.

16. Email correspondence with Margarita Villalba, Mexican historian who worked to reorganize the parish archives in the late 1990s, February 20, 2018.

17. Interview with Aristeo Gutierrez, March 2008.

18. Marco Rangel, cited in *Muerte y vida de Cerro de San Pedro*, dir. Juan Carlos Valladolid.

19. "México, SLP, SubComandante Marcos & EZLN en el Cerro de San Pedro Minera $an Xavier," YouTube, October 19, 2007, https://www.youtube.com/watch?v=dLMp_oOZZLY.

20. Lola Rocha Perez, cited in *Muerte y vida de Cerro de San Pedro*.

21. Septién Prieto, "Borrar de la memoria a Cerro de San Pedro," *La Jornada San Luis*, November 10, 2005; Ruiz Guadalajara, "Vestigios de un prodigio."

22. Ruiz Guadalajara, "Vestigios de un prodigio."

Chapter 19

1. Noveno Tribunal Colegiado en Materia Administrative, June 23, 2004, Expediente D.A. 65/2004-873; Tribunal Federal de Justicia Administrativa, October 5, 2005, Expediente 170/00-05-02-9/634/01-PL-05-04.

2. This unfolded over linked rulings in two Mexican courts: (1) the decision of the Noveno Tribunal Colegiado en Materia Administrativa's sentence of April 17, 2009 (R.A. 59/2009) ruling that previous judgements concerning the inadmissibility of Minera San Xavier's environmental permit was beyond appeal, and (2) the Federal Tribunal of Administrative Justice's sentence of September 24, 2009 (Cumplimiento de ejecutoria exp. 170/00-05-029/634/01-PL-10-QC-DA) declaring the permit null and void and charging the SEMARNAT to advise the company and carry out the sentence.

3. Staff, "Legisladores de México y Canadá piden a Minera San Xavier cumplir con la ley," *La Jornada*, November 13, 2009.

4. Staff, "Mexican Environmental Agency Terminates New Gold's Cerro de San Pedro Mine "Definitively," *The Northern Miner*, November 20, 2009; Frente Amplio Opositor, *Press Release: Canadian Company Disobeys Court, Misleads Shareholders*, November 9, 2009.

5. Angélica Enciso L., "Minera San Xavier está operando legalmente, afirma la Semarnat," *La Jornada*, November 11, 2009; Staff, "SLP acatará disposiciones federales en caso de Minera San Xavier: Gobernador," *La Jornada*, November 20, 2009.

6. Staff, "Burla Minera San Xavier revocación de Semarnat," *La Jornada*, November 15, 2009.

7. Angélica Enciso L., "La delegación de Profepa llega a Cerro de San Pedro para realizar las inspecciones," *La Jornada*, November 19, 2009.

8. Staff, "Profepa coloca sellos de clausura a Minera San Xavier en SLP," *La Jornada*, November 19, 2009; Enciso L., "La delegación de Profepa llega a Cerro de San Pedro para realizar las inspecciones."

9. New Gold Inc., *Press Release: New Gold to Appeal Suspension of Mexican Mining Operation*, November 19, 2009.

10. Staff, "Demandarán penalmente a Minera San Xavier a al delegado de Sedesco," *La Jornada*, November 22, 2009. A week earlier criminal charges were also announced against the former state Governor Marcelo de los Santos, and Minera San Xavier's legal counsel Ángel Candia Pardo, see, Staff, "Obligada, la SEMARNAT revoca permiso a Minera San Xavier," *La Jornada*, November 14, 2009; and Patricia Rodríguez Calva, "Destapan compadrazgos de la Minera San Xavier," *La Jornada de San Luis*, March 2, 2009.

11. "Misa Trabajadores Minera San Xavier," YouTube, December 11, 2009, https://www.youtube.com/watch?v=MZBFYgFGxHs.

12. The first quote is from Carolina Solis; the second from Rosa Flores Cisneros, both recorded in "Misa de Trabajadores Minera San Xavier."

13. Interview with Armando Mendoza, August 2011; see also Amnesty International, "Urgent Action: Environmental Campaigner Threatened," UA:344/09, December 22, 2009.

14. New Gold Inc., *Press Release: New Gold Granted Injunction to Temporarily Lift Shutdown Order at Cerro de San Pedro*, December 14, 2009.

15. Interview with Hector Barri, August 2011.

16. Plan de desarrollo urbano de Cerro de San Pedro, S.L.P.

17. Benson and Kirsch, "Capitalism and the Politics of Resignation"; and Kirsch, *Mining Capitalism*.

18. Oakland Ross, "Tarnish on a Miniature El Dorado," *Toronto Star*, April 16, 2005.

19. Donner, "La dimension immunitaire des enclaves pétroliéres."

20. Peter Benton and Stuart Kirsch speak of the production of a structure of feeling, specifically resignation, that blocking efforts to defend against corporate harms. Benson and Kirsch, "Capitalism and the Politics of Resignation."

21. Danielson, *An Account of the Mining, Minerals and Sustainable Development Projects*, 26, 52.

22. Kirsch, "Sustainable Mining"; Whitmore, "The Emperor's New Clothes: Sustainable Mining?"

23. Dashwood, "Canadian Mining Companies and Corporate Social Responsibility"; Bridge and McManus, "Sticks and Stones"; Szablowski, *Transnational Law and Local Struggles*.

24. For overview and synthesis, see Studnicki-Gizbert, "Canadian Mining in Latin America."

25. Noyes and Young, "The Road to Summitville"; Mitchell and Warhurst. "Corporate Social Responsibility and the Case of the Summitville Mine."

26. Cambior was the operating partner with Metallica Resources in the early years of the CSP project. It ran the Omai mine in Guyana when its tailings dam

breached. Glamis Gold was Metallica Resources next. Among its corporate officers was Chester Miller, previously of Chemgold, operator of the Picacho mine also cited for a toxic spill in California, Office of Compliance, *Profile of the Metal Mining Industry*, 50.

27. Allan Robinson, "Canadian miners strike it rich. Hunt for gold in Latin America grabs attention, cash of speculators worldwide," *The Globe and Mail*, February 14, 1994.

28. Interview with Padre Marguerito Sánchez Grimaldo, May 2008.

29. Herman, "Extracting Consent."

30. Interview Jorge Mendizabal with T. Herman April–May 2008.

31. New Gold, "New Gold Corporate Social Responsibility," YouTube, June 21, 2010. https://www.youtube.com/watch?v=AvJzM_g8tEc; Interview Jorge Mendizabal with T. Herman, April–May 2008; Christmas gifting scene recorded in the Radio-Canada documentary *Les nouveaux conquistadores*, dir. Hélène Pichette.

32. The mine's fuel bill was an estimated USD 8.4 million per year.

33. Heather Scoffield, "Cambior Mine Fuels Controversy," *The Globe and Mail*, November 18, 1999.

34. Jorge Mendizabal interview with T. Herman, May 2008; Peter Foster, "Canada Funds NGOs to Push Suspect Views and Information," *Vancouver Sun*, August 15, 2007; Peter Foster, "New Gold Forced to Suspend Mine; Activist Opposition," *Financial Post*, November 20, 2009.

35. Jorge Mendizabal interview with T. Herman, May 2008; "Canada Funds NGOs to Push Suspect Views and Information."

36. Interview with Aristeo Gutierrez Chávez, March 2008.

37. Jorge Mendizabal interview with T. Herman ,May 2008; "They say we are the enemies of Progress," she [Ana Maria Alvarado] says, "Tarnish on a miniature El Dorado."

38. Angélica Enciso L., "Crimen organizado, involucrado en minería," *La Jornada*, November 26, 2014.

39. Jorge Mendizabal interview with T. Herman, May 2008.

40. Jorge Mendizabal interview with T. Herman, May 2008.

41. Juan Guerrero Peralta, interview with Radio Canada International, November 26, 2009.

42. "Cambior Mine Fuels Controversy."

43. Jorge Mendizabal interview with T. Herman, May 2008.

44. Studnicki-Gizbert, "Canadian Mining in Latin America."

45. Interview Mario Martinez, December 2014.

46. Office of the CSR Counsellor, *Field Visit Report 2013-04-MEX*.

Chapter 20

1. Interview with Enrique Rivera Sierra, May 2018.

2. Calculation by Ugo Lapointe, February 2007.

3. Kempton et al., "Managing Perpetual Environmental Impacts from New Hardrock Mines"; for the "Clock of the Long Now," http://longnow.org/clock/.

4. "Se desbordan piletas de MSX," *Pulso*, May 27, 2014.

5. "MSX cerraría por completo hasta el 2021," *El Expres*, August 29, 2015.

6. Interview Sergio Serrano, December 2014.

7. Interview Enrique Rivera Sierra, May 2018.

8. Though see Dawn Paley's *Drug War Capitalism*; Barbara Anderson, "El Crimen Organizado Se Dedica a La Minería," SIPSE.com, May 27, 2013; Staff, "Intrusión De Minera Canadiense Genera Desalojos Violentos En El Altiplano Potosino," *La Jornada*, February 9, 2012.

9. Interview Mario Martinez, December 2014.

10. Danielle Bochove, "Barrick's Digital Reinvention Taking Shape in Nevada Desert," Bloomberg, November 6, 2017

11. Mumford, *Technics and Civilization*, 70–71.

12. Dawdy, "Clockpunk Anthropology"; Gordillo, *The Afterlife of Destruction*; Stoler, "Imperial Debris"; Martina Teaiwa, *Consuming Ocean Island*.

Bibliography

Archival Collections

AGI—Archivo General de Indias
    Escribania
    Mapas y Planos
    Mexico
AHESLP—Archivo Historico del Estado de San Luis Potosí
    AM—Alcaldia Mayor

Dissertations and Theses

Abdulrahman Nzibo, Yusuf. "Relations between Great Britain and Mexico, 1820–1870." PhD diss., University of Glasgow, 1979.

Azucena Rodríguez López, Dulce. "'Andamos en las entrañas de la tierra.' Trabajo, corporabilidad y ritual en el Mineral de La Paz." MA thesis, Colegio de San Luis, 2011.

Chester, Robert N. "Comstock Creations: An Environmental History of an Industrial Watershed." PhD diss., University of California at Davis, 2009.

Dobado González, Rafael. "El trabajo en las minas de Almadén, 1750–1855." PhD diss. Universidad Complutense de Madrid, 1989.

Gámez Rodríguez, Moisés. "Organización y movilización de los mineros en San Luis Potosí, 1900–1913." MA thesis, Universidad Iberoamericana, 1996.

Herman, Tamara. "Extracting Consent or Engineering Support? An Institutional Ethnography of Mining, 'Community Support' and Land Acquisition in Cerro De San Pedro, Mexico." MA thesis, University of Victoria, 2010.

Jacobo-Marín, Daniel. "Agua para San Luis Potosí: una mirada desde el derecho humano al agua en dos sectores del ámbito urbano." MA thesis, Colegio de San Luis, 2013

Kortheuer, Dennis. "Santa Rosalia and Compagnie du Boleo: The making of a town and company in the Porfirian frontier, 1885–1900." PhD diss., University of California Irvine, 2001.

Luna Morales, Saul. "La fuerza de trabajo en la mineria de Chihuahua, 1880–1910." MA thesis, UNAM, 2000.

Marucci, Horace. "American Smelting and Refining Company in Mexico, 1900–1925." PhD diss., Rutgers University, 1995.

Medina Esquivel, René. "Sobrevivir en un pueblo minero. Vida cotidiana en Cerro de San Pedro, San Luis Potosí durante la posrevolución." MA thesis, Colegio de San Luis, 2008.

Menéndez Taboada, María Teresa. "Minería y enfermedad en Zimapán, Hidalgo. Estudio patológico de la colección esquelética del panteón de Santiago Apóstol." MA thesis, UNAM, 2009.

Montoya, Alejandro. "Población y sociedad en un real de minas de la frontera note Novohispana: San Luis Potosí de finales del siglo XVI a 1810." PhD diss., Université de Montréal, 2004.

Perales, Monica. "Smeltertown: A Biography of a Mexican American Community, 1880–1973." PhD diss., Stanford University, 2003

Pérez Navarro, Mónica. "Litigios y tumultos. Cultura política en Cerro de San Pedro y los ranchos de Soledad, 1760–1767." MA thesis, Colegio de San Luis, 2008.

Schnoor, Steven. "Governmentality and the New Spirit of Extraction. The Politics of Legitimacy and Resistance to Canadian Mining in Guatemala and Honduras." PhD diss., York University, 2013.

Serrano Hernández, Sergio Tonatiuh. "Minería: La articulación económica de San Luis Potosí en el siglo XVII." MA thesis, UNAM, 2014.

Silva Ontiveros, Letizia O. "Transformaciones del paisaje desde la explotación minera en la región central de San Luis Potosí." PhD diss., UNAM, 2014.

Winterbourne, David. "Alteration and Mineralization of the San Pedro Porphyry, Cerro San Pedro, San Luis Potosí, Mexico." PhD diss., Colorado School of Mines, 2000.

Interviews

Interview with Ana Maria Alvarado García, Cerro de San Pedro, March 2008.

Interview with Aristeo Gutierrez Chavéz, Cerro de San Pedro, March 2008.

Interview with Armando Mendoza, Cerro de San Pedro, March 2008; July 2008; August 2011.

Interview with Gabriel Muñoz, February 2006.

Interview Jorge Mendizabal with T. Herman, April—May 2008. https://archive .org/details/MineraSanXavierJorgeMendizabel.

Interviews with Marcos Rangel, Cerro de San Pedro. February 2006; February 2008.

Interviews with Mario Martínez, Cerro de San Pedro, February 2006; February 2008; July 2008; December 2014.

Interview with Padre Marguerito Sánchez Grimaldo, Cerro de San Pedro, May 2008.

Interview with Pedro Rangel, Cerro de San Pedro, March 2008.

Interview with Sergio Serrano Soria, San Luis Potosí, December 2014.

Interview with Tonantzín Mendoza, Cerro de San Pedro, March 2008.

R. S. Shoemaker, *Recollections of Frank Woods McQuiston Jr.* Interview conducted by Eleanor Swent, Western Mining in the Twentieth Century. Regional Oral History Office—Bancroft Library, University of California at Berkeley. https:// www.lib.berkeley.edu/libraries/bancroft-library/oral-history-center/projects /wm.

Media

Annand, Amanda. *[Mine] How One Mexican Town Is Navigating a Canadian Mine.* https://amandaannand.atavist.com/mine.
Frigon, Marin, dir. *Mirages d'un El Dorado.* Difusion Multimonde. 2008.
"Misa de Trabajadores Minera San Xavier." YouTube, December 11, 2009. https://www.youtube.com/watch?v=MZBFYgFGxHs.
New Gold. "New Gold Corporate Social Responsibility." YouTube, June 21, 2010. https://www.youtube.com/watch?v=AvJzM_g8tEc.
Pichette, Hélène, dir. *Les nouveaux conquistadores.* Société Radio-Canada. 2010.
Portillo, Oliva, dir. *La mutilación de San Pedro según San Xavier.* UNAM. 2007.
Valladolid Bocanegra, Juan Carlos, dir. *Muerte y vida de Cerro de San Pedro.* 2014. https://www.youtube.com/watch?v=nlufzXlIBjk.

Newspapers

El Colmillo Público
Cultural Survival QuarterlyDaily Mail
The Dominion
The Economist
The Globe and Mail
El Heraldo
El Informador

Journal of Commerce
Multinational Monitor
El Pais (Mexico City)
Proceso
Pulso
Vancouver Sun

Unpublished Sources

Alvarado García, Ana María. *Mí pueblo viejo—Cerro de San Pedro.* Unpublished manuscript, 2009.
Alvarez Land and Timber Co. *General Ledger, 1898–1953, Compañía Metalúrgica Mexicana and Related Enterprises: Financial Records.* Benson Latin American Collection, University of Texas at Austin.
Anonymous. *Ciudades, Villas y Lugares, reales de Minas y Congregaciones de Espanoles de el Obispado de Mechoacan,* 1649 Newberry Library Ayer MS 1106 A.
———. *Descripción de los Lugares, Confines, y Rentas del Obispado de Mechoacan* [Michoacan] 1639. Newberry Library. Ayer Ms collection, 1106 C3.
———. *Informe del mineral de San Pedro, 1881.* Yale Collections of Latin American Manuscripts. Folder 1289.
Asamblea El Algarrobo. *Comunicado De los vecinos por la vida de Andalgalá.* n.p., December 16, 2009.
Basalenque, Diego. "Historia de la Provincia de San Nicolas de Tolentino de Michoacan." 1673.
Behre Dolbear de Mexico. *Manifestación de Impacto Ambiental, Modalidad General: Proyecto de Explotación Minera San Pedro, Cerro de San Pedro, S.L.P.* May 1997.

Bonney, Wilbert L. *San Luis Potosí—Supplement to Commerce Reports No 32A*. April 30, 1915.

Bureau of Mines. "Production Potential of Known Gold Deposits in the United States." Bureau of Mines Information Circular no. 8331. Washington, DC: US Government Printing Office, 1967.

Camilo Bros. "Sobre el mineral de Cerro de San Pedro." *Minero Mexicano*, no. 30, April 24, 1879.

Chamberlain, Peter, and Michael Pojar. "Gold and Silver Leaching Practices in the United States." Bureau of Mines Information Circular no. 8969. Washington, DC: US Government Printing Office, 1984.

Charles I. "La pregmatica que su magestad ha mandado hazer este ano de MDLII de la pena que ban de aver los Ladrones y rufianes y Vagamundos." Alcala de Henares: Salzedo, 1552.

CIPER. "'El MINEM tiene registrados 6847 pasivos ambientales mineros' Pasivos Ambientales Mineros en Perú: bombas de tiempo de las que nadie se hace responsable." http://ciperchile.cl/multimedia/ciper-presenta-serie-especial-de-reportajes-sobre-la-industria-minera-en-chile-peru-y-colombia/.

Comisión de Medio Ambiente de la USALP. "Opinión técnico-científica sobre los componentes ambientales del proyecto Cerro de San Pedro de Minera San Xavier." San Luis Potosí, December 1998.

del Campo, J.M.G. *Reseña del Mineral del Cerro de San Pedro Octubre 31, 1881, San Luis Potosi*. Yale University Collection of Latin American Manuscripts. Folder 1289.

Dirección General de Minas (DGM). "Proyectos mineros operados por compañías de capital extranjero." 2010.

Domínguez de la Fuente, Manuel. "Leal informe político legal de reflejas y observaciones deducidas de la físico-práctico—mecánico maniobra de las minas." Guanajuato, 1774

Estado de San Luis Potosí. "Plan de desarrollo urbano de Cerro de San Pedro, S.L.P." Periódico oficial del estado de San Luis Potosí, March 26, 2011.

———. "Plan de Ordenación de San Luis Potosí y su Zona Conurbana." Publicado en el periódico oficial del estado el 24 de junio de 1993.

Fabry, Jose Antonio, "Compendiosa demonstracion de los crecidos adelantamientos que pudiera lograr la Real Hacienda de Su Magestad mediante la rebaja en el precio del azogue." Mexico City, 1743.

Foro Biregional Huancavelica—Junín de Comunidades Afectadas por la Mineria. "Declaración de Huancayo." Huancayo, Peru, July 23, 2011.

Frente Amplio Opositor. "Comunicado: La coordinacion de la consulta ciudadana Cerro de San Pedro—MSX: Publica informe y los resultados definitivos del evento." October 30, 2006.

Gálvez, José de. "Informe sobre el estado de Mexico, California, Sonora y Provincias Remotas de Nueva Espana, 1768–1778." Newberry Library, Ayer ms. 1091.

González Márquez, José Juan, and Ivett Montelongo Buenavista. "Minera San Xavier. Grado de cumplimiento con la legislación ambiental mexicana."

Unpublished report. Universidad Autonoma Metropolitano—Azcapotzalco, 2010.

Heffernan, Virginia. "Viola Heffernan: From the Ground Up: An Autobiography" (Afterword) Republic of Mining (blog), January 18, 2011. https://republicof mining.com/2011/01/18/viola-macmillan-from-the-ground-up-an-autobiograpy -afterword-by-virginia-heffernan-part-1-of-2/.

Kelly, T. D., and G. R. Matos, comps. "Historical Statistics for Mineral and Material Commodities in the United States (2018 Version): U.S. Geological Survey Data Series 140." https://minerals.usgs.gov/minerals/pubs/historical-statistics/.

Manahan, R. F. "Mining Operations of Mining Operations of ASARCO in Mexico." n.p., 1948.

Manjón, Lucas Fernandez. "Memorial de Lucas Fernandez Manjon, vezino del pueblo y minas de San Luis Potosi." British Library 725.k.18 (7). Madrid, 1627.

Noveno Tribunal Colegiado en Materia Administrative. "Expediente D.A. 65/2004-873." June 23, 2004.

Office of Compliance—U.S. Environmental Protection Agency. "EPA Office of Compliance Sector—Profile of the Metal Mining Industry." Washington, DC: Office of Enforcement and Compliance Assurance, 1995).

Office of the Extractive Sector, Corporate Social Responsibility (CSR) Counsellor. "Field Visit Report 2013-04-MEX." June 2013. http://www.international.gc.ca /csr_counsellor-conseiller_rse/publications/2013-04-MEX_interim_rep_01-rap _provisoir_01.aspx?lang=eng.

Palacios Garcia, Raúl. "La Fundición de Morales. Símbolo y Motor del Desarrollo Industrial en la Capital Potosina, 1890–2010." n.p., n.d.

Pellicer y Osau, José. "El Comercio Impedido." Bib Nac Madrid VE 35–86. 1640.

"Reales Ordenanzas Para la Dirección, Régimen y Gobierno del Importante Cuerpo de la Minería de Nueva España." Guanajuato: Juan E. Oñate, [1784] 1845.

Shapleigh, Eliot. "Towards a Brighter Future—Away from a Polluted Past." Austin TX: Senate of the State of Texas, 2007.

Thompson, E. G. "A Brief History of the PDAC, 1932–2000." Unpublished manuscript, 2002.

Tribunal Federal de Justicia Administrativa (Mexico). "Expediente 170/00-05-02-9/634/01-PL-05-04." October 5, 2005.

Published Sources

Edited Collections

Cedulario de tierras: Compilación de legislación agraria colonial (1497–1820). Editor Francisco Solano. Mexico DF: UNAM, 1991.

Colección de documentos ineditos para la historia de España (CODOIN). Editor Martín Fernández de Navarrete (1842–95; repr. Vaduz, Liechtenstein: Kraus, 1965).

Documentos para la historia de San Luis Potosí. Editor Primo Feliciano Velásquez. San Luis Potosí: Imprenta del Editor, 1897.

Documentos Sobre el Conflicto de Trabajo con los mineros de Real del Monte. Año de 1766. Editor Luis Chavez Orozco. Mexico: Instituto de Nacional de Estudios Historicos de la Revolucion Mexicana, 1960.

Documentos y grabados para la historia de San Luis Potosí. Editor María Isabel Monroy. Mexico: Archivo Historico del Estado de San Luis Potosí-Casa de la Cultura, 1991.

Las Minas de Nueva España en 1774. Editor Alvaro López Miramontes and Cristina Urrutia de Stebelski. Mexico: SEP—INAH, 1980.

Recopilación De Las Leyes De Los Reynos De Las Indias (1681). Madrid: Por la viuda de D. Joaquin Ibarra, 1791).

Los Virreyes Españoles en America Durante el Gobierno de la Casa de Austria. Editor Lewis Hanke. Madrid: Atlas, 1976.

Monographs and Articles

Abbott, Mark, and Alexander Wolfe. "Intensive Pre-Incan Metallurgy Recorded by Lake Sediments from the Bolivian Andes." *Science* 301, no. 5641 (September 2003): 1893–95.

Acosta, Alberto. "Extractivism and Neo-Extractivism: Two Sides of the Same Curse." In *Beyond Development: Alternative Visions from Latin America,* ed. M. D. Lang and D. Mokrani, 61–86. Amsterdam: Transnational Institute, 2013.

Acosta, José de. *The naturall and morall historie of the East and West Indies.* Translated by E. G. London: Val, 1604.

Alamán, Lucas. *Autobiografía, Documentos Diversos (Inéditos y muy raros) Tomo 4.* Mexico: Editorial Jus, 1947.

———. *Memoria presentada a las dos cámaras del Congreso General de la Federación. 1825. Documentos Diversos (Inéditos y muy raros) Tomo 1.* Mexico: Editorial Jus, 1947.

Altman, Ida. *The War for Mexico's West: Indians and Spaniards in New Galicia, 1524–1550.* Albuquerque: University of New Mexico Press, 2010.

Álvarez, W. C. "Ankylostomiasis in Mexico and its Diagnosis." *Journal of the American Medical Association* 52 (1909): 1388–90.

Álvarez Nogal, Carlos, and Christophe Chamley. "Debt Policy under Constraints: Philip II, the Cortes, and Genoese Bankers." *Economic History Review* 67, no. 1 (2014): 192–213.

Anderson, Rodney D. *Outcasts in Their Own Land. Mexican Industrial Workers, 1906–1911.* DeKalb: Northern Illinois University Press, 1976.

Anonymous. "Contaminación de la aguas de corrientes nacionales por desechos de las haciendas beneficiadores de metales." *Boletin Minero* (1927): 202–10.

———. "El texto de las Leyes de Burgos de 1512," edited by Rafael Altamira. *Revista de Historia de América* 4 (December 1938): 5–79.

———. *Reales Ordenanzas Para la Dirección, Régimen y Gobierno del Importante Cuerpo de la Minería de Nueva España* (1784). Guanajuato: Juan E. Oñate, 1845.

———. "Relación de Las Minas de Çimapan." In *Relaciones Geograficas del Siglo XVI: Mexico*, ed. Rene Acuña, vol. 6, 95–109. Mexico: Instituto de Investigaciones Antropologicas—UNAM, 1984.

———. "Special Map Supplement: The Mines and Railways of Mexico." *Transactions of the American Institute of Mining Engineers* 32 (1903).

Aragón-Piña, Antonio et al. "Influencia de Emisiones Industriales en el polvo historico de la ciudad de San Luis Potosí, Mexico." *Revista Internacional de Contaminación Ambiental* 22, no. 1 (2006): 5–15.

Arnaiz y Freg, Arturo. "Don Fausto de Elhuyar y de Zubice." *Revista de Historia de Américas* 6 (1939): 75–96.

Aschmann, Homer. "The Natural History of a Mine." *Economic Geography* 46, no. 2 (1970): 172–89.

Assadourián, Carlos Sempat. "La producción de la mercancia dinero en la formación del mercado interno colonial." In *Ensayos sobre el desarrollo económico de México y América Latina, 1500–1975*, ed. Enrique Florescano, 223–92. Mexico DF: Fondo de Cultura Económica, 1979.

———. *El sistema de la economía colonial. El mercado interior, regiones y espacio económico*. México, Nueva Imagen, 1983.

Bakewell, Peter. "La periodización de la producción minera en el norte de la Nueva España durante la Època colonial." *Estudios de Historia Novohispana*, 10 (1991): 31–43.

———. *Silver Mining and Society in Colonial Mexico. Zacatecas, 1546–1700*. New York: Cambridge University Press, 1971.

Balliet, Letson. "Wanted—Light and Air." *Mexican Mining Journal* 18 (January 1914): 7–8.

Barba, Alvaro Alonso. *El Arte de los Metales*. Translated by Ross E. Douglass and E. P. Mathewson. New York: John Wiley and Sons, 1923.

Barca, Stefania. "Bread and Poison. Stories of Labor Environmentalism in Italy, 1968–1998." In *Dangerous Trade: Histories of Industrial Hazard across a Globalizing World*, ed. Christopher Sellers and Joseph Selling, 125–39. Philadelphia: Temple University Press, 2011.

Barragán, Rossana. "Working Silver for the World: Mining Labor and Popular Economy in Colonial Potosí." *Hispanic American Historical Review* 97, no. 2 (2017): 193–222.

Barrera de la Torre, Gerónimo. "El paisaje de Real de Catorce: un despojo histórico." *Investigaciones Geográficas (UNAM)* 81 (2013): 110–125.

Barrera-Osorio, Antonio. *Experiencing Nature: The Spanish American Empire and the Early Scientific Revolution*. Austin: University of Texas Press, 2006.

Bauer, Ralph. *The Alchemy of Conquest: Science, Religion, and the Secrets of the New World*. Charlottesville: University of Virginia Press, 2019.

Bebbington, Anthony. "The New Extraction: Rewriting the Political Ecology of the Andes?" *NACLA Report on the Americas* 42, no. 5 (2009): 12–20, 39–40.

Bebbington, Anthony, and Jeffrey Bury. *Subterranean Struggles: New Dynamics of Mining, Oil, and Gas in Latin America*. Austin: University of Texas Press, 2013.

Behar, Ruth. "The Visions of a Guachichil Witch in 1599: A Window on the Subjection of Mexico's Hunter-Gatherers." *Ethnohistory* 34, no. 2 (1987): 115–38.

Bellamy Foster, John. "Marx's Theory of Metabolic Rift: Classical Foundations for Environmental Sociology." *American Journal of Sociology* 105, no. 2 (1999): 366–405.

Benson, Peter, and Stuart Kirsch. "Capitalism and the Politics of Resignation." *Current Anthropology* 51, no. 4 (2010): 459–86.

Benton, Lauren. *A Search for Sovereignty: Law and Geography in European Empires, 1400–1900*. New York: Cambridge University Press, 2009.

Bernstein, Marvin D. *The Mexican Mining Industry, 1890–1950: A Study of the Interaction of Politics, Economics, and Technology*. Albany: State University of New York, 1965.

Berthelot, Jean. "L'exploitation des metaux précieux au temps des Incas." *Annales. Economies Sociétés, Civilizations* 33, no. 5–6 (1978): 948–66.

Betancor, Orlando. *The Matter of Empire. Metaphysics and Mining in Colonial Peru*. Pittsburgh: University of Pittsburgh Press, 2017.

Bigelow, Allison Margaret. *Mining Language: Racial Thinking, Indigenous Knowledge, and Colonial Metallurgy in the Early Modern Iberian World*. Williamsburg, VA and Chapel Hill NC: Omohundro Institute of Early American History and Culture and the University of North Carolina Press, 2020.

———. "Women, Men, and the Legal Languages of Mining in the Colonial Andes." *Ethnohistory* 63, no. 2 (2016): 351–80.

Birrichaga Gardida, Diana. *Agua e Industria en Mexico. Documentos sobre impacto ambiental y contaminación, 1900–1935*. Mexico: CIESAS, 2008.

———. "El dominio de las 'aguas ocultas y descubiertas': Hidraulica colonial en el centro de Mexico, siglos XVI–XVII." In *Mestizajes tecnologicos y cambios culturales en Mexico*, ed. Enrique Florescano and Virginia Garcia Acosta, 94–130. Mexico: CIESAS, 2004.

Bogitsh, Burton J., Thomas N. Oeltmann, and Clint E. Carter. *Human Parasitology*. 3rd ed. New York: Academic Press, 2005.

Bond Head, Francis. *Rough Notes Taken euring Some Rapid Journeys across the Pampas and Among the Andes*. London: J. Murray, 1826.

Boorstein Couturier, Edith. *The Silver King: The Remarkable Life of the Count of Regla in Colonial Mexico*. Albuquerque: University of New Mexico Press, 2003.

Borah, Woodrow. "Un gobierno provincial de frontera en San Luis Potosí (1612–1620)." *Historia Mexicana* 13, no. 4 (1964): 537–38.

Brading, David A. "Mexican Silver Mining in the Eighteenth Century: The Revival of Zacatecas." *Hispanic American Historical Review* 50, no. 4 (1970): 665–81.

Braudel, Fernand. *Civilization and Capitalism. Vol. I: The Structures of Everyday Life*. Translated by Siân Reynolds. Berkeley: University of California Press, 1992.

Braun, Bruce. "Producing Vertical Territory: Geology and Governmentality in Late Victorian Canada." *Ecumene* 7, no. 1 (2000): 7–46.

Bridge, Gavin. "Mapping the Bonanza: Geographies of Mining Investment in an Era of Neoliberal Reform" *The Professional Geographer* 56, no. 3 (2004): 401–21.

———. "Resource Triumphalism: Postindustrial Narratives of Primary Commodity Production." *Environment and Planning A* 33 (2001): 2149–73.

Bridge, Gavin, and Phil McManus. "Sticks and Stones: Environmental Narratives and Discursive Regulation in the Forestry and Mining Sectors." *Antipode* 32, no. 1 (2000): 10–47.

Bros, Camilo. "Sobre el mineral de Cerro de San Pedro." *El Minero Mexicano* 6, no. 30 (1879).

Brown, Kendall W. *A History of Mining in Latin America: From the Colonial Era to the Present.* Albuquerque: University of New Mexico Press, 2012.

———. "Workers' Health and Colonial Mercury Mining at Huancavelica, Peru." *The Americas* 57, no. 4 (April 2001): 467–96.

Bunker, Stephen. *Underdeveloping the Amazon: Extraction, Unequal Exchange. And the Failure of the Modern State.* Urbana: University of Illinois Press, 1985.

Bureau of Labor Statistics, US Department of Labor. "Industrial Accidents [notes]." *Monthly Labor Review* 25, no. 6 (1927): 75–76.

———. "Industrial Accidents [notes]." *Monthly Labor Review* 36, no. 5 (1933): 1070–71.

Bureau of Mines. *Production Potential of Known Gold Deposits in the United States.* Bureau of Mines Information Circular no. 8331. Washington, DC: US Government Printing Office, 1967.

Bury, Jeffrey. "Livelihoods in Transition: Transnational Gold Mining Operations and Local Change in Cajamarca, Peru." *Geographical Journal* 170 (2004): 78–91.

———. "Mining Mountains: Neoliberalism, Land Tenure, Livelihoods and the New Peruvian Mining Industry in Cajamarca" *Environment and Planning A* 37, no. 2 (2005): 221–39.

Bustamante, Margarita Villalba. "El trabajo en las minas de Guanajuato durante la segunda mitad del siglo XVIII." *Estudios de Historia Novohispana* 48 (2013): 35–83.

Calvillo Unna, Tomás. *La fundición de Morales: Una inversión norteamericana durante los gobiernos de la Revolución en San Luis Potosí.* San Luis Potosí: El Colegio de San Luis A.C., 2010.

Camacho Bueno, Anagricel. "'El trabajo mata': Los mineros-metalúrgicos y sus enfermedades en el Primer Congreso Nacional de Higiene y Medicina del Trabajo, México, 1937." *Trashumante. Revista Americana de Historia Social* 7 (2016): 152–71.

Cañete y Domínguez, Pedro Vicente. *Guía Histórica, geográfica, física, política, civil y legal del gobierno e intendencia de la Provincia de Potosí.* Potosí: Editorial Potosí, [1787]1952.

Capoche, Luis. *Relación General de la Villa Imperial de Potosi.* Ed. Lewis Hanke Biblioteca de Autores Espanoles, vol. 122. Madrid: Atlas, [1585] 1959.

Carbajal López, David. *La minería en Bolaños, 1748–1810. Ciclos productivos y actores económicos.* Zamora: Colegio de Michoacan, 2002.

Cárdenas García, Nicolas. *Empresas y trabajadores en la gran minería mexicana (1900–1929). La Revolución y el nuevo sistema de relaciones laborales.* Mexico: Inst. Nacional de Estudios Historicos de la Revolución Mexicana, 1988.

———. "Proceso de Trabajo y Resistencia Obrera. Los mineros mexicanos en los años veinte," *Argumentos* (Mexico), 10/11 (December 1990): 32–46.

Calderón, J., M. E. Navarro, M. E. Jimenez-Capdeville, M. A. Santos-Díaz, A. Golden, I. Rodríguez-Leyva, V. Borja-Aburto, and F. Díaz-Barriga. "Exposure to Arsenic and Lead and Neuropsychological Development in Mexican Children." *Environmental Research Studies, Section A* 85 (2001): 69–76.

Carrizales, L., I. Razo, J. I. Téllez-Hernández, R. Torres-Nerio, A. Torres, et al. 2006. "Exposure to Arsenic and Lead of Children Living Near a Copper-Smelter in San Luis Potosi, Mexico: Importance of soil Contamination for Exposure of Children." *Environmental Research* 101 (1): 1–10.

Carruthers, David. *Environmental Justice in Latin America.* Cambridge: MIT Press, 2008.

Castañeda Delgado, Paulino. "Un capitulo de ética Indiana española: Los trabajos forzados en las minas." *Anuario de Estudios Americanos* 27 (1970): 815–916.

Castro Gutiérrez, Felipe. *Nueva Ley y Nuevo Rey. Reformas borbónicas y rebelión popular en Nueva España.* Zamora: El Colegio de Michoacan, 1996.

Cavioto, Francisco. *La legislación minera vigente en México.* Mexico: FUNDAR—Brot für die Welt, 2013.

Chamberlain, Peter, and Michael Pojar. *Gold and Silver Leaching Practices in the United States.* Bureau of Mines Information Circular no. 8969. Washington, DC: US Government Printing Office, 1984.

Change, L. W., and R. B. Tjalkens. "Neurotoxicology of Metals." In *Comprehensive Toxicology. Vol. 13: Nervous System and Behavioral Toxicology,* 483–97. Oxford: Elsevier, 2010.

Chism, Richard. "Sierra Mojada, Mexico." *Transactions of the American Institute of Mining Engineers* 15 (1886–87): 542–88.

———. "A Synopsis of the Mining Laws of Mexico." *Transactions of the American Institute of Mining Engineers* 22 (1902): 3–55.

Chiu Leung, Chi, Ignatius Tak Sun Yu, and Weihong Chen. "Silicosis." *The Lancet* 379 (2012): 2008–18.

Clark, Brett, and John Bellamy Foster. "Ecological Imperialism and the Global Metabolic Rift. Unequal Exchange and the Guano/Nitrates Trade." *International Journal of Comparative Sociology* 50, no. 3–4 (2009): 311–34.

Cobo, Bernabé. *Historia del Nuevo Mundo.* Seville: E. Rasco, 1890–95.

Colmenares, Germán. *Cali: Terratenientes, mineros y comerciantes, siglo XVIII.* 4th ed. Bogotá: TM Editores, 1997.

Colonel Tepetate (pseudonym). "Safety First." *Mexican Mining Journal* 18 (September 1915).

Columbus, Christopher. *The Diario of Christopher Columbus's First Voyage to America, 1492–93.* Abstracted by Bartolomé de Las Casas, transcribed and translated by Oliver Dunn and James E. Kelley, Jr. Norman: University of Oklahoma Press, 1989.

———. *Select letters of Christopher Columbus: With other original documents relating to his four voyages to the New World.* London: Printed for the Hakluyt Society, 1870.

Cook, Sherburne F. *The Historical Demography and Ecology of the Teotlalpán.* Berkeley: University of California Press, 1949.

Coope, J. Alan. *Carlin Trend Exploration History: Discovery of the Carlin Deposit.* Reno: MacKay School of Mines—University of Nevada, 1991.

Corbeil, Laurent. *The Motions Beneath: Indigenous Migrants on the Urban Frontier of New Spain.* Tucson: University of Arizona Press, 2018.

Cornblit, Oscar. *Power and Violence in the Colonial City: Oruro from the Mining Renaissance to the Rebellion of Tupac Amaru (1740–1782).* Translated by Elizabeth Ladd Glick. New York: Cambridge University Press, 1995.

Coronil, Fernando. *The Magical State: Nature, Money, and Modernity.* Chicago: University of Chicago Press, 1997.

Cortés, Hernán. "De los muchos descubrimientos de minas y las expediciones que envio Hernan Cortes." In *Antologia minera de Mexico. Primera estación, siglo XVI,* ed. Javier Moctezuma Barragan and Sergio Pelaez Parell, 35–37. Mexico DF: Sec. de Energia, Minas e Indutria Paraestatal, 1994.

Cosío Villegas, Daniel. *Historia Moderna de Mexico. El Porfirato. Vol. 8, part 2. La Vida Económica.* Mexico: Editorial Hermes, 1965.

Cox, P. A. *The Elements: Their Origin, Abundance and Distribution.* New York: Oxford University Press, 1989.

Craig, Alan K. "The Ingenious Ingenios: Spanish Colonial Water Mills at Potosi." In *Culture, Form and Place. Essays in Cultural and Historical Geography,* ed. Kent Mathewson, 125–56. Baton Rouge, LA: Geoscience and Man, 1993.

Cramaussel, Chantal. "Sociedad colonial y depredacion ecologica: Parral en el siglo XVII." In *Estudios sobre historia y ambiente en America I: Argentina, Bolivia, Mexco, Paraguay,* ed. Bernardo Garcia Martinez and Alba Gonzalez Jacome, 93–107. Mexico DF: Colegio de Mexico—Instituto Panamericano de Geografia e Historia, 1999.

Cruise, David, and Alison Griffiths. *Fleecing the Lamb: The Inside Story of the Vancouver Stock Exchange.* Vancouver: Douglas and McIntyre, 1987.

Cruz, Pablo, and Pascale Absi. "Cerros ardientes y huayras calladas: Potosí antes y durante el contacto." In *Minas y metalúrgica en los Andes del Sur: Desde la época prehispánica hasta el siglo XVII,* 91–120. La Paz: Institut de Recherche pour le Développement—Representación en Bolivia, 2008.

Cumberland, Charles Curtis. *Mexican Revolution. Genesis Under Madero.* Austin: University of Texas Press, 1952.

Dahlgren, Charles. *Historic Mines of Mexico. A Review of the Mines of That Republic for the Past Three Centuries.* New York: Printed for the Author, 1883.

Danielson, Luke. *Architecture for Change: An Account of the Mining, Minerals and Sustainable Development Projects.* Berlin: Global Public Policy Institute, 2006.

Danks, Noblet Barry. "The Labor Revolt of 1766 in the Mining Community of Real del Monte." *The Americas* 44, no. 2 (1987): 143–65.

Dashwood, Helvina. "Canadian Mining Companies and Corporate Social Responsibility: Weighing the Impact of Global Norms." *Canadian Journal of Political Science* 40 (2007): 129–56.

Davidson, Todd, et al. "Selected Molecular Mechanisms of Metal Toxicity and Carginogenicity." In *Handbook on the Toxicology of Metal*, 79–84. Amsterdam: Academic Press, 2007.

Dawdy, Shannon Lee. "Clockpunk Anthropology and the Ruins of Modernity." *Current Anthropology* 51, no. 6 (2010): 761–93.

Dawson, Frank G. *The First Latin American Debt Crisis: The City of London and the 1822–25 Loan Bubble*. New Haven: Yale University Press, 1990.

Del Rio, Ignacio. "Auge y decadencia de los placeres y del real de la Cieneguilla, Sonora (1771–1783)." *Estudios de Historia Novohispana* 8 (1975): 81–98.

Deneault, Alain, and William Sacher. *Imperial Canada Inc: Legal Haven of Choice for the World's Mining Industries*. Vancouver: Talonbooks, 2012.

Dobado, Rafael, and Gustavo A. Marrero, "The Role of the Spanish Imperial State in the Mining-Led Growth of Bourbon Mexico's Economy." *Economic History Review* 64, no. 3 (2011): 855–84.

Dominguez, Norberto. "The District of Hidalgo del Parral, Mexico in 1820." *Transactions of the American Institute of Mining Engineers* 23 (1903): 396–443.

Donner, Nicolas. "Notes sur la dimension immunitaire des enclaves pétroliéres." *EchoGéo* 17 (2011). https://journals.openedition.org/echogeo/12555#quotation.

Dore, Elizabeth. "Environment and Society—Long-Term Trends in Latin American Mining." *Environment and History* 6 (2000): 1–29.

Dumett, Raymond. *Mining Tycoons in the Age of Empire, 1870–1945: Entrepreneurship, High Finance, Politics and Territorial Expansion*. Farnham, England: Ashgate, 2009.

Dupont, Saint Clair. *De la production des Métaux Precieux au Mexique considerée dans ses rapports avec la Geologie, la Métallurgie et l'Economie Politique*. Paris: Firmin Didot Frères, 1843.

Eggert, Roderick. *Metallic Mineral Exploration: An Economic Analysis*. Washington, DC: Resources for the Future, 1987.

Elhuyar, Fausto de. *El Influjo de la Mineria en la Agricultura, Industria, Poblacion y Civilizacion de la Nueva Espanna en sus diferentes épocas, con varias disertaciones relativas a puntos de economia publica conexos con el propio ramo*. 2nd ed. Mexico DF: Tip. Literaria de F. Mata, [1825] 1883.

Emmons, Samuel Franklin. *Geology and Mining Industry of Leadville, Colorado. With Accompanying Atlas*. Washington, DC: USGS—Government Printing Office, 1886.

Escobar, Arturo. *Encountering Development: The Making and Unmaking of the Third World*. Princeton: Princeton University Press, 1994.

Espinosa Baquero, Armando. "Datos sobre la explotación y el beneficio de los metales preciosos en Nueva Granada en la epoca colonial." In *Mineria y metalurgia. Intercambio tecnologico y cultural entre America y Europa durante el periodo colonial espanol*, ed. Manuel Castillo Martos, 483–503. Sevilla-Bogota: Munoz Moya y Montraveta editores, 1994.

Estrada, Adriana, and Helena Hofbauer. *Impactos De La Inversion Minera Canadiense En Mexico: Una Primera Aproximacion*. Mexico: FUNDAR, 2001.

Estrada, Roque. *La revolución y Francisco I. Madero*. Mexico DF: Instituto de Estudios Históricos de la Revolución Mexicana, 1985.

Ettling, John. "Hookworm Infection." In *The Cambridge Historical Dictionary of Disease*, ed. Kenneth F. Kiple, 784–88. Cambridge: Cambridge University Press, 2003.

Fabry, Jose Antonio. *Compendiosa demonstracion de los crecidos adelantamientos que pudiera lograr la Real Hacienda de Su Magestad mediante la rebaja en el precio del azogue*. Mexico City, 1743.

Fay, Albert H. *A Glossary of the Mining and Mineral Industry*. Washington, DC: Government Printing Office, 1920.

Feliciano Velázquez, Primo. *Historia de San Luis Potosí*. San Luis Potosí: Casa del autor, 1947.

Fernández de Oviedo y Valdés, Gonzalo. *Historia general y natural de las Indias*. Edited by Jose Amador de los Rios, 1851–55. Madrid: Imprenta de la Real academia de la historia.

Ferry, Elizabeth E., and Mandan E. Limbert. *Timely Assets: The Politics of Resources and Their Temporalities*. Santa Fe, NM: School for Advanced Research Press, 2008.

Flores Clair, Eduardo. "Trabajo, salud y muerte: Real del Monte, 1874." *Siglo XIX Cuadernos de Historia* 3, no. 3 (1992): 9–28.

Flynn, Dennis, and Arturo Giráldez. "Born with a 'Silver Spoon': The Origins of World Trade in 1571." *Journal of World History* 6, no. 2 (1995): 201–21.

Fors, Hjalmar. *The Limits of Matter: Chemistry, Mining, and Enlightenment*. Chicago: University of Chicago Press, 2014.

Fowler, Bruce A., et al. "Arsenic, Antimony and Bismuth." In *Patty's Toxicology*, ed. Eula Bingham, Barbara Cohrssen, and F. A. Patty, 475–510. Hoboken, NJ: John Wiley & Sons, 2012.

Francaviglia, Richard V. *Hard Places. Reading the Landscape of America's Historic Mining Districts*. Iowa City: University of Iowa Press, 1991.

Franck, Harry A. *Tramping through Mexico, Guatemala, and Honduras*. New York: Century Company, 1916.

French, William E. *A Peaceful and Working People: Manners, Morals and Class Formation in Northern Mexico*. Albuquerque: University of New Mexico Press, 1996.

Gade, Daniel. "Landscape, System, and Identity in the Post-Conquest Andes." *Annals of the Association of American Geographers* 82, no. 3 (1992): 460–77.

Gálvez, José de. *Informe sobre las rebeliones populares de 1767 y otros documentos ineditos*. Editor Felipe Castro Gutiérrez. Mexico: UNAM, 1990.

———. "Instrucción particular para el restablecimiento y gobierno de las minas de Real del Monte y demás comprendidas en el dsitrito de las Cajas Reales de Pachuca (1771)." In *Los salarios y el trabajo en México durante el siglo XVIII*, ed. Luis Chávez Orozco, 48–58. México: Centro de Estudios Históricos del Movimiento Obrero Mexicano, 1978.

Gamboa, Francisco Xavier de. *Commentaries on the Mining Ordinances of New Spain.* Translated by Richard Heathfield. 2 vols. London: Longman, Rees, Orme, Brown & Green, [1761] 1830.

Garcia, Trinidad. *Los mineros mexicanos. Colección de documentos.* Editor José A. García. Mexico: Editorial Porrua, S.A. 1970.

Garibay, Claudio, and Alejandra Balzaretti Camacho. "Goldcorp y la reciprocidad negativa en el paisaje minero de Mezcala, Guerrero." *Desacatos* 30 (2009): 91–110.

Garner, Richard L. "Long-Term Silver Mining Trends in Spanish America: A Comparative Analysis of Peru and Mexico." *American Historical Review* 93, no. 4 (October 1988): 898–935.

de la Garza Brito, Angel. "La anquilostomasia en el estado de Hidalgo." *Boletin de industria, comercio y trabajo* 1 (1918): 138–41.

Gavira Márquez, and María Concepción. "Disciplina laboral y códigos mineros en los virreinatos del Río de la Plata y Nueva España a fines de la época colonial." *Relaciones* 102 (2005): 201–32.

Gerbi, Antonello. *Nature in the New World. From Christopher Columbus to Gonzalo Fernandez de Oviedo.* Translated by Jeremy Moyle. Pittsburgh: University of Pittsburgh Press, 1985.

Gillespie, Richard. "Accounting for Lead Poisoning: The Medical Politics of Occupational Health." *Social History* 15, no. 3 (1990): 303–31.

Gilliam, Albert M. *Travels in Mexico during the Years 1843 and 44.* Ipswich: J. M. Burton, 1847.

Gilmore, N. Ray. "Henry George Ward, British Publicist for Mexican Mines." *Pacific Historical Review* 32, no. 1 (1963): 35–47.

Gleim, E. J. *Notes Pertaining to Safety Inspections of Permissible Electric Mine Equipment.* Bureau of Mines Circular 6584. Washington, DC: Government Printing Office, 1932.

Gómez Del Campo, José Maria. "El Mineral de San Pedro." *Minero Mexicano,* no. 16 (June 14, 1883) and no. 17 (June 21, 1883).

———. *Noticia del Socavon Aventurero de la Victoria en el Mineral de San Pedro.* San Luis Potosi: Tip. S.M. Velez, 1878.

Gómez Serrano, Jesús, and Enrique Rodríguez Varela. *Aguascalientes: Imperio de los Guggenheim. Estudio sobre la minería y metalurgia en Aguascalientes 1890–1930. El caso Guggenheim-ASARCO.* Mexico DF: Fondo de Cultura Economica, 1982.

González, Michael J. "United States Copper Companies, the State and Labour Conflict in Mexico, 1900–1910." *Journal of Latin American Studies* 26, no. 3 (1994): 651–81.

Goodrich, Carter Lyman. *The Miner's freedom: A Study of the Working Life in a Changing Industry.* Boston: Marshall Jones, 1925.

Gordillo, Gastón R. *Rubble: The Afterlife of Destruction.* Durham NC: Duke University Press, 2014.

Granados y Gálvez, and José Joaquín. *Tardes americanas. Gobierno gentile y catolico y particular noticia de toda la historia indiana.* Mexico: D.F. de Zúñiga y Ontiveros, 1778.

Grandjean, Philippe, and Takashi Yorifuji. "Mercury." In *Patty's Toxicology*, ed. Eula Bingham, Barbara Cohrssen, and F. A. Patty, 213–28. Hoboken, NJ: John Wiley & Sons, 2012.

Grossman, Sarah E. M. *Mining the Borderlands. Industry, Capital, and the Emergence of Engineers in the Southwest Territories, 1855–1910*. Reno: University of Nevada Press, 2018.

Gudynas, Eduardo. "Diez tesis urgentes sobre el nuevo extractivismo." *Extractivismo, política y sociedad* (Quito: Centro Andino de Acción Popular–Centro Latinoamericano de Ecología Social, 2009): 187–225.

———. "Estado compensador y nuevos extractivismos. Las ambivalencias del progresismo sudamericano." *Nueva Sociedad* 237 (2012): 128–46.

———. *Extractivisms: Politics, Economy and Ecology*. Black Point, NS and Winnipeg MB: Fernwood Publishing, 2020.

Guerra, François-Xavier. "La révolution mexicaine: d'abord une révolution minière?" *Annales. Histoire, Sciences Sociales* 36, no. 5 (1981): 785–814.

Guerrero, Saul. *Silver by Fire, Silver by Mercury: A Chemical History of Silver Refining in New Spain and Mexico, 16th and 19th Centuries*. Leiden: Brill, 2017.

Guha, Ramachandra, and Joan Martínez-Alier. *Varieties of Environmentalism: Essays North and South*. London: Earthscan, 1997.

Gunder Frank, André. *ReOrient: Global Economy in the Asian Age*. Berkeley: University of California Press, 1998.

Haber, Stephen H., Noel Maurer, and Armando Razo. *The Politics of Property Rights: Political Instability, Credible Commitments, and Economic Growth in Mexico, 1876–1929*. New York: Cambridge University Pres, 2003.

Hahn, Otto H. "On the Development of Silver Smelting in Mexico," *Transactions of the Institution of Mining and Metallurgy* 8 (1899–1900): 231–303.

Hall, Linda B., and Don M. Coerver. "La frontera y las minas en la revolución mexicana (1910–1920)." *Historia Mexicana* 32, no. 3 (January–March 1983): 389–421.

Hammersley, George. "The Charcoal Industry and Its Fuel, 1540–1750." *Economic History Review* 26, no. 4 (1973): 593–613.

Hampton, Elaine, and Cynthia Ontiveros. *Copper Stain. ASARCO's Legacy in El Paso*. Norman: University of Oklahoma Press, 2019.

Hanke, Lewis. "Viceroy Francisco de Toledo and the Just Titles of Spain to the Inca Empire." *The Americas* 3 (1946): 3–19.

Hardesty, David. *Mining Archaeology in the American West: A View from the Silver State*. Lincoln: University of Nebraska Press and the Society for Historical Archaeology, 2010.

Harvey, David. "Neoliberalism as Creative Destruction." *Annals of the American Academy of Political and Social Science* 61, no. 1 (2007): 21–44.

———. *The New Imperialism*. New York: Oxford University Press, 2005.

Heitt, Dean G. "Newmont's Reserve History on the Carlin Trend." In *Deep Deposits along the Carlin Trend*, ed. T. B. Thompson, Lewis Teal, and R. O. Meeuwig. Nevada Bureau of Mines and Geology Bulletin 111 (2002): 35–45.

Hernández, Raúl Asensio. "Disciplinar la frontera. El juez Francisco Gordillo y el motín de Tumaco de 1709." *Fronteras de la Historia* 13, no. 1 (2008): 15–36.

Higgins, Kathleen J. *"Licentious Liberty" in a Brazilian Gold-Mining Region: Slavery, Gender, and Social Control in Eighteenth-Century Sabará, Minas Gerais.* University Park: Pennsylvania State University Press, 1999.

Himley, Matthew. "Extractivist Geographies. Mining and Development in Late-Nineteenth and Early Twentieth-Century Peru." *Latin American Perspectives* 46, no. 2 (March 2019): 27–46.

Hotchkiss, S. C. "Occupational Diseases in the Mining Industry." *American Labor Legislation Review* 131 (1912): 131–39.

Hughes, Edwin. "Accidents from Falls of Rock or Ore." *Mexican Mining Journal* 18 (October 1914): 263–65.

INEGI (Instituto Nacional de Estadística Geografia e Informatica). *Estadísticas Históricas de México.* Vol. 1. Mexico DF: INEGI, 1986.

Ingersoll, Ralph. *In and Under Mexico.* New York: Century Company, 1924.

Instituto Medico Nacional. *Informe del director sobre la distribucion geografica de la uncinariasis en la Republica.* Mexico DF: Tip. Economica, 1913.

Isenberg, Andrew. *Mining California: An Ecological History.* New York: Hill & Wang, 2005.

Jacobs, David. "Lead." In *Patty's Toxicology,* ed. Eula Bingham, Barbara Cohrssen, and F. A. Patty, 381–426. Hoboken, NJ: John Wiley & Sons, 2012.

Jara, Álvaro. *Guerra y Sociedad.* Santiago: Editorial Universitaria S.A., 1971.

Jonsson, Fredrik Albritton. *Enlightenment's Frontier: The Scottish Highlands and the Origins of Environmentalism.* New Haven: Yale University Press, 2013.

———. "The Origins of Cornucopianism: A Preliminary Genealogy." *Critical Historical Studies* 1, no. 1 (Spring 2014): 151–68.

Junho Anastasia, and Carla Maria. *Vassalos Rebeldes: Violência Coletiva nas Minas na Primeira Metade do Século XVIII.* Belo Horizonte: Coleçào Horizontes Históricos, 1998.

Katz, Friedrich. *The Life and Times of Pancho Villa.* Stanford: Stanford University Press, 1998.

Keeling, Arn, and John Sandlos. "Ghost Towns and Zombie Mines. The Historical Dimensions of Mine Abandonment, Reclamation and Redevelopment in the Canadian North." In *Environmental Histories of Mining in North America,* ed. J. R. McNeill and George Vrtis, 377–420. Berkeley: University of California Press, 2017.

Keen, Benjamin. "The Black Legend Revisited: Assumptions and Realities." *Hispanic American Historical Review* 49, no. 4 (1969): 703–19.

Kelly, T. D., and G. R. Matos. *Historical Statistics for Mineral and Material Commodities in the United States.* U.S. Geological Survey Data Series 140, https://minerals.usgs.gov/minerals/pubs/historical-statistics/.

Kirsch, Stuart. *Mining Capitalism: The Relationship between Corporations and Their Critics.* Berkeley: University of California Press, 2014.

———. "Sustainable Mining." *Dialectical Anthropology* 34, no. 1 (2010): 87–93.

Klubock, Thomas M. *Contested Communities. Class, Gender and Politics in Chile El Teniente Copper Mine*. Durham: Duke University Press, 1998.

———. *La Frontera: Forests and Ecological Conflict in Chile's Frontier Territory*. Durham NC: Duke University Press, 2014.

Knight, Alan. *The Mexican Revolution: counter-revolution and reconstruction. Vol. 1. Porfirians, Liberals, and Peasants*. Lincoln: University of Nebraska Press, 1986.

———. "La Révolution mexicaine: révolution minière ou révolution serrano?" *Annales, Économies, Sociétés, Civilisations* 38, no. 2 (1983): 449–59.

Kryzanowski, Lawrence. "Misinformation and Security Markets." *McGill Law Journal* 24 (1978): 123–35.

Ladd, Doris. *The Making of a Strike. Mexican Silver Workers' Struggles in Real del Monte 1766–1775*. Lincoln: University of Nebraska Press, 1988.

Lane, Kris. *Quito 1599. City and Colony in Transition*. Albuquerque: University of New Mexico Press, 2002.

Langue, Frédérique. "Bibliografía minera colonial." *Suplemento de Anuario de Estudios Americanos* 65 (1988): 137–62.

———. "Problemas y perspectivas de la minería en Sonora, 1770–1780." In *Memoria del X Simposio de Historia de Sonora*, 119–34. Hermosillo, 1986.

———. Trabajadores y formas de trabajo en las minas zacatecanas del siglo XVIII." *Historia Mexicana* 40, no. 3 (1991): 463–506.

Langue, Frédérique, and Carmen Salazar-Soler. "Bibliografia minera hispanoamericana. Siglos XVI-XIX." *Nuevo Mundo—Mundos Nuevos*. 2001. https://doi.org/10.4000/nuevomundo.566.

Las Casas, Bartolome de. *The Devastation of the Indies: A Brief Account*. Translated by Herman Briffault. Baltimore: Johns Hopkins University Press, 1991.

———. *Historia de las Indias*. Edited by Agustín Millares. Mexico: Fondo de Cultura Económica, 1951.

Lasky, S. G. "How Tonnage and Grade Relations Helps Predict Ore Reserves." *Engineering and Mining Journal* 151, no. 4 (April 1950): 81–85.

Leal, Juan Felipe. *Del mutualismo al sindicalismo en México, 1843–1911*. Mexico DF: Juan Pablos Editor, 2012.

———. *En la Revolución (1910–1917)*. Mexico DF: Siglo XXI, 1988.

LeCain, Tim J. *Mass Destruction: The Men and Giant Mines That Wired America and Scarred the Planet*. New Brunswick, NJ: Rutgers University Press, 2009.

Leff, Enrique. *Saber Ambiental. Sustentabilidad, racionalidad, complejidad, poder*. Mexico DF: Siglo XXI, 2004.

LeGoff, Jacques. *Saint Louis*. Paris: Éditions Gallimard, 1996.

Lemieux, André. "Canada's Global Mining Presence." In *Canadian Minerals Yearbook, 1995*, 59–71. Ottawa: Natural Resources of Canada, 1995.

Leung, Chiu, and Chen Tak Sun Yu. "Silicosis." *The Lancet* 379 (2012): 2008–18.

Li, Fabiana. "Relating Divergent Worlds: Mines, Aquifers, and Sacred Mountains in Peru." *Anthropologica* 55 (2013): 399–411.

———. *Unearthing Conflict: Corporate Mining, Activism, and Expertise in Peru*. Durham, NC: Duke University Press, 2015.

Linebaugh, Peter. "Enclosures from the Bottom Up." *Radical History Review* 108 (Fall 2010): 11–27.

Lynch, John. *Spain 1516–1598: From Nation State to World Empire*. Oxford: Blackwell, 1988.

MacLachlan, Colin M. *Spain's Empire in the New World: The Role of Ideas in Institutional and Social Change*. Berkeley: University of California Press, 1988.

Málaga Medina, Alejandro. "Las reducciones en el Virreinato de Perú." *Revista de Historia de America* 80 (1975): 8–42

Malcolmson, James W. "The Sierra Mojada, Coahuila, Mexico, and Its Ore-Deposits." *Transactions of the American Institute of Mining Engineers* 21 (1901): 100–139.

Malm, Andreas. *Fossil Capital: The Rise of Steam Power and the Roots of Global Warming*. London: Verso, 2016.

——. "The Origins of Fossil Capital: From Water to Steam in the British Cotton Industry." *Historical Materialism* 21 (2013): 15–68.

Manuel, Jeffrey T. *Taconite Dreams: The Struggle to Sustain Mining on Minnesota's Iron Range, 1915–2000*. Minnesota: University of Minnesota Press, 2015.

Manuell, Ricardo E. "Anchylostomiasis in Mexico." In *Public Health Papers and Reports Presented at the Annual Meeting of the American Public Health Association, Mexico City* 32, no. 1 (1906): 202–4.

Marichal, Carlos. "The Spanish-American Silver Peso: Export Commodity and Global Money of the Ancient Regime, 1550–1800." In *From Silver to Cocaine: Latin American Commodity Chains and the Building of the World Economy, 1500–2000*, ed. Steven Topik, Carlos Marichal, and Zephyr L Frank, 25–62. Durham, NC: Duke University Press, 2006.

Martinez, Juan. "Minas de la unidad San Pedro, de la Compañia Metalurgica Nacional, en San Luis Potosi." *Boletín Minero*, 23 (1927): 373–76.

Martínez Alier, Joan. "Ecology and the Poor: A Neglected Dimension of Latin American History." *Journal of Latin American Studies* 23, no. 3 (October 1991): 621–39.

——. "The Environmentalism of the Poor: Its Origins and Spread." In *A Companion to Global Environmental History*, ed. J. R. McNeill and Erin Steward Mauldin, 513–29. Chichester, West Sussex: Wiley, 2012.

——. "Mining Conflicts, Environmental Justice, and Valuation." *Journal of Hazardous Materials* 86 (2001): 153–70.

Martínez Baca, Eduardo. "Historical Sketch of Mining Legislation in Mexico." *Transactions of the American Institute of Mining Engineers* 22 (1902): 520–65.

Martínez Chaves, Paula Andrea, Alexander Betancourt Mendieta, Miguel Nicolás Caretta, and Miguel Aguilar Robledo. "Procesos historicas y ambientales en Cerro de San Pedro, San Luis Potosí, México, 1948–1997." *Region y Sociedad* 22, no. 48 (2010): 211–48.

Martínez Ortiz, Juan José, and Adela Tarifa Fernández. *Medicina social, demografía y enfermedad en la minería giennense contemporánea. El centenillo: 1925–1964*. Jaén: Diputación provincial de Jaén, 1999.

Martínez de Viedsma, Joseph. "Informe sobre las minas de Taxco." In *Las minas de Nueva España en 1753*, ed. Alvaro López Miramontes, 33–41. Mexico City: INAH, 1975.

Marx, Karl. *The Capital. A Critique of Political Economy*. Moscow: Progress Publishers, 1965.

Mason Hart, John. *The Silver of the Sierra Madre. John Robinson, Boss Shepherd, and the People of the Canyons*. Tucson: University of Arizona Press, 2008.

McCulloch, Jock. "Air Hunger: The 1930 Johannesburg Conference and the Politics of Silicosis," *History Workshop Journal* 72, no. 1 (October 1, 2011): 118–37.

McEvoy, Arthur F. "Working Environments: An Ecological Approach to Industrial Health and Safety." *Technology and Culture* 36, no. 2 (Supplement to April 1995): s145–s173.

McNeill, John Robert. *Mosquito Empires: Ecology and War in the Greater Caribbean, 1620–1914*. New York: Cambridge University Press, 2010.

Medina Esquivel, René. "Testimonios gráficos de fiestas y diversiones en Cerro de San Pedro, San Luis Potosí." *Vetas. Revista de El Colegio de San Luis* 24, nos. 8–9 (2007): 1–26.

Mexico. *Constitución de 1917—Promulgada en Diario Oficial—Organo del Gobierno Provisional de la Republica Mexicana 5 de Febrero, 1917*. http://www.diputados .gob.mx/LeyesBiblio/ref/cpeum/CPEUM_orig_05feb1917.pdf.

Minaudier, Jean-Pierre. "Une région minière de la colonia à l'Indépendance: Barbacoas 1750–1830 (Economie, société, vie politique locale)." *Bulletin de l'Insitut Français des Etudes Andines* 17, no. 2 (1988): 81–104.

Mining Association of Canada. *Facts and Figures—2007. A Report on the State of the Canadian Mining Industry*. Ottawa: Mining Association of Canada, 2007.

———. *Facts and Figures—2010. A Report on the State of the Canadian Mining Industry*. Ottawa: Mining Association of Canada, 2010.

Mitchell, Paul, and Alyson Warhurst. "Corporate Social Responsibility and the Case of the Summitville Mine." *Resources Policy* 26, no. 2 (2000): 91–102.

Moffitt Watts, Pauline. "Prophesy and Discovery: On the Spiritual Origins of the Enterprise of the Indies." *American Historical Review* 90, no. 102 (1985): 73–102.

Molina Martínez, Miguel. "Legislación minera colonial en tiempos de Felipe II." In *Coloquio de Historia Canario-Americana; VIII Congreso Internacional de Historia de America*, ed. Francisco Morales Padrón, 1014–29. Las Palmas de Gran Canaria, 2000.

Montejano y Aguiñaga, Rafael. *El Real de Minas de la Purisima Concepcion de los Catorce, S.L.P.* Mexico: CONACULTA, 1993.

Montes de Oca, Genaro. "La Fundición de San Luis Potosí." *Boletin Minero* 13 no. 4 (1922): 463–75.

Moore, Jason W. *Capitalism in the Web of Life. Ecology and the Accumulation of Capital*. London: Verso Books, 2015.

——. "Silver, Ecology, and the Origins of the Modern World, 1450–1640."
In *Rethinking Environmental History. World-System History and Global
Environmental Change*, ed. Alf Hornborg, J. R. McNeil, and Joan Martinez-
Alier, 123–42. New York: Altamira Press, 2007.

——. "Sugar and the Expansion of the Early Modern World-Economy:
Commodity Frontiers, Ecological Transformation, and Industrialization."
Review (Fernand Braudel Center) 23, no. 3 (2000): 409–33.

Mota y Escobar, Alonso de la. *Descripción geografica de los reinos de Nueva Galicia,
Nueva Vizcaya y Nuevo Leon*. Editors Jose Parres Arias et al. Guadalajara:
Instituto Jalisciense de Antropologia e Historia, [1605] 1966.

Mourelle, Francisco Antonio. "Viaje a las minas de Guanajuato, noviembre de
1790." In *El ocaso Novohispano: Testimonios documentales*. Translated by
Antonio Saborit, edited by David Brading, 23–76. Mexico DF: INAH—CNCA,
1996.

Mukerji, Chandra. "The Great Forest Survey of 1669–1671. The Use of Archives for
Political Reform." *Social Studies of Science* 37, no. 2 (April 2007): 227–53.

Müller, J., and H. E. Frimmel. "Numerical Analysis of Historic Gold Cycles and
Implications for Future Sub-Cycles." *Open Geology Journal* 4 (2010): 29–34.

Mumford, Lewis. *Technics and Civilization*. London: George Routledge & Sons,
Ltd., 1934.

Muñoz Camargo, Diego. *Historia de Tlaxcala*. Edited by Alfredo Chavero. Mexico:
Oficina Tipografica de la Secretaria de Fomento, [1585] 1892.

Muradian, Roldan, Mariana Walter, and Joan Martinez-Alier. "Hegemonic
Transitions and Global Shifts in Social Metabolism: Implications for Resource-
Rich Countries. Introduction to the Special Section." *Global Environmental
Change* 22, no. 3 (2012): 559–67.

Nash, June. *We Eat the Mines and the Mines Eat Us. Dependency and Exploitation
in Bolivian Tin Mines*. New York: Columbia University Press, 1993.

Nixon, Rob. *Slow Violence and the Environmentalism of the Poor*. Cambridge, MA:
Harvard University Press, 2011.

Nolin, Catherine, and J. P. LaPlante. "Consultas and Socially Responsible
Investing in Guatemala: A Case Study Examining Maya Perspectives on the
Indigenous Right to Free, Prior, and Informed Consent." *Society and Natural
Resources* 27 (2014): 231–48.

Office of Compliance—U.S. Environmental Protection Agency. *EPA Office of
Compliance Sector—Profile of the Metal Mining Industry*. Washington, DC:
Office of Enforcement and Compliance Assurance, 1995.

O'Hara, T. Alan, and Stanley C. Suboleski. "Costs and Cost Estimation." In *Mining
Engineering Handbook*, ed. Howard L. Hartman et al., 405–24. Littleton, CO:
Society for Mining, Metallurgy, and Exploration, Inc., 1992.

Orwell, George. *The Road to Wigan Pier*. New York: Penguin Classics, 2001.

Paley, Dawn. *Drug War Capitalism*. Oakland: AK Press, 2014.

Pagden, Anthony. "Dispossessing the Barbarian: Rights and Property in Spanish
America." In *Spanish Imperialism and the Political Imagination. Studies in*

European and Spanish-American Social and Political Theory, 1513–1830, 13–36. New Haven: Yale University Press, 1990.

Parsons, James J. "Gold Mining in the Nicaragua Rain Forest." *Yearbook of the Association of Pacific Coast Geographers* 17, no. 1 (1955): 49–55.

———. "The Settlement of the Sinú Valley of Colombia." *American Geographical Society* 42, no. 1 (1952): 67–86.

Pastor Bodmer, Beatriz. *The Armature of Conquest. Spanish Accounts of the Discovery of America, 1492–1589.* Stanford: Stanford University Press, 1992.

Peck, Gunther. "The Nature of Labor: Fault Lines and Common Ground in Environmental and Labor History." *Environmental History* 11, no. 2 (2006): 212–38.

Perales, Monica. *Smeltertown—Making and Remembering a Southwest Border Community.* Chapel Hill: University of North Carolina Press, 2010.

Perreault, Tom. "Dispossession by Accumulation? Mining, Water and the Nature of Enclosure on the Bolivian Altiplano." *Antipode* 45 (2013): 1050–69.

Peset, Mariano, and Margarita Menegus. "Rey proprietario o rey soberano." *Historia Mexicana,* 43, no. 4 (1994): 563–99.

Petersen, Mark A., Michelle della Libera, Raymond R. Jannes, and Stephen Maynard. "Geology of the Cerro San Pedro Porphyry-Related Gold-Silver Deposit, San Luis Potosí, Mexico." *Society of Economic Geologists,* Special publication, no. 8 (2001): 217–41.

Powell, Philip W. "Caldera of New Spain: Frontier Justice and Mestizo Symbol." *The Americas* 17, no. 4 (1961): 325–42.

———. "The Chichimecas: Scourge of the Silver Frontier in Sixteenth-Century Mexico." *Hispanic American Historical Review* 25, no 3 (1945): 315–38.

———. "Genesis of the Frontier Presidio in North America." *Western Historical Quarterly* 13, no. 2 (1982): 125–41.

———. *Mexico's Miguel Caldera. The Taming of America's First Frontier.* Tucson: University of Arizona Press, 1977.

———. "Peacemaking on North America's First Frontier." *The Americas* 16, no. 3 (1960): 221–50.

Pratt, Mary Louise. *Imperial Eyes. Travel Writing and Transculturation.* New York: Routledge 1992.

Radding, Cynthia. "Colonial Spaces in the Fragmented Communities of Northern New Spain." In *Contested Spaces of Early America,* ed. Juliana Barr and Edward Countryman, 115–41. Philadelphia: University of Pennsylvania Press, 2014.

Ramírez Reynoso, B. "El conflicto de Pachuca y Real del Monte, un caso 'huelguístico' en el siglo XVIII" *Estudios Históricos—Inst. De Investigaciones Jurídicas* 17 (1984): 553–70.

Ramírez, Santiago. *Apuntes para un proyecto de Código de Minería. Presentado al Sr. Ministro de Fomento, D[irector] General D. Cárlos Pacheco.* Mexico: Tip. de la Secretaria de Fomento, 1884.

———. *Riqueza Minera de Mexico y de su actual estado de exploitacion.* Mexico: Officina tipografica de la Secretaria de Fomento, 1884.

Ramírez, Susan. *The World Upside Down: Cross-Cultural Contact and Conflict in Sixteenth-Century Peru.* Stanford: Stanford University Press, 1996.

Ramos Pérez, Demetrio. "Ordenacion de la mineria en hispanoamerica durante la epoca provincial (siglos XVI, XVII, XVIII)." In *La minería hispana e iberoamericana. Ponencias del I coloquio internacional sobre historia de la minería,* 373–97. Leon: Catedra de San Isidoro, 1970.

Randolph, William E. "The Lakes of Potosi." *Geographical Review* 26, no. 4 (1936): 529–54.

Rankine, Margaret. "The Mexican Mining Industry in the Nineteenth Century with Special Reference to Guanajuato." *Bulletin of Latin American Research* 11, no. 1 (1992): 29–48.

Rawson, William. *The Present Operations and Future Prospects of the Mexican Mine Associations.* London: J. Hatchard and Son, 1825.

Real Academia de Historia. *Diccionario de autoridades.* Madrid, 1726–39.

Réau, Louis. *Iconografía del Arte Cristiano. Tomo 2. Vol. 4: Iconografía de los Santos G-O,* Trad. Daniel Alcoba. Barcelona: Ediciones del Serbal, 1997.

Rebollar Leopol. "La campaña en contra del mal que aflige a los mineros." *Mexican Mining Journal* 18 (1914): 49–50.

Reed, John. *Insurgent Mexico.* New York: Clarion, 1969.

Restrepo, Vicente. *Estudio sobre las minas de oro y plata de Colombia.* Bogotá: Imprenta de Silvestre, 1888.

Reygadas, Pedro. *Las vetas del lenguaje minero: viaje al centro del inframundo.* Colegio de San Luis, 2010.

Rhodes, Nelson. "Barrick Gold Corporation." In *International Directory of Company Histories,* 34. Detroit: St. James Press, 2000.

Rickard, Thomas A. *A History of American Mining.* New York: McGraw-Hill, 1932.

———. *Interviews with Mining Engineers.* San Francisco: Mining and Scientific Press, 1922.

———. *Recent Cyanide Practice.* San Francisco: Mining and Scientific Press, 1907.

Richards, John F. *The Unending Frontier. An Environmental History of the Early Modern World.* Berkeley: University of California Press, 2003.

Rippy, J. Fred. "Latin America and the British Investment 'Boom' of the 1820's." *Journal of Modern History* 19, no. 2 (1947): 122–29.

Rivera Villanueva, Jose Antonio. *Los Tlaxcaltecas: Pobladores de San Luis Potosí.* San Luis Potosí: El Colegio de San Luis, 1999.

Roberts, Ralph J. *A Passion for Gold: An Autobiography.* Reno: University of Nevada Press, 2002.

Roberts Poinsett, Joel. *Notes on Mexico made in the autumn of 1822: Accompanied by an historical sketch of the revolution and translations of official reports on the present state of that country.* London: John Miller, 1825.

Robins, Nicholas. *Mercury, Mining, and Empire: The Human and Ecological Cost of Colonial Silver Mining in the Andes.* Bloomington: Indiana University Press, 2011.

Rodriguez Loubet, Francoise. *Les Chichimeques. Archeologie et Ethnohistoire des Chasseurs-Collectuers de San Luis Potosi, Mxique*. Mexico: Centre d'etudes mexicaines et centriamericains, 1985.

Rogers, Allen H. "Character and Habits of the Mexican Miner." *Engineering & Mining Journal* 85 (1908): 700–702.

Rogers, Thomas D. *The Deepest Wounds. A Labor and Environmental History of Sugar in Northeast Brazil*. Chapel Hill: University of North Carolina Press, 2010.

Romero, Matías. *Geographical and Statistical Notes on Mexico*. New York: G.P. Putman's Sons, 1898.

Romero Gil, Juan. "Sonora: La Revolución en el socavón, 1910–1918." *Signos historicos* 21 (January–June 2009): 14–38.

Ruiz Guadalajara, Juan Carlos. "Capitán Miguel Caldera y la frontier Chichimeca: Entre el mestizo historiográfico y el soldado del Rey." *Revista de Indias* 70, no. 248 (2010): 23–58.

———. "Vestigios de un prodigio: el culto a San Luis de la Paz y el caso del Potosí novohispano." In *Alardes de armas y festividades. Valoración e identificación de elementos de patrimonio histórico*, ed. Ana Díaz Serrano, José Javier Ruiz Ibáñez y Óscar Mazín, 95–113. Murcia, Universidad de Murcia/Red Columnaria/Fundación Séneca, 2008.

Salazar, Luis. "Mexican Railroads and the Mining Industry." *Transactions of the American Institute of Mining Engineers* 32 (1901): 303–34.

Salazar González, Guadelupe. *Las haciendas en el siglo XVII en la región minera de San Luis Potosí: Su espacio, forma, función, material, significado y estructuración regional*. San Luis Potosí: Facultad del Hábitat—UASLP, 2000.

Salazar-Soler, Carmen. *Anthropologie des mineurs des Andes*. Paris: L'Harmattan, 2002.

Sánchez Albavera, Fernando, Georgina Ortiz, and Nicole Moussa, *Mining in Latin America in the Late 1990s*. Santiago: Economic Commission for Latin America and the Caribbean, 2001.

Sánchez Crispín, Álvaro. "The Territorial Organization of Metallic Mining in New Spain," In *In Quest of Mineral Wealth: Aboriginal and Colonial Mining and Metallurgy in Spanish America*, ed. Alan K. Craig and Robert C. West, 155–70. Baton Rouge, LA: Geoscience and Man—Louisiana State University, 1994.

Santa Maria, Fray Guillermo de. *Guerra de los Chichimecas* (Mexico 1575–Zirosto 1585). Edited by Alberto Carrillo Cazares. Zamora: El Colegio de Michoacan—Universidad de Guanajuato, 1999.

Santiago, Myrna. *The Ecology of Oil: Environment, Labor, and the Mexican Revolution, 1900–1938*. New York: Cambridge University Press, 2006.

Sariego, Juan Luis. *Enclaves y minerales en el norte de México: historia social de los mineros de Cananea y Nueva Rosita 1900–1970*. Mexico: CIESAS, 2010.

Sauer, Carl O. *The Early Spanish Main*. Berkeley: University of California Press, 1966.

Sawyer, Suzana. *Crude Chronicles: Indigenous Politics, Multinational Oil and Neoliberalism in Ecuador*. Durham, NC: Duke University Press, 2004.

Sawyer, Suzana, and Edmund T. Gomez. *The Politics of Resource Extraction. Indigenous Peoples, Multinational Corporations, and the State*. New York: Palgrave Macmillan, United Nations Research Institute for Social Development, 2012.

Schmitz, Christopher. "The Rise of Big Business in the World Copper Industry, 1870–1930." *Economic History Review* 39, no. 3 (August 1986): 392–410.

Scott, Heidi. "The Contested Spaces of the Subterranean: Colonial Governmentality, Mining and the Mita in Early Spanish Peru." *Journal of Latin American Geography* 11 (2012): 7–33.

———. "Contested Territories: Arenas of Geographic Knowledge in Early Colonial Peru." *Journal of Historical Geography* 29, no.2 (2003): 166–88.

Sego, Eugene D. *Aliados y adversarios: Los Colonos Tlaxcaltecas en la frontera septentrional de Nueva Espana*. San Luis Potosi: El Colegio de San Luis, 1998.

Semo, Enrique. *The History of Capitalism in Mexico: Its Origins, 1521–1763*. Austin: University of Texas Press, 1993.

Sempat Assadourian, Carlos. "La producción de la mercancia dinero en la formación del mercado interno colonial." In *Ensayos sobre el desarrollo económico de México y América Latina, 1500–1975*, ed. Enrique Florescano, 223–92. Mexico DF: Fondo de Cultura Económica, 1979.

———. *El sistema de la economía colonial: mercado interno, regiones y espacio económico*. Lima: Instituto de Estudios Peruanos, 1982.

Serrano Hernández, Sergio Tonatiuh. "'. . . ¡Hay oro y no nos avisan a los amigos!' Contrabando y evasión fiscal en el Cerro de San Pedro Potosí durante la primera mitad del siglo VII," *Vetas* (San Luis Potosí) 10, no. 29 (2008): 37–62.

Silva Ontiveros, and Garza Merodio. "Neocolonialismo y minería: El ocaso de Cerro de San Pedro, México." *Revista Latino-Americana de História* 6, no. 17 (2017): 14–34.

Smale, Robert L. "I Sweat the Flavor of Tin: Labor Activism in Early Twentieth-Century *Bolivia*." Pittsburgh: University of Pittsburgh Press, 2010.

Smil, Vaclav. *Energy in Nature and Society. General Energetics of Complex Systems*. Cambridge: MIT Press, 2008.

Snoke, A. W. "North America—Southern Cordillera." In *Encyclopedia of Geology*, ed. Richard C. Selley, L.R.M. Cocks, and I. R. Plimer, 48–61. Amsterdam: Elsevier Academic, 2005.

Soluri, John. *Banana Cultures: Agriculture, Consumption, and Environmental Change in Honduras and the United States*. Austin: University of Texas Press, 2005.

———. "Labor, Rematerialized: Putting Environments to Work in the Americas." *International Labor and Working-Class History* 85 (2014): 162–76.

Southworth, John R., and Percy G. Holmes. *The Official Mining Directory of Mexico*. Mexico City: n.p., 1908.

Stein, Stanley J., and Barbara H. Stein. *Apogee of Empire: Spain and New Spain in the Age of Charles III, 1759–1789*. Baltimore: Johns Hopkins University Press, 2003.

Stern, Steven J. "Feudalism, Capitalism and the World-System in the Perspective of Latin America and the Caribbean." *American Historical Review* 93 (4): 829–72.

Stoll, Steven. "A Metabolism of Society: Capitalism for Environmental Historians." In *The Oxford Handbook of Environmental History,* ed. Andrew C. Isenberg, 369–90. New York: Oxford University Press, 2014.

Studnicki-Gizbert, Daviken. "Canadian Mining in Latin America. A Provisional History." *Canadian Journal of Latin American and Caribbean Studies* 41, no. 1 (2016): 95–113.

Studnicki-Gizbert, Daviken, and David Schecter. "The Environmental Dynamics of a Colonial Fuel-Rush: Silver Mining and Deforestation in New Spain, 1522 to 1810." *Environmental History* 15 (January 2010): 94–119.

Sutter, Paul. "Nature's Agents or Agents of Empire? Entomological Workers and Environmental Change during the Construction of the Panama Canal." *ISIS* 98 (2007): 724–54.

Svampa, Maristella. "Resource Extractivism and Alternatives: Latin American Perspectives on Development." In *Beyond Development: Alternative Visions from Latin America,* ed. M. Lang and D. Mokrani, 117–43. Amsterdam: Transnational Institute, 2013.

Svampa, Maristella, and Enrique Viale. *Maldesarrollo. La Argentina del extractvismo y el despojo.* Buenos Aires: Katz Editores, 2014.

Swann, Michael M. *Tierra Adentro: Settlement and Society in Colonial Durango.* Boulder, CO: Westview Press, 1982.

Szablowski, David. *Transnational Law and Local Struggles: Mining, Communities and the World Bank.* Oxford: Hart, 2007.

Tandeter, Enrique. "Forced and Free Labour in Late Colonial Potosi." *Past & Present* 93, no. 1 (1981): 98–136.

Tardy, M., et al. "Observaciones generales sobre la estructura de la Sierra Madre Oriental: La oloctonia del conjunto Alta-Altiplano Central." *Revista del Instituto de Geología—UNAM* 75, no.1 (1975): 1–11.

Taussig, Michael T. *The Devil and Commodity Fetishism in South America.* Chapel Hill: University of North Carolina Press, 1980.

Tays, E.A.H. "Present Labor Conditions in Mexico." *Engineering and Mining Journal* 84, no. 14 (1907): 621–24.

Teisch, Jessica B. *Engineering Nature: Water, Development and the Global Spread of American Environmental Expertise.* Chapel Hill: University of North Carolina Press, 2011.

TePaske, John J., and Kendall W. Brown. *A New World of Gold and Silver.* Leiden: Brill, 2010.

Thompson, Edward P. "The Moral Economy of the English Crowd in the Eighteenth Century." *Past & Present* 50 (1971): 76–136.

de Toledo, Francisco. *Memorial y Ordenanzas in Relaciones de los Vireyes y Audiencias que han Gobernado el Perú.* Tomo 1. Lima: Imprenta del Estado por J.E. del Campo, 1867.

Torres Pares, Javier. *La Revolución Sin Frontera.* Mexico DF: UNAM, 1990.

Truett, Samuel. *Fugitive Landscapes: The Forgotten History of the U.S.-Mexico Borderlands.* New Haven: Yale University Press, 2006.

Tuffnell, Stephen. "Engineering Inter-Imperialism: American Miners and the Transformation of Global Mining, 1871–1910." *Journal of Global History* 10, no. 1 (2015): 53–76.

Tutino, John. *Making a New World. Founding Capitalism in the Bajío and Spanish North America*. Durham, NC: Duke University Press, 2011.

———. *The Mexican Heartland: How Communities Shaped Capitalism, a Nation, and World History, 1500–2000*. Princeton NJ: Princeton University Press, 2017.

Ulloa, Antonio de. *Noticias Americanas: entretenimientos phisicos-historicos, sobre la América meridional, y la septentrianal*. F. Manuel de Mena, 1772.

Urkidi, Leire, and Mariana Walter. "Dimensions of Environmental Justice in Anti-Gold Mining Movements in Latin America." *Geoforum* 42 (2011): 683–95.

Urquiola Permisan, Jose Ignacio. *Agua para los ingenios. San Luis Potosi y el valle de San francisco a inicios de la epoca colonial. Estudio introductoria y documentos sobres ingenios de beneficios de metales*. San Luis Potosi: El Colegio de San Luis, 2004.

Van Buren, Mary, and Ana María Presta. "The Organization of Inka Silver Production in Porco, Bolivia." In *Distant Provinces in the Inka Empire: Toward a Deeper Understanding of Inka Imperialism*, ed. Sonia Alconini and Michael A. Malpass, 173–92. Iowa City: University of Iowa Press, 2010.

Van Law, W. "Aerial Tramway of the Real del Monte Company, in Pachuca." *Informe y Memorias del Instituto Mexicano de Minas y Metalurgia* 2, no. 6 (1910–11): 214–17.

Varela, Consuelo, and Isabel Aguirre. *La caída de Cristóbal Colón. El juicio de Bobadilla*. Madrid: Marcial Pons, 2006.

Vázquez de Espinosa, Antonio. *Compendium and Description of the West Indies*. Edited and translated by Charles U. Clark. Washington, DC: Smithsonian, [1628] 1948.

Velasco Ávila, Cuahtémoc, Eduardo Flores Clair, A. Parra Campos, and Edgar Omar Gutiérrez López. *Estado y Minería en México (1767–1910)*. Mexico DF: Fondo de Cultura Económica, 1988.

Velasco Murillo, Dana. *Urban Indians in a Silver City. Zacatecas, Mexico, 1546–1810*. Stanford: Stanford University Press, 2016.

Vergara, Germán. *Fueling Mexico: Energy and Environment, 1850–1950*. New York: Cambridge University Press, 2021.

———. "How Coal Kept My Valley Green: Forest Conservation, State Intervention, and the Transition to Fossil Fuels in Mexico." *Environmental History* 23, no. 1 (2018): 82–105.

Vicuña Mackenna, Benjamín. *La Edad del Oro en Chile*. Santiago: Imprenta Cervantes, 1881.

Villarello Vélez, Ildefonso. *Historia de la Revolución mexicana en Coahuila*. Mexico: Talleres Gráficos de la Nación, 1970.

Villaseñor y Sánchez, Joseph Antonio de. *Theatro Americano: Descripción General de los Reynos y Provincias de la Nueva España y sus jurisdicciones*. 2 vols. Mexico: Imprenta de la viuda de D/Joseph Bernardo de Hogal, 1748.

Vitoria, Francisco de. "On the American Indians." In *Political Writings*, ed. Anthony Pagden and Jeremy Lawrence, 231–92. New York: Cambridge University Press, 1991.

von Humboldt, Alexander. *Political Essay on the Kingdom of New Spain.* Translated by John Black. London: Longman, Hurst, Rees, Orme & Brown, 1814.

Wakild, Emily. "Environmental Justice, Environmentalism, and Environmental History in Twentieth-Century Latin America." *History Compass* 11, no. 2 (2013): 163–76.

Walker, Brett. *Toxic Archipelago. A History of Industrial Disease in Japan.* Seattle: University of Washington Press, 2010.

Wallerstein, Immanuel Maurice. *The Modern World-System I. Capitalist Agriculture and the Origins of the European World-Economy in the Sixteenth Century.* New York: Academic Press, 1974.

Ward, H. G. *Mexico in 1827.* 2 vols. London: H. Colburn, 1828.

Warde, Paul. "Fear of Wood Shortage and the Reality of the Woodland in Europe, c. 1450–1850." *History Workshop Journal* 62, no. 1 (Fall 2006): 28–57.

Warsh, Molly. "A Political Ecology in the Early Spanish Caribbean." *William and Mary Quarterly* 71, no. 4 (October 2014): 517–48.

Wasserman, Mark. "The Social Origins of the 1910 Revolution in Chihuahua." *Latin American Research Review* 15, no. 1 (1980): 15–38.

Wernke, Steven A. "A Reduced Landscape: Toward a Multi-Causal Understanding of Historic Period Agricultural Deintensification in Highland Peru." *Journal of Latin American Geography* 9, no. 3 (2010): 51–83.

West, Robert C. *Colonial Placer Mining in Colombia.* Baton Rouge: Louisiana State University Press, 1952.

———. *The Mining Community in Northern New Spain: The Parral Mining District.* Berkeley: University of California Press, 1949.

———. *Sonora: Its Geographical Personality.* Austin: University of Texas Press, 1993.

Wey Gómez, Nicolás. *The Tropics of Empire. Why Columbus Sailed South to the Indies.* Cambridge: MIT Press, 2008.

Weyl, Walter E. "Labor Conditions in Mexico." *Bulletin of the Department of Labor* 7, no. 38 (1902): 1–4.

Whitehead, Neil L. "The Crises and Transformations of Invaded Societies: The Caribbean (1492–1580). In *The Cambridge History of the Native Peoples of the Americas. South America, Part 1*, ed. Frank Salomon and Stuart B. Schwartz, 864–903. New York: Cambridge University Press, 1999.

Whitmore, Andrew. "The Emperor's New Clothes: Sustainable Mining?" *Journal of Cleaner Production* 14, no. 3–4 (2006): 309–14.

Wing, John T. "Keeping Spain Afloat: State Forestry and Imperial Defense in the Sixteenth Century." *Environmental History* 17 (2012): 116–45.

———. *Roots of Empire. Forests and State Power in Early Modern Spain, c. 1500–1750.* Leiden: Brill, 2015.

Worstell, Jonathan H. "Precious Metal Heap Leaching in North America." *Mining Magazine* 154, no. 5 (1986): 405–7.

Wyman, Mark. *Hard Rock Epic: Western Miners and the Industrial Revolution, 1860–1910.* Berkeley: University of California Press, 1979.

——. "Industrial Revolution in the West: Hard-Rock Miners and the New Technology." *Western Historical Quarterly* 5, no. 1 (1974): 39–57.

Ye, Jongzhong, Jan Douwe van der Ploeg, Sergio Schneider, and Tedor Shanin. "The Incursions of Extractivism: Moving from Dispersed Places to Global Capitalism." *Journal of Peasant Studies* 47, no. 1 (2020): 155–83.

Zulawski, Ann. *They Eat from Their Work: Work and Social Change in Colonial Bolivia.* Pittsburgh: University of Pittsburgh Press, 1995.

Index

constitution of 1917, 162–63
cooperatives, miners, 161, 165, 171
copper mining, open-pit, 179
cornucopianism, 105, 113
Cortes, Hernán, 33
cottage miners and smelters, 62, 88–89, 99; at Cerro de San Pedro, 92, 103, 122, 170–71, 195; environmental impacts, 130; persistence, 104; suppression, 94. *See also* rescate
credit, 40
cuadrillas, 29, 30, 63, 87
cumulative impacts, 133, 135, 144
cyanidation, 116, 171. *See also* heap leaching
cyanide, 220, 224, 225, 227
cycles, mining, 7, 8; Cerro de San Pedro, 1, 4–5, 62, 179, 215, 227, 229, 232; Latin America, 2, 9, 32; Mexico, 5, 118; open-pit mining, 177, 193, 197; revival, 104, 119, 129, 193

Dahlgren, Charles, 116
deep sea mining, 232
Deneault, Alain, 185
deregulation, in financial sector, 186
Díaz, Porfirio, 153
diesel, 127, 184, 195, 197
disease, 31, 35, 131, 135, 145
drills, mechanical, 115, 140
dust, 132, 140, 141, 147, 150
dynamite, 116, 148

de Eguía, José Joaquín, 109, 112
Ejército Zapatista de Liberación Nacional (EZLN), 213, 216, 233
ejidatarios, 14
ejido of Cerro de San Pedro, 170, 172; rental agreement with company, 207, 208, 222, 223
El Dorado, icon and myth, 188, 189, 190, 193
electricity, 115, 125, 126, 131, 223
de Elhuyar, Fausto, 107–8, 109
El Paso, Texas, 121

energy, 1, 10, 74, 77, 78; colonial period, 80, 96, 98; geography of, 131; industrial period, 115, 125, 126, 129; open-pit mines, 184, 195, 197
entradas, 32
Environmental Impact Assessment (EIA), 205, 206, 208, 213
environmentalism of the poor, 15
environmental justice, 141, 143, 149, 151, 164, 203, 213, 226
environmental justice movements. *See* socio-ecological movements
environmental toxicology, 132–35
Española, 26, 27, 28, 30, 31, 32
Essequibo River, 206
exhaustion, 6, 8, 9, 25, 35, 165, 187; bodily, 91; resolving, 1, 7, 10, 179, 183
extractivism, 24, 204; assemblage, 39, 103; critiques of, 38, 85, 157–58, 204; definition, 3, 24–25; ideological aspects, 23, 38, 189–91; liberalism, 109, 110, 111, 157; Neoliberal, 189–91, 199; in place, 219; political aspects, 3, 16–18, 24, 28, 38, 85, 103, 199; social and ecological aspects, 37

Fabry, José, 85, 105, 106
Fernández de Oviedo, Gonzalo, 30
Fernández Leal, Manuel, 114
fetishism of commodities, 73
floods, 80, 172, 206, 229
Flores Magón, Ricardo, 158, 160
flow-through shares, 187
fomento, 112
forests: afforestation, 100, 170, 232; deforestation, 10, 34, 80, 98; at Guanajuato, 126
fossil fuels, 10, 184, 195, 197
Fox, Vicente, 209
Franck, Harry, 146–47
Franco-Nevada company, 187, 194
French, William, 151
Frente Amplio Opositor a la Minera San Xavier (FAO-MSX), 212, 216, 231
Friedland, Robert, 184, 186, 206, 221

and organization, 181–83; US West, 184, 187. *See also* Carlin, Nevada

opposition to Cerro de San Pedro project, 200, 202, 211, 226, 230; blockades, 213; community plebiscite, 213; composition, 202, 211–12; demonstrations and marches, 205, 214, 224; lawsuits, 208–9, 217, 218, 219; mobilization of the past, 214–16; praxis, 202; targeted with violence, 219, 223

opposition to large-scale mining in Latin America, 201, 203, 211, 231; community plebiscites, 214, 231; direct action, 213; praxis, 202

oral history, 147

ores: chemical composition, 97, 103; colonial period, 75, 76, 103; grades, 6, 7, 43, 97, 113, 164, 167, 181, 197; industrial period, 115, 118, 125, 128; open-pit period, 183, 195; processing rates, 7, 11, 181

Orwell, George, 156

de Ovando, Nicolás, 28, 29, 31

Pacheco, Carlos, 112, 121

Pachuca, 87, 91, 116, 136

Panama, 32, 33, 34, 89, 201, 203, 211, 213

Panama Canal, 136, 137, 138

parasites, 11, 131, 135

Parral, 151

Partido, 2, 13, 75, 86–88; abolished by Gálvez, 94; across Spanish and Portuguese Americas, 89; ASARCO period, 171; nineteenth century, 108, 112, 114; Real del Monte, 91

peace by purchase, 51, 58–60

pension funds, 186

Peru: anti-extractivism in, 203; Canadian mining projects in, 192; community-based referenda, 214; deforestation in, 80; huaqueria, 33; mining law reforms, 188, 191; open-pit mining in, 187, 188, 189, 190

pillage, 4, 33, 40, 190, 202, 215

pillaring, 98, 100

plate tectonics, 44

Poinsett, Joel Roberts, 108, 126

Poma de Ayala, Guaman, 38

Porfiriato, 104, 113, 114

porphyry, 42, 46, 49, 183, 226, 227, 229

Potosí, 33, 35, 37, 38, 50, 85; *kajcheo* (partido) at, 90

Procuraduría Federal de Protección al Ambiente (Profepa), 218

Pro Defensa de Cerro de San Pedro, 204

Pro San Luis Ecologico, 204, 205, 212, 218, 219

prospecting, 26, 33, 49, 50, 51, 52, 172, 180

protected areas: ecological and patrimonial designation for Cerro de San Pedro, 177, 208, 219; Guadalcazar, 204

providentialism, 27, 38, 103, 110, 190

provisioning of mines, 31; at Cerro de San Pedro, 52, 57, 69, 73, 75, 78, 96

Québec City, protests against Summit of the Americas, 205

racialism, 146, 149

railways, 121, 123, 126, 159

Ramírez, Santiago, 112, 113, 122

Rangel, Marcos, 4, 128, 172

raubwirtshaft, 33

Rawson, William, 110

Real de Catorce, 100, 157

Real del Monte, 87, 91, 123, 165

real de minas, 54

Rebolloso, Pedro, 224

Red Mexicana de Afectados por la Minería (REMA), 231

Regla, Conde de, first, 91; second, 111

Reid, John, 159

rescate, 88

reserves, mineral, as socio-technological constructs, 180, 183–84; at the Carlin mine, 182–83; at Cerro de San Pedro, 194, 195

www.ingramcontent.com/pod-product-compliance
Lightning Source LLC
Chambersburg PA
CBHW031356270326
41929CB00010BA/1207